Very High Resolution (VHR) Satellite Imagery

Very High Resolution (VHR) Satellite Imagery

Processing and Applications

Special Issue Editors

Francisco Eugenio
Javier Marcello

MDPI • Basel • Beijing • Wuhan • Barcelona • Belgrade

MDPI

Special Issue Editors

Francisco Eugenio
University of Las Palmas of Gran Canaria
(ULPGC)
Spain

Javier Marcello
University of Las Palmas of Gran Canaria
(ULPGC)
Spain

Editorial Office
MDPI
St. Alban-Anlage 66
4052 Basel, Switzerland

This is a reprint of articles from the Special Issue published online in the open access journal *Remote Sensing* (ISSN 2072-4292) from 2018 to 2019 (available at: https://www.mdpi.com/journal/remotesensing/special_issues/VHR_Satellite_Imagery).

For citation purposes, cite each article independently as indicated on the article page online and as indicated below:

LastName, A.A.; LastName, B.B.; LastName, C.C. Article Title. *Journal Name* **Year**, *Article Number*, Page Range.

ISBN 978-3-03921-756-4 (Pbk)
ISBN 978-3-03921-757-1 (PDF)

Cover image courtesy of Francisco Eugenio and Javier Marcello.

Contents

About the Special Issue Editors vii

Preface to "Very High Resolution (VHR) Satellite Imagery" ix

Kui Jiang, Zhongyuan Wang, Peng Yi, Junjun Jiang, Jing Xiao and Yuan Yao
Deep Distillation Recursive Network for Remote Sensing Imagery Super-Resolution
Reprinted from: *Remote Sensing* **2018**, *10*, 1700, doi:10.3390/rs10111700 1

Yun Ren, Changren Zhu and Shunping Xiao
Deformable Faster R-CNN with Aggregating Multi-Layer Features for Partially Occluded
Object Detection in Optical Remote Sensing Images
Reprinted from: *Remote Sensing* **2018**, *10*, 1470, doi:10.3390/rs10091470 24

Yao Yao and Shixin Wang
Evaluating the Effects of Image Texture Analysis on Plastic Greenhouse Segments via
Recognition of the OSI-USI-ETA-CEI Pattern
Reprinted from: *Remote Sensing* **2019**, *11*, 231, doi:10.3390/rs11030231 37

Wei Zhang, Ping Tang and Lijun Zhao
Remote Sensing Image Scene Classification Using CNN-CapsNet
Reprinted from: *Remote Sensing* **2019**, *11*, 494, doi:10.3390/rs11050494 58

Melanie K. Vanderhoof and Clifton Burt
Applying High-Resolution Imagery to Evaluate Restoration-Induced Changes in Stream
Condition, Missouri River Headwaters Basin, Montana
Reprinted from: *Remote Sensing* **2018**, *10*, 913, doi:10.3390/rs10060913 80

Javier Marcello, Francisco Eugenio, Javier Martín and Ferran Marqués
Seabed Mapping in Coastal Shallow Waters Using High Resolution Multispectral and
Hyperspectral Imagery
Reprinted from: *Remote Sensing* **2018**, *10*, 1208, doi:10.3390/rs10081208 108

Wei Wu, Qiangzi Li, Yuan Zhang, Xin Du and Hongyan Wang
Two-Step Urban Water Index (TSUWI): A New Technique for High-Resolution Mapping of
Urban Surface Water
Reprinted from: *Remote Sensing* **2018**, *10*, 1704, doi:10.3390/rs10111704 129

George Marmorino and Wei Chen
Use of WorldView-2 Along-Track Stereo Imagery to Probe a Baltic Sea Algal Spiral
Reprinted from: *Remote Sensing* **2019**, *11*, 865, doi:10.3390/rs11070865 150

Livia Piermattei, Mauro Marty, Wilfried Karel, Camillo Ressl, Markus Hollaus,
Christian Ginzler and Norbert Pfeifer
Impact of the Acquisition Geometry of Very High-Resolution Pléiades Imagery on the Accuracy
of Canopy Height Models over Forested Alpine Regions
Reprinted from: *Remote Sensing* **2018**, *10*, 1542, doi:10.3390/rs10101542 159

Donato Amitrano, Raffaella Guida, Domenico Dell'Aglio, Gerardo Di Martino,
Diego Di Martire, Antonio Iodice, Mario Costantini, Fabio Malvarosa and Federico Minati
Long-Term Satellite Monitoring of the Slumgullion Landslide Using Space-Borne Synthetic
Aperture Radar Sub-Pixel Offset Tracking
Reprinted from: *Remote Sensing* **2019**, *11*, 369, doi:10.3390/rs11030369 181

Angel Garcia-Pedrero, Consuelo Gonzalo-Martín, Mario Lillo-Saavedra and Dionisio Rodriguez-Esparragon
The Outlining of Agricultural Plots Based on Spatiotemporal Consensus Segmentation
Reprinted from: *Remote Sensing* **2018**, *10*, 1991, doi:10.3390/rs10121991 . . . . **194**

Yongfa You, Siyuan Wang, Yuanxu Ma, Guangsheng Chen, Bin Wang, Ming Shen and Weihua Liu
Building Detection from VHR Remote Sensing Imagery Based on the Morphological Building Index
Reprinted from: *Remote Sensing* **2018**, *10*, 1287, doi:10.3390/rs10081287 . . . . **207**

Lipeng Gao, Wenzhong Shi, Zelang Miao and Zhiyong Lv
Method Based on Edge Constraint and Fast Marching for Road Centerline Extraction from Very High-Resolution Remote Sensing Images
Reprinted from: *Remote Sensing* **2018**, *10*, 900, doi:10.3390/rs10060900 . . . . **229**

About the Special Issue Editors

Francisco Eugenio received his B.S., M.S., and Ph.D. degrees in electrical engineering from the Universidad de Las Palmas de Gran Canaria (ULPGC), Las Palmas de Gran Canaria, Spain, in 1986, 1993, and 2000, respectively. In June 1996, he joined the Department of Signal and Communications, ULPGC. From 1998 to December 2000, he was with the Technical University of Catalonia (UPC), Barcelona, Spain, working in image processing. Since 2017, he has been a Full Professor with ULPGC, where he served as the Dean of the Telecommunication School in 2004–2010 and is currently lecturing on the area of remote sensing and radar. His current research interests at the Institute of Oceanography and Global Change (IOCAG, ULPGC), focuses on new methodologies and algorithms for multispectral and hyperspectral high-resolution remote sensing processing for the monitoring of shallow-water environments and fusion of multisensor/multiresolution satellite image data. In these areas, he is the author or coauthor of many publications that have been published in journals, and he has also been a reviewer for more than 15 publications. He is a Guest Editor for the Special Issue in Remote Sensing: Very High Resolution (VHR) Satellite Imagery: Processing and Applications.

Javier Marcello received his M.S. degree in electrical engineering from the Technical University of Catalonia (UPC), Barcelona, Spain, in 1993 and the Ph.D. degree from the Universidad de Las Palmas de Gran Canaria (ULPGC), Las Palmas de Gran Canaria, Spain, in 2006. From 1992 to 2000, he was the Head Engineer at the Spanish Aerospace Defense Administration (Instituto Nacional de Técnica Aeroespacial), where he served different programs at the Canary Space Center (Cospas-Sarsat, MINISAT, Helios, and CREPAD). In January 1994, he joined the Department of Signals and Communications, ULPGC, where he has been an Associate Professor in the Telecommunication School, lecturing on the areas of satellite and radio communications since 2000. His research is carried out at the Institute of Oceanography and Global Change (IOCAG, ULPGC) and includes multisensor remote sensing image processing (image fusion, classification, segmentation, etc.) and the generation of coastal and land products. He has authored 30 papers in remote sensing journals with medium-high impact factors. Additionally, he has been a reviewer in more than 20 remote sensing publications. He is a member of the Editorial Board of Remote Sensing and has also served as Guest Editor for the Special Issue in Remote Sensing: Very High Resolution (VHR) Satellite Imagery: Processing and Applications. Since 2016, he is the vice-president of the IEEE Geoscience and Remote Sensing Spanish Chapter.

Preface to "Very High Resolution (VHR) Satellite Imagery"

Nowadays, optical sensors provide multispectral and panchromatic imagery at much finer spatial resolutions than in previous decades. Ikonos was the first commercial high-resolution satellite sensor. Launched on September 24, 1999, it broke the one meter mark. Since then, Quickbird, Geoeye, Pleiades, Kompsat, and many other very high resolution (VHR) satellites have been launched.

Another important milestone was the 2009 launch of WorldView-2, the first VHR satellite to provide eight spectral channels in the visible to near-infrared range. On the other hand, very high-resolution SAR finally became available in 2007 with the launch of the Italian Cosmo-Skymed and German TerraSAR-X, both providing X band imagery at a 1-m resolution. Following these innovations, the recent advances in sensor technology and algorithm development have enabled the use of VHR remote sensing to quantitatively study the biophysical and biogeochemical processes in coastal and inland waters. Apart from bodies of water, VHR can be fundamental for the monitoring of complex land ecosystems for biodiversity conservation or precision agriculture for the management of soils, crops and pests. In this context, recent very high resolution satellite technologies and image processing algorithms present the opportunity to develop quantitative techniques that have the potential to improve upon traditional techniques in terms of cost, mapping fidelity, and objectivity. Typical applications include multi-temporal classification, recognition and tracking of specific patterns, multisensor data fusion, analysis of land/marine ecosystem processes and environment monitoring, etc. This book aims to collect new developments, methodologies, and applications of very high resolution satellite data for remote sensing. The research works included in this book present the most recent advances on all aspects of VHR satellite remote sensing, including image preprocessing (super-resolution, atmospheric modeling, sunglint correction, feature extraction, etc.), data fusion and integration of multiresolution and multiplatform data, image segmentation and classification, change detection and multi-temporal analysis, vegetation monitoring in complex ecosystems, precision agriculture, urban mapping, shallow waters monitoring, etc.

Francisco Eugenio, Javier Marcello
Special Issue Editors

remote sensing

MDPI

Article

Deep Distillation Recursive Network for Remote Sensing Imagery Super-Resolution

Kui Jiang [1], Zhongyuan Wang [1,*], Peng Yi [1], Junjun Jiang [2], Jing Xiao [1] and Yuan Yao [3]

[1] National Engineering Research Center for Multimedia Software, School of Computer Science, Wuhan University, Wuhan 430072, China; 2017282110506@whu.edu.cn (K.J.); 2017202110008@whu.edu.cn (P.Y.); jing@whu.edu.cn (J.X.)

[2] School of Computer Science and Technology, Harbin Institute of Technology, Harbin 150001, China; junjun0595@163.com

[3] State Key Laboratory of Information Engineering in Surveying, Mapping and Remote Sensing, Wuhan University, Wuhan 430079, China; whyaoyuan@163.com

* Correspondence: wzy_hope@163.com; Tel.: +86-136-2865-2051

Received: 22 September 2018; Accepted: 24 October 2018; Published: 29 October 2018

Abstract: Deep convolutional neural networks (CNNs) have been widely used and achieved state-of-the-art performance in many image or video processing and analysis tasks In particular, for image super-resolution (SR) processing, previous CNN-based methods have led to significant improvements, when compared with shallow learning-based methods. However, previous CNN-based algorithms with simple direct or skip connections are of poor performance when applied to remote sensing satellite images SR. In this study, a simple but effective CNN framework, namely deep distillation recursive network (DDRN), is presented for video satellite image SR. DDRN includes a group of ultra-dense residual blocks (UDB), a multi-scale purification unit (MSPU), and a reconstruction module. In particular, through the addition of rich interactive links in and between multiple-path units in each UDB, features extracted from multiple parallel convolution layers can be shared effectively. Compared with classical dense-connection-based models, DDRN possesses the following main properties. (1) DDRN contains more linking nodes with the same convolution layers. (2) A distillation and compensation mechanism, which performs feature distillation and compensation in different stages of the network, is also constructed. In particular, the high-frequency components lost during information propagation can be compensated in MSPU. (3) The final SR image can benefit from the feature maps extracted from UDB and the compensated components obtained from MSPU. Experiments on Kaggle Open Source Dataset and Jilin-1 video satellite images illustrate that DDRN outperforms the conventional CNN-based baselines and some state-of-the-art feature extraction approaches.

Keywords: remote sensing imagery; super-resolution; ultra-dense connection; feature distillation; video satellite; compensation unit

---

1. Introduction

In recent years, remote sensing imaging technology is developing rapidly and provides extensive applications, such as object matching and detection [1–4], land cover classification [5,6], assessment of urban economic levels, resource exploration [7], etc. [8,9]. In these applications, high-quality or high-resolution (HR) imageries are usually desired for remote sensing image analysis and processing procedure. The most technologically advanced satellites are able to discern spatial within a squared meter on the Earth surface. However, due to the high cost of launch and maintenance, the spatial resolution of these satellite imageries in ordinary civilian applications is often low-resolution (LR). Therefore, it is very useful to construct HR remote sensing images from existing LR observed images [10].

Compared with the general images, the quality of satellite imageries can be subject to additional factors, such as ultra-distanced imaging, atmospheric disturbance, as well as relative motion. All these factors can impair the spatial resolution or clarity of the satellite images, but video satellite imageries are more severely affected due to the over-compression. More specifically, for the video satellite, since it captures continuous dynamic video, in order to improve the temporal resolution, the optical imaging system has to sacrifice spatial resolution. At present, the original data volume of the video satellite has reached to the Gb/s level, but the channel transmission capacity of the spaceborne communication system is only in Mb/s level. To adapt to the transmission capacity of the satellite channel, the video acquisition system has to increase the compression ratio or reduce the spatial sampling resolution. For example, taking the video imagery taken by "Jilin No. 1" launched in China in 2015 as an example, although its frame rate reaches 25 fps, the resolution is only in 2048×960 pixels (equivalent to 1080P), and hence the imagery looks very blurred. Therefore, the loss of high-frequency details caused by excessive compression is a special concern for video satellite imagery SR.

To address the above mentioned problems, a series of SR techniques for the restoration of HR remote sensing images have been proposed [10–14]. For example, Merino et al. proposed the super-resolution with variable-pixel linear reconstruction algorithm, named SRVPLR [15], which recombines a set of LR images in a linear nonuniform optimum manner. In [16], a hidden Markov tree model is proposed to establish a prior model in the wavelet domain to regularize the ill-conditioned problem for remote sensing image SR restoration. To fully use prior knowledge from a given LR image, Gou et al. [17] presented a non-local pairwise dictionary learning (NPDL) based model. In this model, the photometric, geometric, and feature information of the given LR image can be considered to improve the quality of reconstruction.

However, these shallow learning-based frameworks, show poor reconstruction performance when a high object resolution is required in practical applications. Recently, given the strength of deep CNNs, many CNN-based methods have evolved to deal with complex tasks in various applications [18–20], such as medical imaging, satellite imaging and video surveillance [21,22]. In particular, these effective architectures have achieved very good performance in general image SR reconstruction. For example, Dong et al. [23] introduced a three-layer CNN into single image SR (SISR) and achieved considerable improvement. Then, Kim et al. [24] proposed a residual network, called VDSR by using adaptive gradient clipping and skip connection to alleviate training difficulty. More recently, Sheng et al. [25] proposed the deep laplacian pyramid super-resolution network (LapSRN) to reconstruct the sub-band residuals of HR images at multiple pyramid levels. In LapSRN, a weight-sharing mechanism is implemented in the same structure, thus considerably reducing large quantity of parameters. However, the incremental depth in a deep CNN framework causes loss of information, thus weakening the continuity of information propagation. Moreover, these conventional CNN-based or residual-learning-based structures fail to restore fine texture details with simply direct or skip connections under complex imaging conditions. In particular, remote sensing satellite imageries have a complicated degradation process, low ground object resolution, and weak textures, thus posing considerable challenges for SR reconstruction.

Recently, Huang et al. [26] introduced the dense convolutional network (DenseNet) to strengthen feature propagation and encourage feature reuse by connecting each layer to every other layer in a feed-forward manner. Furthermore, in [27], the feature maps of each layer are propagated into all subsequent layers, thus providing an effective method of combining the low- and high-level features to boost reconstruction performance. Tai et al. [28] proposed memory blocks to build MemNet by heavily using long-term dense connections in MemNet to recover more high-frequency information. Although these methods can enforce information propagation by increasing nodes between layers with skip or dense connections, the features are fused in the network with a concatenated manner and will lead to large computational burden and high memory consumption.

Following the idea of sharing weights among recursive nodes, recursive learning networks have been recently used to reduce redundancy parameters of the network. For example, Kim et al. [29]

presented to use more layers to increase the receptive field of the network. It proposes a very deep recursive layer to avoid excessive parameters. In addition, a skip-connection manner is used to mitigate the training difficulty. Tai et al. [30] proposed a deep recursive residual network to address the problems of model parameters and accuracy, which recursively learns the residual unit in a multi-path model. More recently, Yang et al. [31] used the LR image and its edge map to infer sharp edge details of an HR image during the recurrent recovery process. However, the simple-connection manner used in these models [29,30] extremely limits the SR reconstruction performance.

In this study, a novel ultra-dense-connection manner is proposed to improve the reconstruction performance along with recursive strategy to mitigate memory consumption. Compared with the conventional skip- and dense-connection-based networks [24,26], the proposed UDB contains approximately twice as many short and long paths as the conventional dense block given the same convolution layers. Therefore, this will greatly enhance the representational power of the network. In addition, parameters sharing strategy between UDBs can extremely release the memory burden. We also find ferture distillation in different stages leads to better accuracy for deep SR networks. Thus, we distill the feature maps by partly choosing output (with a special ratio) in different stages yet retain its integrity. After getting feature maps in different UDBs, we aggregate these components for gaining more abundant and efficient information in a multi-scale purification unit.

The strategy of feature distillation and compensation is obviously different from the knowledge distillation in these studies [32,33]. They compacted deep networks by letting a small simple network learn from a large complex network. In [34], the authors distilled a multi-model complex network by retaining the necessary network knowledge while keeping close performance. In [35], Pintea et al. showed substantially reduced parameters by recasting multiple residual layers in the large network into a single recurrent simple layer. However, our proposed distillation and compensation strategy is mainly used to compensate for the high-frequency details lost during information propagation rather than model compression.

In summary, the main contributions of this work are as follows:

1. We propose a novel deep distillation recursive network DDRN for remote sensing satellite image SR reconstruction in a convenient and effective end-to-end training manner.
2. We propose a novel multiple-path residual block UDB, which provides additional possibilities for feature extraction through ultra-dense connections, quite agreeing with the uneven complexity of image content.
3. We construct a distillation and compensation mechanism to compensate for the high-frequency details lost during information propagation through the network with a special distillation ratio.

The remainder of this paper is organized as follows. In Section 2, we introduce previous works on CNN-based SR reconstruction algorithms, particularly network structures for feature extraction. Section 3 particularly presents the framework of the proposed DDRN. Section 4 individually presents the design of each key module under the proposed DDRN framework in details, including UDB, MSPU, resolution lifting, and loss function. Experimental results are given in Section 5, and the conclusions of this study are given in Section 6.

2. Related Work

We briefly review previously related works on structure-efficient networks [25,29,36–38], from which our network draws inspiration. These previous deep networks are committed to learning fine detail textures by designing a sophisticated structure. In this section, we focus on recent skip- and dense-connection-based methods.

Skip connection: A skip connection that directly connects input to output through an identity map, as shown in Figure 1b, was pioneered for SISR by Kim et al. [24]. They proposed a 20-layer CNN model known as VDSR. Instead of learning the actual pixel values, VDSR harnesses the global residual learning paradigm to predict the differences between ground truth and bicubic interpolated image.

This learning strategy makes the feature maps very sparse, enabling easy training and convergence. Compared with the traditional methods [39–42], this learning strategy on the benchmark datasets shows a significant superiority on reconstruction performance in terms of visual and quantitative indicators. In addition, DRCN [29] constructes a recursive-supervision structure to alleviate the difficulty in training a deep residual network further. Recently, Sheng et al. [25] proposed a deep Laplacian pyramid super-resolution network (LapSRN) to reconstruct the sub-band residuals of HR images at multiple pyramid levels with skip connection.

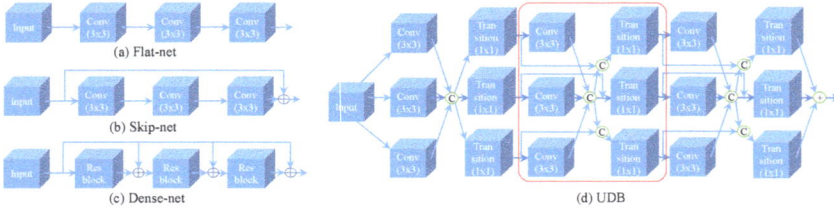

Figure 1. Frameworks of the CNN-based modules. (**a**) Flat-net (e.g., SRCNN [23] and FSRCNN [43]): Direct connections are commonly used to learn the features. (**b**) Skip-net (e.g., VDSR [24]) : An identity map with connecting input to the output is pioneered for SISR. (**c**) Dense-net (e.g., DenseNet [26] and SRDenseNet [27]): The feature maps are directly passed from the preceding layers to the current layers through the identity function with much richer connections. (**d**) UDB: Interacted multiple-path units are embedded for extracting local feature maps with a richer ultra-dense connections. "C" and " + " denote the concatenation and adding operation, respectively.

Dense connection: Enlightened by previous works, Huang et al. [26] recently represented an intensive skip connection called dense connection. As shown in Figure 1c, the feature maps of the current layer are connected to every subsequent layer in a feed-forward manner. With rich local dense connections, the current layer can aggregate the information from all of the preceding layers within the dense block for further selection and fusion. These strategies effectively address the vanishing-gradient problem and enhance information propagation, thus strengthening the feature expression and boosting the convergence. Subsequently, Tong et al. [27] proposed an enhancement version called SRDenseNet. In SRDenseNet, the feature maps obtained from each dense block are propagated into the deconvolution layers to reconstruct SR images, providing an effective way to combine the low-level and high-level features, which further boosts the reconstruction performance. In addition, the dense skip connections in the network enable short paths to be built directly linking to the output from each layer, thus mitigating the vanishing-gradient problem. While considering the research on feature extraction and fusion, the earlier work of Gao et al. [38] is also noteworthy. They proposed a technique called multi-scale dense network for resource-efficient image classification. Their main idea is to train multiple classifiers in different stages using a two-dimensional multi-scale architecture, enabling them to preserve the coarse-and-fine level features all throughout the network.

Ultra-dense connection: These above mentioned strategies have been proven effective in addressing vanishing-gradient problem, guaranteeing accurate feature extraction and fusion. However, the directly concatenated operation on all layers in previous works [27,38] have led to high memory consumption and computation burden. In addition, conventional dense-connection-based networks have to construct a deeper network the more the skip paths required. Moreover, the increasing computational burden and memory consumption are unacceptable.

As shown in Figure 1d, on the basis of the dense network [26], we propose a multiple-path residual block called UDB. Compared with conventional skip or dense networks [24,26,27,29], UDB contains richer short and long paths with the same convolution layers. In particular, given the multiple-path units and transition layer, the feature channels becomes shallower, extremely reducing the parameters and decreasing the computational burden and memory consumption.

3. Network Architecture

As shown in Figure 2, the proposed model is a deep recursive neural network that can be roughly partitioned into three substructures, namely, local feature extraction and fusion, feature distillation, and feature compensation and SR reconstruction. Except for the upsampling operation, motivated by previous works on SISR [24,25,27,43], the entire process of local feature extraction and fusion is in the LR space. I_{LR} and I_{SR} are considered the LR input and HR output of the proposed DDRN, respectively. F_i and B_j refer to the output in the i_{th} layer and the j_{th} block, respectively. In this work, the LR RGB images are directly fed into the network and processed with the initial convolutional layers (two layers with 3×3 kernel) to extract features as follows:

$$F_1 = H(I_{LR}), \tag{1}$$

$$F_2 = H(F_1), \tag{2}$$

where $H(\cdot)$ denotes the convolution operation. F_1 and F_2 represent the shallow feature maps extracted through the initial convolutional layers, served as the input of the UDB. Moreover, the proposed residual block UDB is used as a basic module for local feature extraction in DDRN. For each UDB, the information cannot only be shared among layers and multiple-path units but also be used as the input for the subsequent residual blocks with ultra-dense connections. These strategies enforce information propagation and lead to fine feature expression by combining the multi-scale coarse-and-fine features in different stages. The operation can be defined as follows:

$$B_i = H_{block,i}(B_{i-1}) + B_{i-1}, \tag{3}$$

where $H_{block,i}$ denotes the entire convolution operation in the i_{th} UDB and B_{i-1} refers to the extracted feature maps from the $(i-1)_{th}$ UDB. As shown in Figure 1, compared with the conventional CNN-based modules [24–26,29,30], whose commonly used residual block contains the simply direct or skip connections between layers, the proposed UDB module is composed of several interactive multiple-path units and parametric rectified linear units (PReLU). The dedicated architecture for UDB enjoys more linking paths in the same layers and provides more possibilities for feature extraction than do these previous strategies, thus matching the uneven content complexity of remote sensing imagery. Specifically, the simple links are adapted to smooth areas, whereas complex connections are suited for high-frequency texture details.

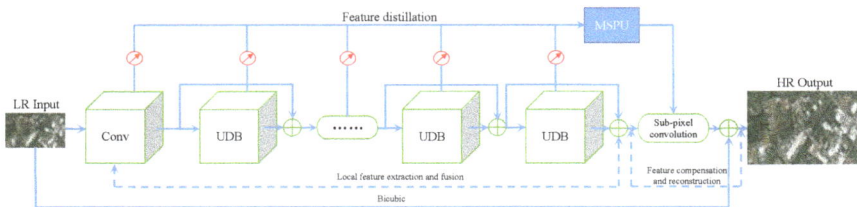

Figure 2. Outline of the proposed deep distillation recursive network (DDRN). The red distillation symbol followed the UDB represents the distillation operation with a special distilled ratio of α.

According to previous SISR algorithms [24,27,29,30], the output of the current stage is directly transmitted to the next stage. Then the final residual maps are obtained at the top layer for SR reconstruction. However, information loss is inevitable during its propagation in the network, thereby weakening the continuity of information propagation. Previous works add a set of nodes to shorten the transmission distance, thus boosting information propagation and reducing information loss during propagation, so-called skip connections [24,29]. However, increasing the nodes between the input and the output cannot only deepen the network but also increase computational burden

5

and memory consumption. Differently, we facilitate information propagation with the multiple-path residual module UDB. Furthermore, we also present a distillation and compensation strategy for fine feature expression by compensating for extra-high-frequency details. As shown in Figure 3, unlike the traditional network, whose output in each block is directly transmitted to the subsequent part, our proposed method can adaptively distill and preserve the feature maps by partly choosing information from the current output yet retain its integrety. Then, these feature maps collected from different stages are aggregated and purified in MSPU to infer and compensate for the high-frequency details before the reconstruction operation.

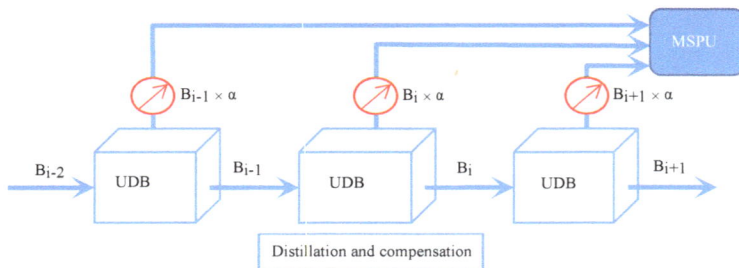

Figure 3. The distillation and compensation mechanism. The red components indicate that the distilled feature maps $B_i \times \alpha$ in current UDB are adaptively preserved. α denotes the distillation ratio for current UDB output B_i. MSPU refers to the further purification operation.

In this study, we denote the preserved part from B_i as the distillation unit (DU) with the ratio of α. At the same time, B_i is used as the input to the subsequent residual block for further extraction. This process can be formulated as follows:

$$DU_i = S(B_i, \alpha), \tag{4}$$

where α refers to the distillation ratio, which indicates that the feature maps in each stage with the ratio of α will be distilled and preserved. In our experiments, we set α to $\{0.0, 0.125, 0.25, 0.5\}$. $S(\cdot)$ represents the distillation operation, and DU_i denotes the distilled information from the i_{th} residual block B_i.

In addition, the reserved feature maps DU_i in different stages are aggregated through a concatenation operation, and then they are fed into the purified unit MSPU, where the HR components lost in the previous blocks are reactivated as a compensation for SR reconstruction. In Equation (5), $H_C(\cdot)$ denotes the concatenation operation adopted to collect the distillation information and $M(\cdot)$ refers to the MSPU. Through the distillation and compensation mechanism, the high-frequency components compensated from MSPU can further promote reconstruction performance.

$$P = M(H_C(DU_0, \cdots, DU_i, \cdots, DU_n)), \tag{5}$$

At the end of the network, the feature maps extracted from the top UDB and the compensated high-frequency details purified from MSPU are combined to infer and restore the HR components by a transition layer with 3×3 kernel. Then, a sub-pixel upsampling operation is used to project these features into HR space to obtain the residual image. The detailed operation is expressed as follows:

$$I_{SR} = PS(H_S(D_n, P)) + I_B, \tag{6}$$

where D_n and P represent the feature maps extracted from the top UDB and the compensated details from MSPU, respectively. H_S denotes a transition function that contains a 3×3 convolution layer to fuse features and infer HR components, adaptively. I_B refers to the bicubic interpolated image.

$PS(\cdot)$ represents the reconstruction operation performing a sub-pixel amplification to obtain the HR residual image in the ending part of the network.

4. Feature Extraction and Distillation

In this section, we present the design of each key module under our DDRN framework in details, including UDB, MSPU, and Resolution Lifting.

4.1. Ultra-Dense Residual Block (UDB)

It is acknowledged that rich dense connections can promote feature expression [26,27]. Therefore, we design a dense connection module for feature extraction. In this study, a multiple-path residual block UDB is constructed to enforce the correlation among layers and blocks with rich dense connections. Compared with existing skip- or dense-connection-based methods, UDB considers diverse short and long linking paths (the multiple-path structure) and exhibits effective information-sharing capability among the layers. Therefore, our network provides additional possibilities for feature extraction, quite agreeing with the uneven complexity of image content. More precisely, simple links are adapted to smooth areas, whereas complex connections are suited for high-frequency texture details. As shown in Figure 1d, UDB includes several interactive multiple-path units, which can fuse the feature maps extracted from parallel multiple convolution paths. The information-sharing mechanism aggregates features in different levels to ensure a rich feature representation further. The function of the i_{th} unit can be formulated as follows:

$$y_i = H_C([F_{i,0}(x_0), F_{i,1}(x_1), \cdots, F_{i,n}(x_n)]), \tag{7}$$

$$s_{i,n} = H_1(H_C(y_i, s_{i-1,n})). \tag{8}$$

Equations (7) and (8) formally show the operation process in a multiple-path unit. In Equation (7), $F_{i,n}(x_n)$ and $H_C([F_{i,0}(x_0), F_{i,1}(x_1), \cdots, F_{i,n}(x_n)])$ refer to the single convolution operation and the feature congregation of multiple convolution layers in each unit, respectively. In Equation (8), y_i denotes feature concatenation in the current unit. $s_{i,n}$ indicates the transition output in the n_{th} path of the i_{th} unit, and $s_{i-1,n}$ represents the output from the n_{th} path of the $(i-1)_{th}$ unit. Functionally, a group of skip connections is used to enforce the correlation among the input and output feature maps, where the transition layers represented as H_1 are embedded to reduce feature channels with 1×1 convolution kernel.

Unlike skip- or dense-connection-based algorithms [26–28], the proposed multiple-path ultra-dense connection block can simultaneously explore and infer local and global features. In particular, the feature maps in the multiple-path unit cannot only be shared among the layers in the current unit through aggregation and dense connections but also be used as the input of other units with skip connections. Given the simplicity, effectiveness, and robustness of this strategy, local features can be well expressed through numerous short and long paths. Furthermore, owing to the effective structure for feature extraction in UDB, the network can become shallow in the channels but wide for the convolution paths, which extremely reduces the parameters and simultaneously boosts the reconstruction performance.

4.2. Multi-Scale Purification Unit (MSPU)

In [44], the authors focused on channels and proposed a novel architectural unit termed "squeeze-and-excitation" (SE) block to recalibrate channel-wise feature responses adaptively by explicitly modeling the interdependencies between channels. The SE block can learn to use global information to emphasise informative features and suppress less useful features selectively. This model won the first place in the classification contest *ILSVRC2017* [45].

In this study, we adopt the SE module because of its promising efficiency and efficacy. On the basis of this finding, we propose an applicable module MSPU for information compensation. The basic structure of MSPU building unit is illustrated in Figure 4. Contrary to the squeeze-and-excitation network (SEN) [44], the redundant residual connections between SE blocks used for features transmission are removed. In addition, given that the full connection layer can destroy the internal structure of the image, we therefore replace it with a 1×1 convolution layer. Moreover, we adopt a robust activation function, e.g., parametric rectified linear unit (PReLU), to replace the previous version rectified linear unit (ReLU).

On the basis of MSPU process, we further propose a distillation and compensation strategy to compensate for lost details. By partially distilling the components from B_i with the distillation ratio of α, as shown in Figure 3, we can obtain feature maps originating from UDB in different stages. Then, these features are aggregated into MSPU to purify and gain more abundant and efficient information. The extraction functions can be defined as follows:

$$MS = H(x), \tag{9}$$

$$P = \sigma(H_1(A_P(MS))) \times MS. \tag{10}$$

In Equation (9), the input x denotes the concatenation of the distilled components in different satges, equivalent to $H_C(DU_0, \cdots, DU_i, \cdots, DU_n)$ in Equation (5), and $H(\cdot)$ represents a group of convolutional operations (with 3×3 kernel) that is adopted to fuse the features distilled from different levels. As expressed in Equation (10), A_P denotes the global average pooling, H_1 refers to the group of transition layers that comprises the bottleneck structure, and σ represents the sigmoid function.

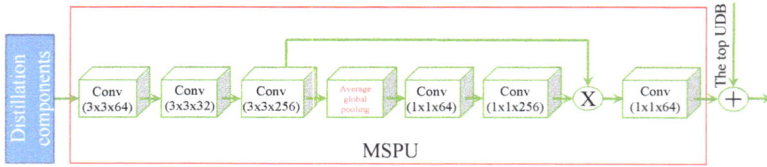

Figure 4. The Multi-scale feature purification unit (MSPU). The distillation components preserved from the different stages are fused to obtain compensation information lost during the information delivery. X denotes the matrix multiplication.

4.3. Resolution Lifting

To project a single LR image into HR space, the resolution of LR image must be increased to match that of the HR image at a certain point. Osendorfer et al. [46] presented a computationally efficient architecture for image SR by leveraging the fast approximate inference to increase the image resolution in the middle of the network gradually. Another well-known approach can also achieve spatial resolution enhancement by linear interpolation [23,24]. They obtained the same image resolution by directly using the common bicubic interpolation before loading the dataset into the network.

In addition, the early work of Shi et al. [47] is noteworthy when considering the upsampling operation. Contrary to authors of previous works, the researchers proposed an efficient sub-pixel convolution layer to increase the image resolution only at the final layer, eliminating the need to perform most of the SR operations in the large HR space. Compared with the transposed convolution and bicubic interpolation, sub-pixel magnification [47] is actually a realignment of feature maps without extra parameters, thus quite decreasing memory consumption and computational cost. These reasons enable the network go deeper and be trained easily.

As expressed in Equation (11), PS is a shuffling operator that rearranges the elements of a $H \times W \times C \cdot r^2$ tensor acquired in the top layer into a $rH \times rW \times C$ tensor (where r is the magnification factor of the network, and C refers to the feature channels of the input image). Mathematically, the upsampling function can be expressed as follows:

$$PS(T)_{x,y,c} = T_{\lfloor x/r \rfloor, \lfloor y/r \rfloor}(\mathrm{mod}\,(x,r), \mathrm{mod}\,(y,r)), \tag{11}$$

where T indicates the output from the final layer with the size of $W \times H \times Cr^2$, (x,y) denotes the output pixel coordinate in the HR space, $(x/r, y/r)$ represents the pixel area of $r \times r$ in the sub-pixel space, and $(\mathrm{mod}\,(x,r), \mathrm{mod}\,(y,r))$ refers to the pixel coordinate in LR space. The Cr^2 channels of each pixel in the same location in the LR space is rearranged into a region of $1r \times 1r \times C$, which corresponds to a subblock in an HR image, and the feature image is rearranged into an HR image of $rW \times rH \times C$.

In this work, as in many CNN-based SISR methods [25,47,48], we adopt the sub-pixel upsampling strategy to reconstruct the HR image at the top layer because of its promising efficiency and efficacy.

4.4. Loss Function

It is well known that SISR is an ill-posed problem whose solution from the reconstruction constraint is not unique because of the insufficient number of LR images, ill-conditioned registration, and unknown degradation process. In previous works, the loss function is commonly used to fit the real target image by minimizing the distance between the reconstructed HR image and the ground truth. The commonly used distance measurements include pixel-based l_1-norm [25] and l_2-norm [23,24,29], and cosine distance based on feature level.

Most of the previous works [23,27,29] constrain the reconstruction image by minimizing the mean squared error (MSE) or maximizing the peak signal to noise ratio (PSNR), which is a common measure used to evaluate SR algorithms [49]. However, the capability of MSE to capture perceptually relevant components, such as high-frequency texture details, is insufficient because they are defined on basis of pixel-wise image differences [50]. For example, the previous works [23,29,43] use MSE loss as the cost function and produce overly smooth reconstruction results that are inconsistent with human vision. In [25,51], the authors proposed a novel optimal function charbonnier loss based on the l_1-norm, which can recover a large amount of realistic details, more faithful to the ground truth. In our work, we therefore introduce the charbonnier penalty function to penalize the deviation of the prediction from the residuals of ground truth. The loss function can be expressed as follows:

$$Loss(I_{SR}, I_{HR}, \theta) = \arg\min_{\theta} \sum \rho(I_{HR} - f(I_{LR}, \theta)), \tag{12}$$

where θ denotes a set of model parameters to be optimized and $\rho(x) = \sqrt{x^2 + \varepsilon^2}$ represents the charbonnier penalty function (a differentiable variant of l_1-norm). We empirically set the compensation parameter ε of 10^{-3}. I_{SR} and I_{HR} refer to the predicted HR image and the ground truth.

5. Experimental Results and Analysis

In this section, first, we describe the experimental settings, including the data collection and model parameters. Then, we assess the effect of the distillation ratio α and the network depth m on the reconstruction performance. Subsequently, we compare our results with these state-of-the-art techniques and provide a thorough analysis. We retrain the comparison algorithms with our training dataset to ensure a fair comparison, including SRCNN [23] and VDSR [24]. Moreover, we directly apply the original models [23–25] trained with general image datasets, as the anchors.

5.1. Data Collection

For general image SR, a large quantity of public training and assessing datasets, such as DIV2K [52], BSD500 [53] and Yang291 [39], are used to evaluate the results. However, few available datasets can be used as the training samples for satellite imagery SR because of the special requirements of ground target resolution. We use two available satellite image datasets, namely, *Kaggle Open Source Dataset* and *Jilin-1* video satellite imagery, to train and evaluate the proposed DDRN method.

1. The first imagery dataset is the *Kaggle Open Source Dataset* (https://www.kaggle.com/c/draper-satellite-image-chronology/data), which contains more than 1000 HR images of aerial photographs captured in southern California. The photographs were taken from a plane and

meant as a reasonable facsimile for satellite images. The images are grouped into five sets, each of which having the same setId. Each scenario in a set contains five images captured on different days (not necessarily at the same time each day). The images for each set cover approximately the same area but are not exactly aligned. Images are named according to the convention (setId-day). In this dataset, the scene has 3099×2329 pixels and 324 different scenarios. A total of 1720 satellite images cover agriculture, airplane, buildings, golf course, forest, freeway, parking lot, tennis court, storage tanks, and harbor. In this study, 30 different categories are selected for the test and 10 for the evaluation. Meanwhile, a total of 350 images are used for the training. Regarding the training dataset, the entire images are cropped into many batches with 720×720 pixels, but only the central area of the testing images with size of 720×720 pixels is cropped for testing and evaluation.

2. The second satellite dataset is from *Jilin-1* video satellite imagery. In 2015, the Changchun Institute of Optics, Fine Mechanics, and Physics successfully launched the *Jilin-1* video satellite which had 1.12 m resolution. To cover the duration of video sequences, we select one for every five frames from each video and crop the central part with the size of 480×204 as test samples. We select several areas in different countries with certain typical surface coverage types, including vegetation, harbor, and a variety of buildings as the test images.

5.2. Model Parameters and Experiment Setup

In our experiments, we use an NVIDIA GTX1080Ti GPU and an Intel I7-8700K CPU for training and testing, respectively. Our model is implemented on TensorFlow with Python3 under Windows10, CUDA8.0, and CUDNN5.1 systems. We mainly focus on the up-scaling factor of 4, which is usually the most challenging case in image SR.

The original HR images are downsized by bicubic interpolation to generate LR images for training. We augment the training patches by horizontal or vertical flipping and rotating 90°. By following the settings presented in [54], we send one batch consisting of 16 LR RGB patches with the size of 32×32 from the training datasets to our network each time. The learning rate is initialized to 10^{-3} for all layers and halved for every 10^4 steps up to 10^{-5}. In our model, each convolution layer contains 64 filters, followed by PReLU. We empirically set the distillation ratio α to $\{0.0, 0.125, 0.25, 0.5\}$ and the number of parallel convolution layers n in each multiple-path unit to 3. For the basic module DDRN, the depth of UDB is 15. In our experiments, training a basic module consumes approximately 20 h under the previously presented experimental settings.

5.3. Quantitative Indicators (QI)

Similar to many previous representative works [23,24,28,29], we also select two commonly used evaluation metrics, i.e., PSNR and structural similarity (SSIM), to evaluate the model performance. These evaluation metrics differ in terms of visual perception but involve reference images for comparison. However, in real SR scenes, we have only LR images to be super-resolved, without the corresponding HR reference image. Therefore, we need to introduce quantitative non-reference image quality assessment methods. Quality with no reference (QNR) [55,56], generalized quality with no reference (GQNR) [57] and average gradient (AG) [58] are commonly used image quality evaluation algorithms without reference, which can reasonably assess the clarity of reconstructed image. Nevertheless, QNR and GQNR are used for multispectral or hyperspectral images rather than ordinary RGB images, which needs to calculate the spectral distortion index and spatial distortion index. Thus, in this study, we propose to alternatively use AG for objective evaluation without reference. This process can be expressed as follows:

$$G(x,y) = \mathrm{d}x_{(i,j)} + \mathrm{d}y_{(i,j)}, \tag{13}$$

$$\mathrm{d}x_{(i,j)} = I_{(i+1,j)} - I_{(i,j)}, \tag{14}$$

$$dy_{(i,j)} = I_{(i,j+1)} - I_{(i,j)}, \tag{15}$$

where dx and dy refer to the horizontal and vertical gradients, respectively, and $I_{(i,j)}$ denotes the pixel value corresponding to the coordinate of (i, j).

The indicator of the AG can reasonably assess image clarity because it sensitively reflects content sharpness, detail contrast, and texture diversity. Generally, the larger the AG, the richer the details. Thus, the AG can be used to evaluate the reconstruction quality of satellite imagery in real-world scenes, such as *Jilin-1* video satellite imageries.

5.4. Validation of the Ultra-Dense Residual Block

We examine the effectiveness of the proposed deep recursive CNN network DDRN and the multiple-path UDB. Given that SRCNN [23] and VDSR [24] are the most representative and most effective deep-learning-based SR methods, in our experiments, we retrain these two models by using the same training datasets and label them as SRCNN* and VDSR*. Figure 5 shows the comparison results according to the iterations of DDRN, SRCNN, and VDSR. Comparatively, our DDRN exhibits faster convergence and higher scores than do direct-connection-based SRCNN and skip-connection-based VDSR. This superiority can be mainly attributed to the proposed multiple-path ultra-dense connections which can readily capture local features. Thus, our framework significantly boosts the SR efficacy of remote sensing imagery.

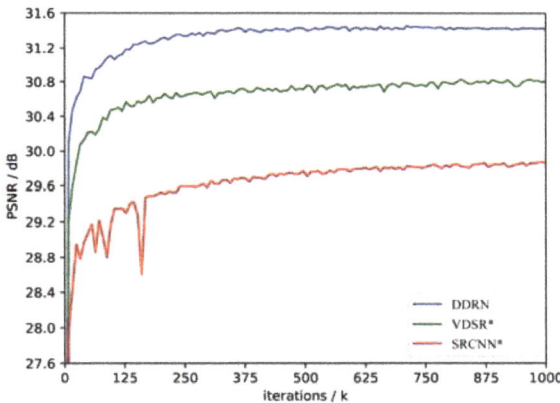

Figure 5. Training process for different models with the scale of 4. On the top, the blue line denotes the convergence process of the basic module DDRN with depth of 15 while the green and the red lines at the bottom refer to the VDSR and SRCNN. The competitive algorithms marked by * denote the retrained versions with our dataset.

In Figure 6, we show the evaluation results of the proposed DDRN method and the comparison algorithms on the *Kaggle Open Source Dataset* to verify the usefulness of the ultra-dense connections strategy further. The test set contains 30 different scenarios, which are labeled 1 to 30 in Figure 6. The figure shows that by using ultra-dense connections, we obtain better reconstruction results than do the conventional CNN-based methods, i.e., SRCNN [23] and VDSR [24]. For the average PSNR, our DDRN shows substantial improvements, surpassing VDSR by 0.92 dB, and SRCNN by 1.94 dB. Similarly, SSIM is also considerably improved.

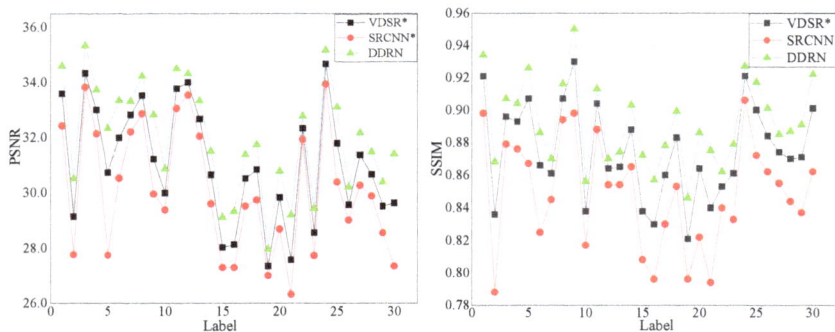

Figure 6. The SR performance comparisons for 30 different scenarios (denoted by label) from *Kaggle Open Source Dataset*. The competitive algorithms marked by * denote the retrained versions with our dataset.

In summary, the proposed residual block UDB effectively captures realistic detail textures. Although SRCNN and VDSR are effective, the well-designed deep recursive framework DDRN is more suitable for satellite image SR reconstruction.

5.5. Influence of Parameters α and m

On the basis of the basic module DDRN, we implement a distillation and compensation mechanism to compensate for the HR components lost during information propagation to infer and restore more realistic high-frequency details. The improved model with MSPU embedment is called DDRN⁻. In particular, a couple of comparison simulation experiments are conducted to analyze the influences of (i) the hyperparameter α in Equation (4) for partial feature maps distillation and preservation, (ii) the depth value m of UDB on the reconstruction performance.

We report the training process of the proposed DDRN⁺ with respect to different distillation ratios to verify the necessity of the proposed distillation and compensation mechanism. When α is set to 0, no components are distilled in the current stage, whereas MSPU does not function. Figure 7 shows the comparison results of the training process under different distillation ratios. From the figure, we learn that the proposed DDRN⁺ exhibits better training performance than the basic module DDRN. In addition, we observe that, with an increase in the distillation ratio α, the module exhibits robust and fast convergence. This result can be attributed to the increasing compensated high-frequency details from the MSPU by an increased distillation ratio. However, we also observe that the performance starts to decline when α is set to a large value, e.g., 0.5. This result can be mainly attributed to the large distillation rate, which may result in information redundancy. In addition, excessive parameters might lead to overfitting. All of these results indicate that the proposed distillation and compensation mechanism show substantial improvements by compensating for high-frequency details. Therefore, embedding MSPU into the basic module for satellite image SR reconstruction is an effective and reliable choice.

In light of the observations in these previous works [26–28], fine features can be well inferred from a deep CNN framework. Thus, we gradually increase the depth of the network by simply adding the number of the UDB (i.e., m is set to 10, 15, 20, 25, 30, and 35). We assess the performance of different values of m. In Figure 8, we show the training details of the proposed DDRN⁺ method with different depths. When simply increasing the value of m to 30, the improvement gradually increases and surpasses the basic module by approximately 0.22 dB in the scale of 4. By contrast, the performance declines when we continue to increase m to 35 and the network exhibits slow convergence. This result can be mainly attributed to the overfitting, and the convergence of the network becomes more difficult in such a depth.

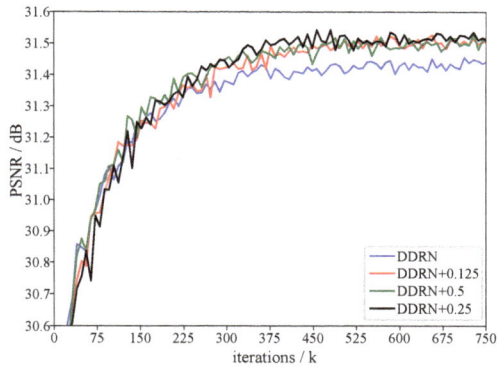

Figure 7. Training process for different distillation ratios by the scale of 4. DDRN$^+$ represents the improved module with MSPU embedded at different ratios on the basis of the basic module. In particular, DDRN denotes the improved module with the distillation ratio α of 0, which is actually the basic module.

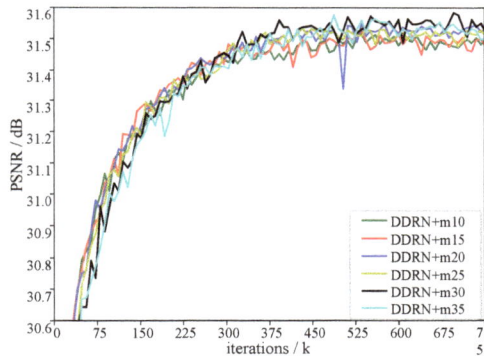

Figure 8. Training process for different depths of DDRN$^+$ with scale of 4 and the distillation ratio α of 0.25. We set UDB number m to 10, 15, 20, 25, 30 and 35 while keeping other parameters consistent.

On the basis of the experiments, we can obtain the optimal distillation ratio α and UDB depth m for satellite image SR reconstruction, which are set to 0.25 and 30, respectively.

5.6. Comparison Results with the State-of-the-Art

We compare our basic model DDRN and the improved version DDRN$^+$ ($\alpha = 0.25$, $m = 30$) with other SISR algorithms, including Bicubic, SRCNN [23], VDSR [24], and LapSRN [25], by the scaling factors of $\times 2$, $\times 3$, and $\times 4$. The implementations of these anchor methods have been released online and can thus be conducted on the same test datasets.

The reconstruction results obtained with above mentioned *Kaggle Open Source Dataset* for the proposed approaches and the comparison methods are shown in Figure 9. We select several different but representative scenarios (i.e., crossroads, factory, freeway, tennis court, and parking lot) to produce a visual presentation. Experimentally, we crop these representative scenarios into a sub-batch with the size of 120×120 pixels from each reconstructed SR image and compute PSNR and SSIM. Notably, the proposed method DDRN and its improved version DDRN$^+$ surpass these state-of-the-art methods by a large margin. Moreover, the modules that we propose exhibit the most accurate and realistic image details from the visual effect. Most of the comparison methods produce noticeable artifacts and blurred edges, whereas the proposed DDRN$^+$ can recover sharper and clearer edges because of successful feature extraction and fusion, more faithful to the ground truth. For example, as shown in Figure 9, only our proposed modules restore the clear court boundary in the tennis court scenario and the accurate and credible car outline in the four other scenarios. Therefore, all of the proposed models exhibit solid performance improvements compared with the conventional direct- or skip-connection-based algorithms [23–25].

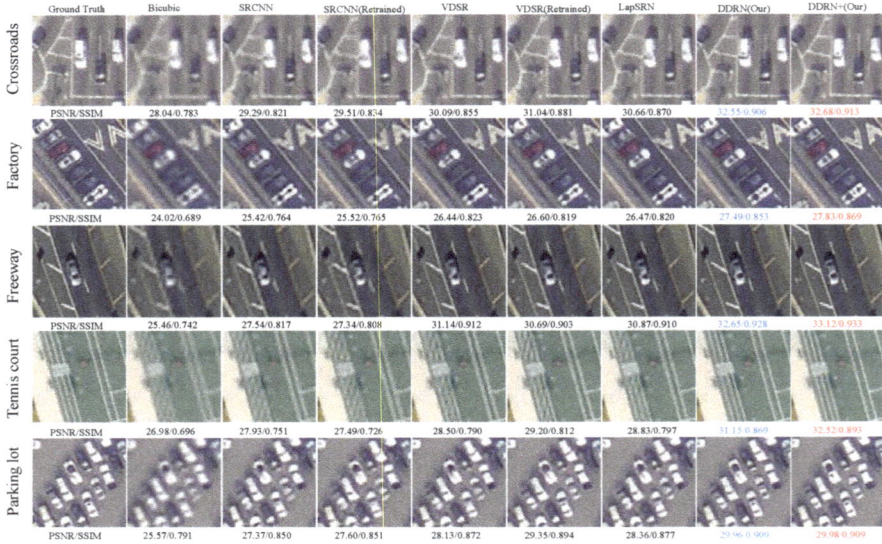

Figure 9. The reconstruction results on *Kaggle Open Source Dataset* and by the scale of 4. We select several different but representative scenarios, i.e., crossroads, factory, freeway, tennis court and parking lot, and then crop them into small image batches in size of 120 × 120 for demonstration. Red and blue indicate the best and the second best performance, respectively.

Objectively, Tables 1, 2 and 3 tabulate the detailed evaluating results in terms of PSNR, SSIM and AG with the magnification scales of ×2, ×3, and ×4, respectively. From these records, we learn that raw CNN-based or skip connection methods, such as SRCNN [23] and VDSR [24], exhibit lower scores than do DDRN-based methods (i.e., in terms of PSNR, the proposed DDRN$^+$ surpasses SRCNN and VDSR retrained by approximately 2.16 and 1.14 dB with the scale of 4 in the first test dataset, respectively.). Among these comparison methods, the basic module DDRN shows the best performances because of its ultra-dense-connection-based effective framework for local spatial information extraction. In addition, through the compensated high-frequency details obtained from the MSPU, the improved version DDRN$^+$ can produce fine detail textures. With regard to PSNR and SSIM, Figure 6 shows an more intuitive result that the proposed modules outperform these state-of-the-art methods [23–25] by a large margin. For the metric AG, the proposed DDRN and DDRN$^+$ are also better than previous works on average. In particular, in the comparison results shown in the three tables, our methods exhibit remarkable advantages when the upsampling factor is large, as reported at the bottom of the three tables. These results indicate the advantages of the proposed ultra-dense-connection manner in modeling the relationship between LR and HR images with lager magnification factors.

Table 1. Quantitative evaluation of the proposed DDRN approach and its improved version DDRN+ against some state-of-the-art SISR algorithms on *Kaggle Open Source Dataset* with 30 different scenarios for the scale factor of ×2. Bold indicates the best performance. Particularly, * refers to the modules retrained by us with *Kaggle Open Source Dataset*.

Labels	Methods Scale	Bicubic PSNR/SSIM/AG	SRCNN [23] PSNR/SSIM/AG	SRCNN* PSNR/SSIM/AG	VDSR [24] PSNR/SSIM/AG	VDSR* PSNR/SSIM/AG	LapSRN [25] PSNR/SSIM/AG	DDRN (Our) PSNR/SSIM/AG	DDRN+ (Our) PSNR/SSIM/AG
(1)	2	36.77/0.960/3.468	39.17/0.973/3.878	39.49/0.974/3.849	40.52/0.978/3.887	40.83/0.979/3.881	40.65/0.979/3.881	41.33/0.980/3.958	**41.38/0.981/3.958**
(2)	2	31.97/0.919/4.729	35.27/0.953/5.848	35.21/0.951/5.791	35.99/0.959/5.907	36.58/0.962/5.838	35.95/0.959/5.879	37.10/0.965/5.915	**37.23/0.965/5.920**
(3)	2	37.42/0.945/2.700	39.20/0.959/3.176	39.31/0.960/3.100	40.20/0.965/3.197	40.36/0.966/3.170	40.34/0.966/3.177	40.77/0.968/3.216	**40.82/0.968/3.220**
(4)	2	36.78/0.953/3.698	39.07/0.968/4.172	38.97/0.966/4.119	39.47/0.969/4.174	39.70/0.970/4.147	39.61/0.970/4.158	39.95/0.971/4.194	**40.05/0.971/4.201**
(5)	2	31.97/0.948/6.149	35.63/0.970/6.808	35.54/0.969/6.846	36.75/0.974/6.821	37.23/0.976/6.860	36.84/0.975/6.808	37.16/0.977/**6.921**	**37.66/0.978/**6.903
(6)	2	33.81/0.913/3.614	35.78/0.936/4.238	35.91/0.935/4.192	37.16/0.944/4.346	37.26/0.945/4.269	37.20/0.945/4.310	37.57/0.946/4.357	**37.73/0.947/4.373**
(7)	2	35.80/0.924/3.474	37.26/0.941/4.020	37.10/0.939/3.908	37.50/0.943/4.037	37.56/0.944/3.986	37.59/0.944/4.022	37.72/0.945/4.054	**37.79/0.945/4.061**
(8)	2	36.66/0.953/2.538	39.05/0.968/3.050	38.88/0.966/3.022	40.00/0.971/3.067	40.02/0.971/3.048	39.96/0.971/3.041	40.66/0.973/3.097	**40.74/0.973/3.104**
(9)	2	33.39/0.962/5.090	37.62/0.982/5.652	38.29/0.982/5.785	39.77/0.987/5.604	40.02/0.988/5.705	39.72/0.987/5.576	40.81/0.989/**5.748**	**41.10/0.990/**5.737
(10)	2	32.91/0.922/3.573	35.15/0.930/4.470	35.35/0.950/4.440	36.29/0.957/4.540	36.90/0.960/4.525	36.25/0.957/4.499	**37.96/0.964/4.622**	37.88/0.964/4.608
(11)	2	37.05/0.951/2.866	39.42/0.966/3.352	39.27/0.964/3.290	39.81/0.967/3.353	40.07/0.968/3.304	39.92/0.968/3.325	40.35/0.969/3.356	**40.38/0.969/3.360**
(12)	2	38.34/0.949/2.916	40.53/0.967/3.486	40.40/0.966/3.422	40.91/0.968/3.510	41.04/0.970/3.499	41.06/0.969/3.497	41.24/0.970/3.543	**41.31/0.971/3.548**
(13)	2	36.20/0.941/3.775	38.55/0.959/4.353	38.51/0.958/4.306	38.93/0.960/4.364	39.07/0.962/4.368	38.99/0.961/4.355	39.33/0.963/4.399	**39.36/0.963/4.405**
(14)	2	33.84/0.945/4.742	36.50/0.964/5.355	36.43/0.963/5.305	37.18/0.967/5.349	37.64/0.969/5.348	37.44/0.968/5.333	38.15/0.970/5.424	**38.17/0.970/5.427**
(15)	2	31.83/0.936/6.327	35.17/0.966/7.572	35.60/0.967/7.550	36.46/0.972/7.548	37.21/0.975/7.534	36.72/0.974/7.532	38.02/0.978/7.652	**38.08/0.978/7.660**
(16)	2	31.26/0.920/5.463	34.63/0.956/6.625	34.61/0.955/6.569	36.14/0.964/6.717	36.44/0.966/6.648	35.99/0.964/6.701	37.19/0.969/6.740	**37.35/0.969/6.747**
(17)	2	33.78/0.933/4.433	36.88/0.959/5.294	36.82/0.958/5.199	37.56/0.963/5.300	37.86/0.964/5.247	37.69/0.964/5.270	38.22/0.966/5.355	**38.35/0.966/5.362**
(18)	2	34.00/0.944/4.304	37.17/0.967/5.066	37.15/0.966/4.986	38.28/0.972/5.065	38.51/0.973/5.022	38.40/0.973/5.029	39.04/0.975/5.085	**39.21/0.975/5.085**
(19)	2	31.33/0.924/6.328	34.07/0.957/7.558	33.80/0.954/7.620	34.77/0.963/7.567	34.76/0.963/7.619	34.72/0.963/7.518	35.15/0.966/7.659	**35.32/0.967/7.664**
(20)	2	32.37/0.926/4.947	35.42/0.956/5.800	35.75/0.959/5.739	37.20/0.968/5.867	37.56/0.970/5.786	37.17/0.968/5.828	38.07/0.972/5.890	**38.19/0.972/5.897**
(21)	2	29.57/0.905/5.137	32.84/0.945/6.269	32.72/0.944/6.186	34.62/0.959/6.412	35.03/0.961/6.308	34.29/0.958/6.324	**36.10/0.967/6.434**	36.08/0.967/6.430
(22)	2	35.46/0.931/3.450	37.54/0.954/4.091	37.45/0.952/3.990	38.33/0.959/4.103	38.41/0.960/4.065	38.34/0.959/4.082	38.72/0.962/4.153	**38.80/0.962/4.157**
(23)	2	31.57/0.934/5.460	35.06/0.964/6.485	34.96/0.963/6.549	36.23/0.970/6.487	36.63/0.972/6.499	36.32/0.971/6.445	37.32/0.975/6.553	**37.47/0.975/6.560**
(24)	2	38.26/0.965/3.085	40.78/0.976/3.375	40.47/0.974/3.367	41.25/0.977/3.362	41.17/0.977/3.391	41.44/0.978/3.358	41.63/0.978/3.437	**41.69/0.979/3.437**
(25)	2	34.75/0.948/3.281	37.61/0.968/3.958	37.70/0.967/3.946	39.04/0.974/4.018	39.37/0.975/4.014	39.12/0.974/3.991	**40.31/0.978/4.097**	40.20/0.978/**4.098**
(26)	2	32.86/0.946/3.699	34.87/0.966/4.312	34.75/0.964/4.242	36.62/0.976/4.480	37.29/0.978/4.310	36.98/0.977/4.314	39.34/0.983/4.514	**39.86/0.984/4.535**
(27)	2	34.43/0.944/3.425	37.35/0.965/4.132	37.36/0.963/4.092	38.35/0.967/4.176	38.80/0.969/4.143	38.37/0.968/4.136	39.14/0.970/4.188	**39.21/0.970/4.193**
(28)	2	33.36/0.930/4.591	36.51/0.959/5.420	36.25/0.957/5.359	37.56/0.965/5.411	37.73/0.967/5.364	37.57/0.966/5.367	38.13/0.968/5.417	**38.19/0.969/5.422**
(29)	2	32.19/0.929/4.881	35.30/0.959/5.782	35.53/0.959/5.813	36.52/0.967/5.819	37.07/0.969/5.804	36.59/0.967/5.778	37.61/0.971/5.881	**37.70/0.972/5.886**
(30)	2	31.26/0.941/5.749	34.69/0.964/6.484	34.43/0.963/6.527	35.69/0.969/6.513	36.54/0.971/6.559	35.76/0.970/6.486	37.10/**0.974/6.605**	**37.23/**0.974/6.604
Avg	2	34.04/0.938/4.263	36.80/0.961/5.004	36.80/0.960/4.970	37.83/0.966/5.033	38.15/0.968/5.008	37.90/0.967/5.000	38.70/0.970/5.080	**38.81/0.970/5.085**

15

Table 2. Quantitative evaluation of the proposed DDRN approach and its improved version DDRN⁺ against some state-of-the-art SISR algorithms on *Kaggle Open Source Dataset* with 30 different scenarios for the scale factor of ×3. Bold indicates the best performance. Particularly, * refers to the modules retrained by us with *Kaggle Open Source Dataset*.

Labels	Methods Scale	Bicubic PSNR/SSIM/AG	SRCNN [23] PSNR/SSIM/AG	SRCNN * PSNR/SSIM/AG	VDSR [24] PSNR/SSIM/AG	VDSR * PSNR/SSIM/AG	DDRN (Our) PSNR/SSIM/AG	DDRN⁺ (Our) PSNR/SSIM/AG
(1)	3	33.06/0.915/3.021	34.58/0.935/3.579	35.07/0.940/3.586	36.22/0.953/3.616	36.65/0.955/3.599	37.63/0.961/3.653	**37.70/0.962/3.663**
(2)	3	28.25/0.821/3.717	30.10/0.871/4.911	30.12/0.870/4.941	31.68/0.903/5.274	31.86/0.903/5.047	32.97/0.919/5.333	**33.12/0.920/5.346**
(3)	3	34.30/0.897/2.256	35.38/0.913/2.796	35.76/0.918/2.737	36.56/0.930/2.812	36.57/0.930/2.798	37.51/0.938/**2.868**	**37.68/0.939**/2.864
(4)	3	32.88/0.901/3.197	34.68/0.926/3.825	34.89/0.925/3.780	35.58/0.934/3.796	35.71/0.935/3.766	36.40/0.940/3.881	**36.49/0.941/3.884**
(5)	3	27.86/0.885/5.309	30.35/0.921/6.439	30.87/0.923/6.403	32.11/0.940/6.345	33.31/0.945/6.531	**34.68/0.953/6.502**	34.43/0.953/6.452
(6)	3	31.17/0.852/3.031	32.44/0.880/3.686	32.38/0.879/3.728	33.72/0.901/3.803	34.03/0.902/3.731	35.13/0.912/3.843	**35.27/0.914/3.870**
(7)	3	32.92/0.870/2.979	34.18/0.893/3.548	34.15/0.891/3.474	34.83/0.902/3.546	34.67/0.900/3.519	35.11/0.905/3.589	**35.25/0.906/3.609**
(8)	3	33.32/0.907/2.070	34.90/0.930/2.593	34.93/0.929/2.566	35.89/0.939/2.611	35.74/0.938/2.573	36.48/0.943/2.668	**36.70/0.944/2.680**
(9)	3	29.16/0.897/4.537	32.80/0.945/5.666	32.79/0.943/5.643	34.18/0.964/5.508	34.90/0.966/5.432	36.07/0.974/5.563	**36.29/0.974/5.578**
(10)	3	29.89/0.843/2.766	31.04/0.878/3.594	31.04/0.877/3.579	31.79/0.894/3.743	32.21/0.897/3.710	**33.06/0.906/3.803**	33.05/0.907/3.793
(11)	3	33.43/0.903/2.384	35.52/0.930/2.971	35.44/0.928/2.961	36.16/0.937/2.988	36.23/0.937/2.927	36.87/0.942/3.050	**37.01/0.943/3.062**
(12)	3	34.62/0.888/2.311	35.95/0.913/2.991	35.97/0.912/2.937	36.47/0.919/3.007	36.50/0.919/2.972	36.86/0.923/3.070	**36.93/0.923/3.078**
(13)	3	32.70/0.881/3.189	34.14/0.908/3.837	34.22/0.908/3.804	35.17/0.916/3.871	35.35/0.917/3.860	35.81/0.920/3.927	**35.88/0.921/3.940**
(14)	3	30.20/0.888/4.099	32.03/0.919/4.961	32.24/0.919/4.888	33.07/0.931/4.898	33.50/0.933/4.900	34.37/0.940/4.947	**34.58/0.942/4.966**
(15)	3	27.84/0.844/5.081	29.70/0.893/7.098	30.40/0.907/6.934	31.27/0.924/7.059	31.69/0.927/7.010	32.46/0.942/**7.138**	**33.49/0.948**/7.104
(16)	3	27.70/0.822/4.472	29.21/0.872/5.548	29.29/0.872/5.605	30.99/0.902/5.901	31.20/0.904/5.744	32.33/0.916/5.947	**32.60/0.918/5.969**
(17)	3	29.95/0.854/3.684	31.87/0.896/4.683	32.18/0.897/4.620	33.26/0.916/4.722	33.22/0.915/4.644	33.98/0.924/4.793	**34.13/0.925/4.807**
(18)	3	30.15/0.875/3.615	32.14/0.913/4.622	32.36/0.913/4.540	33.61/0.933/4.625	33.66/0.931/4.550	34.52/0.940/4.699	**34.68/0.941/4.703**
(19)	3	27.82/0.829/5.063	29.49/0.886/6.717	29.46/0.884/6.934	30.36/0.907/6.705	30.07/0.901/6.727	30.79/0.915/6.951	**30.97/0.918/6.957**
(20)	3	28.97/0.842/4.186	30.99/0.891/5.177	31.20/0.895/5.132	32.50/0.921/5.324	32.68/0.923/5.276	33.62/0.934/5.392	**33.85/0.936/5.397**
(21)	3	26.45/0.808/4.169	28.29/0.865/5.253	28.30/0.863/5.204	29.87/0.900/5.555	30.04/0.901/5.449	31.57/0.920/5.623	**31.59/0.921/5.637**
(22)	3	32.43/0.866/2.898	33.84/0.895/3.522	33.77/0.893/3.453	34.45/0.905/3.524	34.38/0.904/3.488	34.81/0.909/3.572	**34.91/0.911/3.586**
(23)	3	27.88/0.852/4.529	30.12/0.900/5.799	30.11/0.898/5.763	31.24/0.919/5.785	31.40/0.920/5.731	32.41/0.932/5.829	**32.53/0.932/5.833**
(24)	3	34.58/0.927/2.702	36.80/0.948/3.215	36.72/0.947/3.185	37.82/0.956/3.164	37.66/0.955/3.191	38.46/0.960/3.199	**38.64/0.961/3.210**
(25)	3	31.05/0.891/2.640	32.75/0.921/3.440	33.19/0.923/3.448	34.36/0.938/3.549	34.60/0.938/3.467	35.75/0.947/3.575	**35.97/0.949/3.592**
(26)	3	29.70/0.884/3.075	30.66/0.912/3.843	30.90/0.913/3.724	31.01/0.923/3.967	31.55/0.928/3.747	32.31/0.940/3.991	**33.16/0.948/4.009**
(27)	3	30.89/0.880/2.747	32.44/0.909/3.499	32.70/0.909/3.452	33.80/0.922/3.584	34.07/0.923/3.538	34.83/0.929/3.623	**34.92/0.929/3.632**
(28)	3	30.14/0.862/3.958	32.37/0.905/4.850	32.09/0.900/4.838	33.41/0.923/4.877	33.21/0.920/4.801	34.02/0.930/4.731	**34.18/0.932/4.747**
(29)	3	28.67/0.851/4.072	30.74/0.899/5.104	30.92/0.900/5.130	32.09/0.924/5.207	32.33/0.926/5.162	33.23/0.936/5.277	**33.41/0.938/5.294**
(30)	3	27.62/0.876/4.913	29.80/0.916/5.971	29.86/0.915/5.927	31.03/0.933/5.969	31.91/0.937/6.018	**33.74/0.949/6.097**	33.49/0.948/6.065
Avg	3	30.52/0.870/3.555	32.31/0.906/4.457	32.44/0.906/4.430	33.48/0.923/4.511	33.69/0.924/4.463	34.59/0.933/4.571	**34.76/0.935/4.577**

Table 3. Comparison results of the proposed DDRN approach and its improved version DDRN$^+$ with some state-of-the-art algorithms on *Kaggle Open Source Dataset* for the scale factor of 4. Bold indicates the best performance. Particularly, * refers to the modules retrained by us with *Kaggle Open Source Dataset*.

Labels	Methods Scale	Bicubic PSNR/SSIM/AG	SRCNN [23] PSNR/SSIM/AG	SRCNN* PSNR/SSIM/AG	VDSR [24] PSNR/SSIM/AG	VDSR* PSNR/SSIM/AG	LapSRN [25] PSNR/SSIM/AG	DDRN (Our) PSNR/SSIM/AG	DDRN$^+$ (Our) PSNR/SSIM/AG
(1)	4	30.84/0.867/2.664	32.15/0.892/3.234	32.41/0.897/3.150	33.29/0.917/3.281	33.69/0.922/3.311	33.50/0.920/3.263	34.60/0.934/3.343	**34.74/0.936/3.354**
(2)	4	26.41/0.739/2.997	27.80/0.792/4.129	27.77/0.788/3.924	28.92/0.834/4.438	29.27/0.839/4.290	28.96/0.835/4.328	30.51/0.868/4.547	**30.67/0.872/4.680**
(3)	4	32.35/0.852/1.929	33.31/0.873/2.475	33.81/0.878/2.349	34.49/0.895/2.502	34.58/0.898/2.474	34.72/0.899/2.469	35.33/0.907/2.467	**35.67/0.910/2.551**
(4)	4	30.52/0.847/2.809	32.19/0.880/3.432	32.13/0.875/3.385	32.94/0.893/3.452	33.11/0.894/3.387	33.06/0.895/3.404	33.72/0.904/3.362	**33.95/0.906/3.365**
(5)	4	25.31/0.816/4.522	27.18/0.861/5.814	27.75/0.867/5.655	28.30/0.888/5.735	30.92/0.909/6.079	28.39/0.890/5.738	**32.32/0.926/6.258**	31.94/0.924/6.364
(6)	4	29.57/0.799/2.610	30.85/0.836/3.258	30.51/0.825/3.169	31.41/0.856/3.440	32.04/0.865/3.393	32.03/0.865/3.331	33.33/0.886/3.401	**33.55/0.888/3.485**
(7)	4	31.03/0.822/2.635	32.32/0.849/3.171	32.20/0.845/3.105	32.95/0.864/3.205	32.89/0.862/3.177	33.07/0.865/3.173	33.30/0.870/3.137	**33.45/0.873/3.276**
(8)	4	31.58/0.871/1.771	32.81/0.895/2.196	32.84/0.894/2.138	33.66/0.908/2.208	33.63/0.908/2.177	33.76/0.909/2.188	34.22/0.916/2.187	**34.45/0.918/2.189**
(9)	4	26.90/0.831/4.097	30.23/0.904/5.256	29.94/0.898/5.059	31.51/0.935/4.986	31.42/0.933/4.864	31.69/0.938/4.957	32.81/0.950/5.094	**33.18/0.954/5.228**
(10)	4	28.47/0.783/2.246	29.26/0.816/2.947	29.37/0.817/2.864	29.77/0.835/3.061	30.09/0.839/3.004	29.71/0.834/3.013	30.86/0.856/3.153	**30.88/0.856/3.182**
(11)	4	31.32/0.858/2.041	33.19/0.891/2.586	33.03/0.887/2.518	33.81/0.903/2.637	33.87/0.904/2.549	33.94/0.905/2.605	34.49/0.913**2.554**	**34.71/0.915**/2.552
(12)	4	32.49/0.830/1.867	33.53/0.856/2.479	33.53/0.854/2.360	33.99/0.866/2.529	34.05/0.865/2.554	34.01/0.867/2.481	34.31/0.870**2.408**	**34.52/0.873**/2.394
(13)	4	30.75/0.822/2.745	31.95/0.853/3.404	32.04/0.853/3.320	32.59/0.865/3.412	32.68/0.865/3.344	32.62/0.866/3.358	33.32/0.874**3.340**	**33.55/0.876**/3.333
(14)	4	27.94/0.830/3.570	29.52/0.868/4.509	29.60/0.865/4.340	30.29/0.884/4.413	30.72/0.889/4.437	30.51/0.888/4.414	31.49/0.903/4.555	**31.79/0.907/4.601**
(15)	4	25.70/0.744/4.120	27.11/0.805/6.071	27.29/0.808/5.721	28.16/0.842/6.114	28.27/0.845/6.140	28.33/0.850/6.032	29.11/0.872/6.271	**29.37/0.875/6.326**
(16)	4	25.98/0.738/3.809	27.29/0.797/4.809	27.29/0.796/4.712	28.00/0.827/4.964	28.15/0.830/4.890	28.02/0.828/4.894	29.33/0.857/5.089	**29.84/0.863/5.152**
(17)	4	27.91/0.784/3.139	29.49/0.832/4.069	29.51/0.930/3.904	30.56/0.862/4.130	30.60/0.862/3.997	30.63/0.864/4.101	31.37/0.878/4.083	**31.61/0.881/4.100**
(18)	4	28.10/0.810/3.091	29.65/0.855/4.082	29.72/0.853/3.910	30.81/0.884/4.117	30.93/0.885/4.172	30.90/0.886/4.072	31.73/0.899/4.083	**31.89/0.902/4.108**
(19)	4	25.79/0.734/4.064	27.01/0.802/5.738	27.00/0.796/5.497	27.55/0.827/5.677	27.48/0.823/5.552	27.57/0.829/5.684	27.96/0.846/5.910	**28.17/0.852/5.999**
(20)	4	27.06/0.766/3.612	28.50/0.818/4.575	28.69/0.821/4.411	29.57/0.854/4.643	29.90/0.864/4.606	29.64/0.858/4.621	30.77/0.886/4.753	**31.18/0.893/4.787**
(21)	4	24.87/0.733/3.487	26.18/0.792/4.517	26.32/0.794/4.429	27.12/0.831/4.737	27.70/0.841/4.798	27.07/0.832/4.684	**29.20/0.875/4.943**	28.96/0.873/5.042
(22)	4	30.72/0.811/2.541	31.96/0.844/0.050	31.91/0.840/2.941	32.37/0.855/3.066	32.42/0.854/3.003	32.41/0.855/3.025	32.77/0.862**2.945**	**32.87/0.864**/2.935
(23)	4	25.85/0.779/3.834	27.52/0.832/5.104	27.73/0.833/4.950	28.47/0.860/5.037	28.62/0.862/4.988	28.60/0.862/4.991	29.42/0.879/5.049	**29.67/0.882/5.120**
(24)	4	32.16/0.883/2.382	34.09/0.912/2.929	33.91/0.906/2.800	35.09/0.926/2.902	34.82/0.922/2.872	35.18/0.927/2.877	35.16/0.927/2.904	**35.86/0.934/2.909**
(25)	4	29.08/0.839/2.174	30.34/0.872/2.898	30.38/0.871/2.819	31.50/0.898/3.091	31.97/0.901/3.140	31.59/0.901/3.075	33.09/0.917/3.149	**33.39/0.919/3.173**
(26)	4	27.96/0.824/2.586	28.89/0.864/3.433	29.01/0.861/3.194	28.89/0.877/3.787	29.78/0.888/3.629	29.13/0.879/3.489	30.20/0.901/3.626	**30.82/0.910/3.629**
(27)	4	29.06/0.824/2.275	30.30/0.855/2.943	30.25/0.854/2.885	31.15/0.873/3.054	31.45/0.875/2.987	31.23/0.874/2.995	32.17/0.885/**3.012**	**32.35/0.887**/3.001
(28)	4	28.33/0.800/3.517	30.02/0.850/4.333	29.87/0.844/4.237	30.93/0.874/4.340	30.74/0.871/4.310	31.02/0.876/4.289	31.47/0.887/4.352	**31.72/0.891/4.357**
(29)	4	26.80/0.783/3.483	28.46/0.837/4.489	28.56/0.836/4.439	29.41/0.868/4.537	29.61/0.872/4.440	29.47/0.870/4.509	30.38/0.891/4.610	**30.65/0.895/4.709**
(30)	4	25.46/0.810/4.213	27.40/0.863/5.412	27.34/0.862/5.313	28.18/0.888/5.406	30.00/0.904/5.599	28.26/0.891/5.428	**31.39/0.922/5.756**	31.26/0.920/5.729
Avg	4	28.54/0.808/3.028	30.01/0.850/3.911	30.06/0.848/3.783	30.86/0.873/3.963	31.08/0.875/3.60	30.97/0.875/3.916	32.00/0.892/4.018	**32.22/0.895/4.064**

Another group of comparison experiments are conducted with the *Jilin-1* satellite imagery to illustrate the effectiveness and applicability of the proposed ultra-dense strategy and distillation and compensation mechanism further. Compared with the first dataset *Kaggle Open Source Dataset*, the test images obtained from *Jilin-1* show lower quality (small ground objects and weak textures) but more realistic satellite imagery characteristics. Unlike the images in the training dataset, the test images have completely different imaging conditions, including ultra-high imaging distance, atmospheric scattering, relative motion between satellite and moving ground targets, and compression distortion. These severe imaging conditions pose substantial demands to SR networks.

With an operation similar to the previously presented preprocessing of the testing images, we crop the test images with the size of 480×204. The reconstruction results obtained from our proposed approaches and the comparison methods are shown in Figure 10. For the first and second images, most of the comparison methods produce noticeable artifacts and blurred edges. By contrast, the proposed DDRN and DDRN$^+$ can recover sharp and clear edges because of fine feature expression that is faithful to the ground truth. At the bottom of the figure, only our proposed modules can reconstruct a clear outline of the warships and dock, whereas the other conventional methods fail to restore the realistic details. These results further indicate the effectiveness of the proposed method.

Furthermore, we perform a set of realistic SR reconstruction experiments for the unknown real degradation process (i.e., using the observed LR images instead of the downscaled LR images as input). These test images are randomly selected from *Jilin-1* satellite imagery using the same preprocessing to acquire the test images with the size of 480×204. Then, the processed images used as the LR input are directly transmitted to the network to obtain the reconstructed HR images. The comparison results with other state-of-the-art algorithms are shown in Figure 11 (we show only one example due to space constrains). Evidently, most of compared methods [23,24] produce noticeable artifacts and blurred building outlines, whereas the proposed DDRN and DDRN$^+$ yield better results with fewer jagged lines and ringing artifacts. Instead of the commonly used evaluation metrics PSNR and SSIM (because the original HR images are unavailable), we introduce the AG to measure the sharpness of the SR results. As shown in Figure 11, the proposed modules DDRN and DDRN$^+$ enjoy the second and first highest AG scores, respectively. The results for real video satellite imagery indicate that our model is more robust than the comparison methods in super-resolving the image with unknown degradation process.

In brief, the SR reconstruction experiments on different test datasets and magnification scales show the advantages of feature expression and indicate the robustness of our modules against images of unknown degradation models.

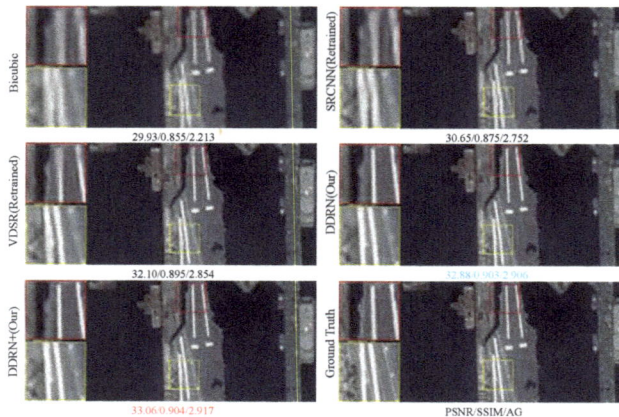

Figure 10. *Cont.*

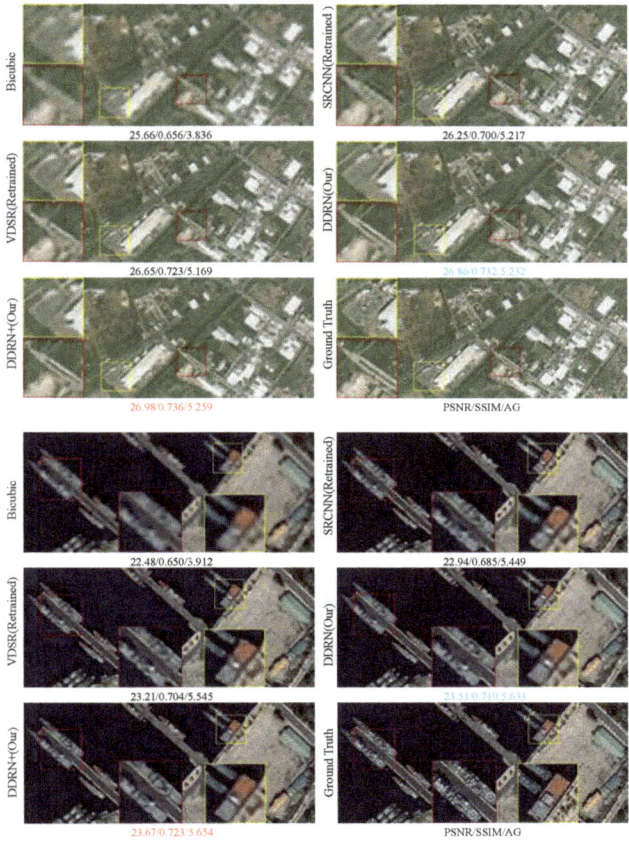

Figure 10. The reconstruction results on *Jilin-1* dataset with the scale of 4. We select several different but representative scenarios, i.e., aircraft carrier, city suburb, and military harbour to make comparisons. Red and blue indicate the best and the second best performance, respectively.

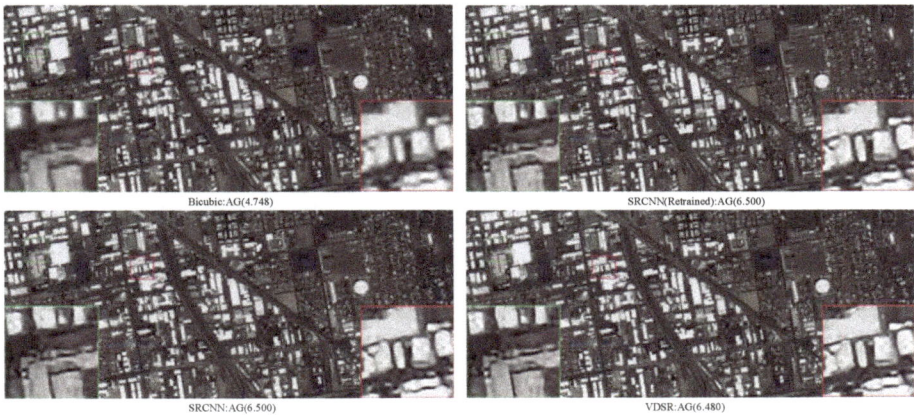

Figure 11. *Cont.*

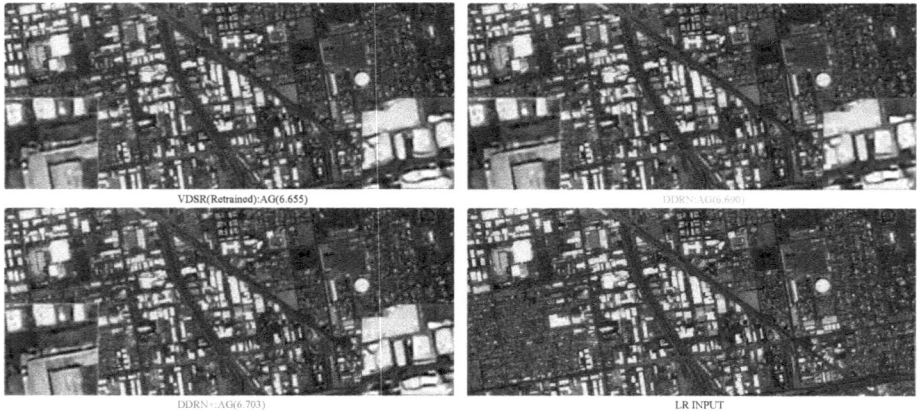

Figure 11. An example for the reconstruction results on *Jilin-1* imagery by the scale of 4. The experiment is performed with real low satellite images rather than simulation degradation. Red and blue respectively indicate the first and the second best performance in terms of AG. Note that the enlarged details are shown in the boxes on the bottom left and bottom right in each image.

6. Conclusions

In this study, we propose a simple but very effective technique for remote sensing image SR reconstruction. In particular, we present a multiple-path UDB for local feature extraction and fusion. Unlike in the conventional methods, rich dense connections between layers and units promote information interaction and improve reutilization. In addition, we further promote feature expression by advocating a distillation and compensation mechanism. The feature maps distilled from different stages with a special distillation ratio α are aggregated to compensate for the high-frequency details lost during information propagation in MSPU. Extensive experiments on the test datasets indicate that the proposed DDRN and its improved version DDRN$^+$ outperform existing state-of-the-art feature extraction techniques, including conventional direct- and skip-connection-based methods. In particular, when the image degradation model is unknown, the proposed algorithm can still obtain competitive reconstruction results compared with the comparison algorithms.

Author Contributions: Conceptualization, K.J. and Z.W.; Data curation, K.J. and P.Y.; Formal analysis, P.Y. and J.J.; Funding acquisition, Z.W.; Investigation, J.X.; Methodology, K.J. and Z.W.; Project administration, K.J.; Resources, Z.W.; Software, P.Y.; Supervision, Y.Y.; Validation, K.J., Z.W. and J.J.; Visualization, P.Y.; Draft Preparation, K.J.; Writing, review and editing, Z.W. and J.J.

Funding: This research was funded by [National Natural Science Foundation of China] grant number [61671332, 41771452, 41771454, 61501413], [National Key R&D Project] grant number [2016YFE0202300], and [Hubei Province Technological Innovation Major Project] grant number [2017AAA123]. (*Corresponding author: Zhongyuan Wang*).

Conflicts of Interest: The authors declare no conflict of interest.

Abbreviations

The following abbreviations are used in this manuscript:

CNNs	Convolutional neural networks
SR	Super-resolution
SISR	Single image super-resolution
LR	Low resolution
HR	High resolution
DDRN	Deep distillation recursive network

UDB   Ultra-dense residual block
MSPU   Multi-scale purification unit

References

1. Xu, Y.; Zhu, M.; Li, S.; Feng, H.; Ma, S.; Che, J. End-to-end airport detection in remote sensing images combining cascade region proposal networks and multi-threshold detection networks. *Remote Sens.* **2018**, *10*, 1516. [CrossRef]

2. Shao, Z.; Wu, W.; Wang, Z.; Du, W.; Li, C. SeaShips: A large-scale precisely annotated dataset for ship detection. *IEEE Trans. Multimed.* **2018**, *20*, 2593–2604. [CrossRef]

3. Ma, J.; Zhao, J.; Tian, J.; Yuille, A.L.; Tu, Z. Robust point matching via vector field consensus. *IEEE Trans. Image Process.* **2014**, *23*, 1706–1721.

4. Ma, J.; Jiang, J.; Zhou, H.; Zhao, J.; Guo, X. Guided locality preserving feature matching for remote sensing image registration. *IEEE Trans. Geosci. Remote Sens.* **2018**, *56*, 4435–4447. [CrossRef]

5. Jiang, J.; Ma, J.; Chen, C.; Wang, Z.; Cai, Z.; Wang, L. SuperPCA: A superpixelwise PCA approach for unsupervised feature extraction of hyperspectral imagery. *IEEE Trans. Geosci. Remote Sens.* **2018**, *56*, 4581–4593. [CrossRef]

6. Jiang, J.; Ma, J.; Wang, Z.; Chen, C.; Liu, X. Hyperspectral image classification in the presence of noisy labels. *IEEE Trans. Geosci. Remote Sens.* **2018**. [CrossRef]

7. Li, C.; Ma, Y.; Mei, X.; Liu, C.; Ma, J. Hyperspectral unmixing with robust collaborative sparse regression. *Remote Sens.* **2016**, *8*, 588. [CrossRef]

8. Ma, J.; Ma, Y.; Li, C. Infrared and visible image fusion methods and applications: A survey. *Inf. Fus.* **2019**, *45*, 153–178. [CrossRef]

9. Ma, J.; Yu, W.; Liang, P.; Li, C.; Jiang, J. FusionGAN: A generative adversarial network for infrared and visible image fusion. *Inf. Fus.* **2019**, *48*, 11–26. [CrossRef]

10. Dong, W.; Fu, F.; Shi, G.; Cao, X.; Wu, J.; Li, G.; Li, X. Hyperspectral image super-resolution via non-negative structured sparse representation. *IEEE Trans. Image Process.* **2016**, *25*, 2337–2352. [CrossRef] [PubMed]

11. Yang, S.; Sun, F.; Wang, M.; Liu, Z.; Jiao, L. Novel super resolution restoration of remote sensing images based on compressive sensing and example patches-aided dictionary learning. In Proceedings of the International Workshop on Multi-Platform/Multi-Sensor Remote Sensing and Mapping, Xiamen, China, 10–12 January 2011; pp. 1–6.

12. Jiang, K.; Wang, Z.; Yi, P.; Jiang, J. A progressively enhanced network for video satellite imagery superresolution. *IEEE Signal Process. Lett.* **2018**, *25*, 1630–1634. [CrossRef]

13. Luo, Y.; Zhou, L.; Wang, S.; Wang, Z. Video satellite imagery super resolution via convolutional neural networks. *IEEE Geosci. Remote Sens. Lett.* **2017**, *14*, 2398–2402. [CrossRef]

14. Xiao, A.; Wang, Z.; Wang, L.; Ren, Y. Super-resolution for "Jilin-1" satellite video imagery via a convolutional network. *Sensors* **2018**, *18*, 1194. [CrossRef] [PubMed]

15. Merino, M.T.; Nunez, J. Super-resolution of remotely sensed images with variable-pixel linear reconstruction. *IEEE Trans. Geosci. Remote Sens.* **2007**, *45*, 1446–1457. [CrossRef]

16. Li, F.; Jia, X.; Fraser, D. Universal HMT based super resolution for remote sensing images. In Proceedings of the 15th IEEE Conference on ICIP, San Diego, CA, USA, 12–15 October 2008; pp. 333–336.

17. Gou, S.; Liu, S.; Yang, S.; Jiao, L. Remote sensing image super-resolution reconstruction based on nonlocal pairwise dictionaries and double regularization. *IEEE J. Sel. Top. Appl. Earth Observ.* **2014**, *7*, 4784–4792. [CrossRef]

18. Li, Y.; Zhang, Y.; Huang, X.; Zhu, H.; Ma, J. Large-scale remote sensing image retrieval by deep hashing neural networks. *IEEE Trans. Geosci. Remote Sens.* **2018**, *56*, 950–965. [CrossRef]

19. Shao, Z.; Cai, J. Remote sensing image fusion with deep convolutional neural network. *IEEE J. Sel. Top. Appl. Earth Observ.* **2018**, *11*, 1656–1669. [CrossRef]

20. Zhou, W.; Newsam, S.; Li, C.; Shao, Z. Learning low dimensional convolutional neural networks for high-resolution remote sensing image retrieval. *Remote Sens.* **2017**, *9*, 489–508. [CrossRef]

21. Lu, T.; Ming, D.; Lin, X.; Hong, Z.; Bai, X.; Fang, J. Detecting building edges from high spatial resolution remote sensing imagery using richer convolution features network. *Remote Sens.* **2018**, *10*, 1496. [CrossRef]

22. Zhang, W.; Witharana, C.; Liljedahl, A.K.; Kanevskiy, M. Deep convolutional neural networks for automated characterization of arctic ice-wedge polygons in very high spatial resolution aerial imagery. *Remote Sens.* **2018**, *10*, 1487. [CrossRef]

23. Dong, C.; Loy, C.C.; He, K.; Tang, X. Image super-resolution using deep convolutional networks. *IEEE Trans. Pattern Anal. Mach. Intell.* **2016**, *38*, 295–307. [CrossRef] [PubMed]

24. Kim, J.; Lee, J.K.; Lee, K.M. Accurate image super-resolution using very deep convolutional networks. In Proceedings of the IEEE Conference on CVPR, Las Vegas, NV, USA, 27–30 June 2016; pp. 1646–1654.

25. Lai, W.S.; Huang, J.B.; Ahuja, N.; Yang, M.H. Deep laplacian pyramid networks for fast and accurate super-resolution. In Proceedings of the IEEE Conference on CVPR, Honolulu, HI, USA, 21–26 July 2017; pp. 5835–5843.

26. Huang, G.; Liu, Z.; van der Maaten, L.; Weinberger, K.Q. Densely connected convolutional networks. In Proceedings of the IEEE Conference on CVPR, Honolulu, HI, USA, 21–26 July 2017; pp. 2261–2269.

27. Tong, T.; Li, G.; Liu, X.; Gao, Q. Image super-resolution using dense skip connections. In Proceedings of the IEEE Conference on ICCV, Venice, Italy, 22–29 October 2017; pp. 4809–4817.

28. Tai, Y.; Yang, J.; Liu, X.; Xu, C. MemNet: A persistent memory network for image restoration. In Proceedings of the IEEE Conference on ICCV, Venice, Italy, 22–29 October 2017; pp. 4549–4557.

29. Kim, J.; Lee, J.K.; Lee, K.M. Deeply-recursive convolutional network for image super-resolution. In Proceedings of the IEEE Conference on CVPR, Las Vegas, NV, USA, 27–30 June 2016; pp. 1637–1645.

30. Tai, Y.; Yang, J.; Liu, X. Image super-resolution via deep recursive residual network. In Proceedings of the IEEE Conference on CVPR, Honolulu, HI, USA, 21–26 July 2017; pp. 2790–2798.

31. Yang, W.; Feng, J.; Yang, J.; Zhao, F.; Liu, J.; Guo, Z.; Yan, S. Deep edge guided recurrent residual learning for image super-resolution. *IEEE Trans. Image Process.* **2017**, *26*, 5895–5907. [CrossRef] [PubMed]

32. Yim, J.; Joo, D.; Bae, J.; Kim, J. A gift from knowledge distillation: Fast optimization, network minimization and transfer learning. In Proceedings of the IEEE Conference on CVPR, Honolulu, HI, USA, 21–26 July 2017; pp. 7130–7138.

33. Gupta, S.; Hoffman, J.; Malik, J. Cross modal distillation for supervision transfer. In Proceedings of the IEEE Conference on CVPR, Las Vegas, NV, USA, 27–30 June 2016; pp. 2827–2836.

34. Hinton, G.; Vinyals, O.; Dean, J. Distilling the knowledge in a neural network. *Comput. Sci.* **2015**, *14*, 38–39.

35. Pintea, S.L.; Liu, Y.; van Gemert, J.C. Recurrent knowledge distillation. In Proceedings of the 25th IEEE Conference on ICIP, Athens, Greece, 7–10 October 2018; pp. 3393–3397.

36. Zhao, P.; Liu, K.; Zou, H.; Zhen, X. Multi-stream convolutional neural network for SAR automatic target recognition. *Remote Sens.* **2018**, *10*, 1473. [CrossRef]

37. Gu, S.; Zuo, W.; Xie, Q.; Meng, D.; Feng, X.; Zhang, L. Convolutional sparse coding for image super-resolution. In Proceedings of the IEEE Conference on ICCV, Santiago, Chile, 7–13 December 2015; pp. 1823–1831.

38. Huang, G.; Chen, D.; Li, T.; Wu, F.; van der Maaten, L.; Weinberger, K.Q. Multi-scale dense networks for resource efficient image classification. *arXiv* **2018**, arXiv:1703.09844.

39. Yang, J.; Wright, J.; Huang, T.S.; Ma, Y. Image super-resolution via sparse representation. *IEEE Trans. Image Process.* **2010**, *19*, 2861–2873. [CrossRef] [PubMed]

40. Timofte, R.; De, V.; Gool, L.V. Anchored neighborhood regression for fast example-based super-resolution. In Proceedings of the IEEE Conference on ICCV, Sydney, Australia, 8 April 2013; pp. 1920–1927.

41. Yang, J.; Wright, J.; Huang, T.; Ma, Y. Image super-resolution as sparse representation of raw image patches. In Proceedings of the IEEE Conference on CVPR, Anchorage, Alaska, 23–28 June 2008; pp. 1–8.

42. Yang, C.Y.; Yang, M.H. Fast direct super-resolution by simple functions. In Proceedings of the IEEE Conference on ICCV, Sydney Australia, 8 April 2013; pp. 561–568.

43. Dong, C.; Chen, C.L.; Tang, X. Accelerating the super-resolution convolutional neural network. In Proceedings of the IEEE Conference on ECCV, Amsterdam, The Netherlands, 8–16 October 2016; pp. 391–407.

44. Hu, J.; Shen, L.; Sun, G. Squeeze-and-excitation networks. *arXiv* **2018**, arXiv:1709.0150.

45. Russakovsky, O.; Deng, J.; Su, H.; Krause, J.; Satheesh, S.; Ma, S.; Huang, Z.; Karpathy, A.; Khosla, A.; Bernstein, M.; et al. ImageNet large scale visual recognition challenge. *Int. J. Comput. Vis.* **2015**, *115*, 211–252. [CrossRef]

46. Osendorfer, C.; Soyer, H.; Smagt, P.V.D. Image super-resolution with fast approximate convolutional sparse coding. In Proceedings of the International Conference on Neural Information Processing, Kuching, Malaysia, 3–6 November 2014; pp. 250–257.

47. Shi, W.; Caballero, J.; Huszár, F.; Totz, J.; Aitken, A.P.; Bishop, R.; Rueckert, D.; Wang, Z. Real-time single image and video super-resolution using an efficient sub-pixel convolutional neural network. In Proceedings of the IEEE Conference on CVPR, Las Vegas, NV, USA, 27–30 June 2016; pp. 1874–1883.
48. Lim, B.; Son, S.; Kim, H.; Nah, S.; Lee, K.M. Enhanced deep residual networks for single image super-resolution. In Proceedings of the IEEE Conference on CVPRW, Honolulu, HI, USA, 21–26 July 2017; pp. 1132–1140.
49. Yang, C.Y.; Ma, C.; Yang, M.H. Single-image super-resolution: A nenchmark. *Lect. Notes Comput. Sci.* **2014**, *8692*, 372–386.
50. Wang, Z.; Bovik, A.C.; Sheikh, H.R.; Simoncelli, E.P. Image quality assessment: from error visibility to structural similarity. *IEEE Trans. Image Process.* **2004**, *13*, 600–612. [CrossRef] [PubMed]
51. Lai, W.; Huang, J.; Ahuja, N.; Yang, M. Fast and accurate image super-resolution with deep laplacian pyramid networks. *IEEE Trans. Pattern Anal. Mach. Intell.* **2018**, 1. [CrossRef] [PubMed]
52. Timofte, R.; Agustsson, E.; Gool, L.V.; Yang, M.H.; Zhang, L.; Limb, B.; Som, S.; Kim, H.; Nah, S.; Lee, K.M.; et al. NTIRE 2017 challenge on single image super-resolution: Methods and results. In Proceedings of the IEEE Conference on CVPRW, Honolulu, HI, USA, 21–26 July 2017; pp. 1110–1121.
53. Huang, J.B.; Singh, A.; Ahuja, N. Single image super-resolution from transformed self-exemplars. In Proceedings of the IEEE Conference on CVPR, Boston, MA, USA, 7–12 June 2015; pp. 5197–5206.
54. Tao, X.; Gao, H.; Liao, R.; Wang, J.; Jia, J. Detail-revealing deep video super-resolution. In Proceedings of the International Conference on ICCV, Venice, Italy, 22–29 October 2017; pp. 4482–4490.
55. Loncan, L.; de Almeida, L.B.; Bioucas-Dias, J.M.; Briottet, X.; Chanussot, J.; Dobigeon, N.; Fabre, S.; Liao, W.; Licciardi, G.A.; Simoes, M.; et al. Hyperspectral pansharpening: A review. *IEEE Geosc. Remote Sens. Mag.* **2015**, *3*, 27–46. [CrossRef]
56. Vivone, G.; Alparone, L.; Chanussot, J.; Mura, M.D.; Garzelli, A.; Licciardi, G.A.; Restaino, R.; Wald, L. A critical comparison among pansharpening algorithms. *IEEE Trans. Geosci. Remote Sens.* **2015**, *53*, 2565–2586. [CrossRef]
57. Kwan, C.; Budavari, B.; Bovik, A.C.; Marchisio, G. Blind quality assessment of fused WorldView-3 images by using the combinations of pansharpening and hypersharpening paradigms. *IEEE Geosci. Remote Sens. Lett.* **2017**, *14*, 1835–1839. [CrossRef]
58. Chen, A.; Chen, B.; Chai, X.; Bian, R.; Li, H. A novel stochastic stratified average gradient method: Convergence rate and its complexity. *arXiv* **2017**, arXiv:1710.07783.

remote sensing

MDPI

Letter

Deformable Faster R-CNN with Aggregating Multi-Layer Features for Partially Occluded Object Detection in Optical Remote Sensing Images

Yun Ren [1] [iD], Changren Zhu [1,*] and Shunping Xiao [2]

[1] ATR National Lab, National University of Defense Technology, Changsha 410073, China;
    renyun_nudt@163.com
[2] State Key Lab of Complex Electromagnetic Environment Effects on Electronics and Information System,
    National University of Defense Technology, Changsha 410073, China; shun_ping_xiao@163.com
* Correspondence: changrenzhu@nudt.edu.cn; Tel.: +86-139-7495-8436

Received: 24 August 2018; Accepted: 12 September 2018; Published: 14 September 2018

Abstract: The region-based convolutional networks have shown their remarkable ability for object detection in optical remote sensing images. However, the standard CNNs are inherently limited to model geometric transformations due to the fixed geometric structures in its building modules. To address this, we introduce a new module named deformable convolution that is integrated into the prevailing Faster R-CNN. By adding 2D offsets to the regular sampling grid in the standard convolution, it learns the augmenting spatial sampling locations in the modules from target tasks without additional supervision. In our work, a deformable Faster R-CNN is constructed by substituting the standard convolution layer with a deformable convolution layer in the last network stage. Besides, top-down and skip connections are adopted to produce a single high-level feature map of a fine resolution, on which the predictions are to be made. To make the model robust to occlusion, a simple yet effective data augmentation technique is proposed for training the convolutional neural network. Experimental results show that our deformable Faster R-CNN improves the mean average precision by a large margin on the SORSI and HRRS dataset.

Keywords: Deformable CNN; Faster R-CNN; data augmentation; occluded object detection

1. Introduction

Recently, Convolutional Neural Networks (CNNs) [1] have achieved flourishing success for visual recognition tasks, such as image classification [2], semantic segmentation [3], and object detection [4]. With the powerful feature representation capability of Deep CNNs, object detection has witnessed a quantum leap in the performance on benchmark datasets. Within the last five years, there have been massive improvements on standard benchmarks such as PASCAL and COCO by the family of region-based CNNs. However, little effort has been made towards occluded object detection in optical remote sensing images. Besides, modeling geometric variations or transformations in the scale of objects, pose, viewpoint, and part deformations is a key challenge in optical remote sensing visual recognition.

Object detection in optical remote sensing images often suffers from several increasing challenges including the large variations in the visual appearance of objects caused by viewpoint variation, occlusion, resolution, background clutter, illumination, shadow, etc. In the past few decades, various methods have been developed for the detection of different types of objects in satellite and aerial images, such as buildings [5], storage tanks [6], vehicles [7], and airplanes [8]. In general, they can be divided into four main categories: Template matching-based methods, knowledge-based methods, OBIA-based methods, and machine learning-based methods. According to the selected template type, template matching-based methods could be further subdivided into two classes, as rigid template matching and deformable template matching [5,9]. For knowledge-based object detection methods, there are two kinds of the most widely used, which used prior knowledge involved geometric information and context information [10–12]. In general, OBIA-based object detection methods include two steps: Image segmentation and object classification [13]. With regard to machine learning-based methods, three crucial steps, which include feature extraction, feature fusion dimension reduction, and classifier training, play important roles in the performance of object detection. Many recent approaches have formulated object detection as feature extraction and classification problems and have achieved significant improvements.

With the prosperity and rapid development of CNNs, object detection tasks have been formulated as feature extraction and classification problems, whose results have been shown to be promising with the help of the powerful feature representation capability of advanced CNN architecture. Currently, the most popularly CNN-based object detection algorithms could be roughly divided into two streams: The region-based methods and the region-free methods. The region-based methods firstly generate about 2000 category-independent region proposals for the input image, extract a fixed-length feature vector from each proposal using a CNN, and then classify those regions and refine their spatial locations. As a ground-breaking work, R-CNN [4] consists of three modules. The first module generates category-independent region proposals that are fed into the second module. It is a large CNN to extract a fixed-length feature vector from each region, while the third module is a set of class-specific linear SVMs. Compared to traditional R-CNN and its accelerated version SPPnet [14], Fast R-CNN [15] trains networks using a multi-task loss in a single training stage, which simplifies learning and tremendously increases runtime efficiency. Merging the proposed RPN and Fast R-CNN into a single network by sharing their convolutional features, Faster R-CNN [16] enables a unified, deep-learning-based object detection system to run at near real-time frame rates. In contrast, the region-free methods frame object detection as a regression problem and directly estimates the objects region, which truly enables real-time detection. YOLO [17] is extremely fast because it utilizes a single convolutional network to simultaneously predict bounding boxes and class probabilities directly from full images in one evaluation. Using a single CNN as well, SSD [18] discretizes the output space of bounding boxes into a set of default boxes over different aspect ratios and scales per feature map location. Additionally, the network combines predictions from multiple feature maps with different resolutions to naturally handle objects of various sizes, which improves the accuracy on high-speed detection. What is noteworthy is that the above-mentioned CNN-based object detection algorithms are designed somewhat specially for general object detection benchmarks, which is not suitable for object detection in optical remote sensing images because the object instances occupy a minor portion of the image that usually have the characteristic of small size in the optical remote sensing images. Furthermore, to deal with the problem of small objects, some methods like Fast R-CNN and Faster R-CNN achieve this by directly up-sampling the input image at the training phase or testing phase. It significantly increases the memory usage and processing time.

However, CNNs are inherently limited to model geometric transformations shown in visual appearance. The limitations derive from the fixed geometric structures of CNN modules: A convolution operation samples the input feature map at fixed locations. As long as a standard CNN architecture is adopted, the only method available to model geometric transformations are artificially generating sufficient complete training samples with various deformations. As said by Cheng et al. [19], it is problematic to directly use it for object detection in optical remote sensing images because it is difficult to effectively handle the problem of object rotation variations. Rotation Invariant CNN (RICNN) augments training objects by rotating them 360 degrees by a step of 10 degrees, which does not actually solve the inherent limitation in CNN. The emergence of deformable convolution overcomes the mapping limitations in CNN [20]. By adding 2D offsets to the regular convolution grid in the standard convolution, deformable convolution sample features from flexible locations instead of fixed locations, allowing for the free deformation of the sampling grid. In other words, deformable convolution refines standard convolution by adding learned offsets. The deformable convolution modules can readily replace the convolution layer in standard CNN and form deformable ConvNet. The spatial sampling locations in deformable convolution modules are augmented with additional offsets, which are learned from data and driven by the target task. Deformable ConvNet is a simple, efficient, deep, and end-to-end solution to model dense spatial transformations. We believe that it is feasible and effective to learn dense spatial transformation in CNNs for object detection in optical remote sensing images.

In this paper, we present a deformable Faster R-CNN with aggregating multi-layer features for partially occluded object detection in optical remote sensing images. In other words, Deformable ConvNet, embedded within Faster R-CNN, is introduced in the field of optical remote sensing for object detection. The main contributions of this paper are summarized as follows:

➢ A unified deformable Faster R-CNN is introduced for object detection in optical remote sensing images. Geometric variation modeling is completed within the deformable convolution layers. Feature maps extracted by deformable ConvNet contain more information about various geometric transformations.

➢ A modified backbone network is specially designed for small object to generate more abundant feature maps with high semantic information at low layer. Therefore, a Transfer Connection Block (TCB) adopting top-down and skip connections is presented to produce a single high-level feature map of a fine resolution.

➢ A simple, yet effective, data augmentation technique named Random Covering is proposed for training CNN. In training phase, it randomly selects a rectangle region in a region of interest and covers its pixels with random values. Hence, we can obtain augmented training samples with random levels of occlusion, which are fed into the model to enhance the generalization ability of the CNN model.

The rest of this paper is organized as follows. Section 2 introduces the methodology of our deformable Faster R-CNN with the transfer connection block. The last subsection of Section 2 proposes the data augmentation technique, namely the Random Covering. Section 3 presents the datasets and experimental settings. The results of our methodology and other approaches in the SORSI and HRRS dataset are presented in Section 4, while Section 5 gives our conclusion and the future work.

2. Methodology

Figure 1 presents a roundup of our deformable Faster R-CNN with three transfer connection blocks. Deformable Faster R-CNN is constructed by substituting the standard convolution layer with a deformable convolution layer in the fifth network stage. The proposed network consists of a deformable proposal network and a deformable object detection network, both of which share a deformable backbone network with three transfer connection blocks for feature map generation. More details are provided in the following content.

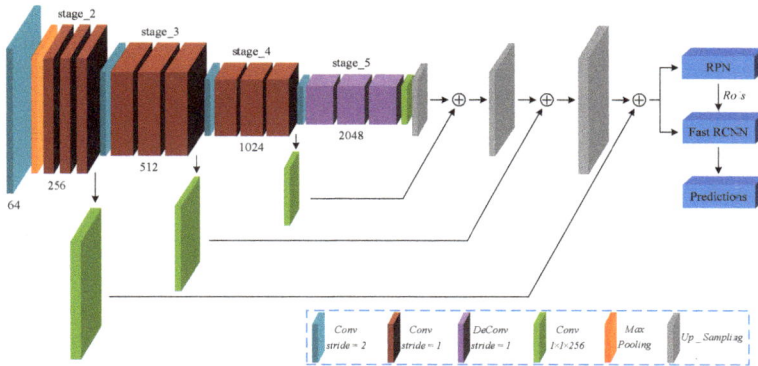

Figure 1. Architecture of the deformable Faster CNN with three TCBs.

2.1. Deformable Convolution

While convolution in CNNs can be regarded as 3D spatial sampling, deformable convolution operates on the 2D spatial domain and remains the same across the channel dimension. In general, they are explained in 2D here. Extending the equations to 3D should be straightforward and omitted for notation clarity.

A standard 2D convolution consists of two steps: (1) Sampling using a regular grid $\mathcal{R}$ over the input feature map $\mathbf{X}$; and (2) summation of sampled values weighted by $\mathbf{W}$. The grid $\mathcal{R}$ defines the convolution kernel by size and dilation. For example, $\mathcal{R} = \{(-1,1),(-1,0), \dots ,(0,1),(1,1)\}$ defines a 3×3 kernel with dilation 1. We can derive the standard convolution output of each position $\mathbf{p}_0$ on the output feature map $\mathbf{Y}$, according to the following formula:

$$\mathbf{Y}(\mathbf{p}_0) = \sum_{\mathbf{p}_i \in \mathcal{R}} \mathbf{W}(\mathbf{p}_i) \cdot \mathbf{X}(\mathbf{p}_0 + \mathbf{p}_i) \tag{1}$$

In Dai et al. [20], deformable convolution was defined by augmenting the regular grid $\mathcal{R}$ with 2D offsets $\{\Delta\mathbf{p}_i | i = 1, \dots, N\}$, where $N = |\mathcal{R}|$. Then the deformable convolution output of each position $\mathbf{p}_0$ on the output feature map $\mathbf{Y}$ can be formulized as follows:

$$\mathbf{Y}(\mathbf{p}_0) = \sum_{\mathbf{p}_i \in \mathcal{R}} \mathbf{W}(\mathbf{p}_i) \cdot \mathbf{X}(\mathbf{p}_0 + \mathbf{p}_i + \Delta\mathbf{p}_i) \tag{2}$$

Obviously, the sampling is over the unfixed positions $\mathbf{p}_i + \Delta\mathbf{p}_i$ of the input feature grid. As the offset $\Delta\mathbf{p}_i$ might be non-integer, Equation (2) is implemented by bilinear interpolation to obtain the fractional position. As we know, the bilinear interpolation can be formulated as

$$\mathbf{X}(\mathbf{p}) = \sum_{\mathbf{q}} \mathbf{G}(\mathbf{q}, \mathbf{p}) \cdot \mathbf{X}(\mathbf{q}) \tag{3}$$

where $\mathbf{p}$ denotes an arbitrarily fractional position ($\mathbf{p} = \mathbf{p}_0 + \mathbf{p}_i + \Delta\mathbf{p}_i$ for Equation (2)), $\mathbf{q}$ enumerates four integral spatial positions nearest to the position $\mathbf{p}$, and $\mathbf{G}(\cdot, \cdot)$ indicates the bilinear interpolation kernel. Note that $\mathbf{G}$ can be decomposed into two 1D kernels as

$$\mathbf{G} = g(q_x, p_x) \cdot g(q_y, p_y) \tag{4}$$

where the 1D bilinear interpolation kernel is defined as $g(a, b) = max(0, 1 - |a - b|)$.

As illustrated in Figure 2, the additional offsets are learned by adding a standard convolutional layer branch whose convolution kernel is the same spatial resolution as the current convolutional layer. Additionally, the output offset fields have the same spatial resolution with the input feature map.

The output channel dimension is set at $2N$ to encode N 2D offset vectors. During training, both the convolutional kernels for producing the output features and for generating offsets can be learned. The gradients enforced on the deformable convolution layer can be back-propagated through the bilinear operations in Equations (3) and (4).

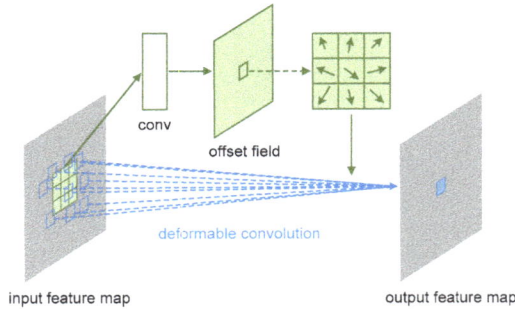

Figure 2. Illustration of 3×3 deformable convolution.

2.2. Transfer Connection Block

Generally, the objects have the characteristics of small size in the optical remote sensing images. The region-based methods consist of a region proposal network and an object detection network, both of which share a backbone network to generate feature representation. However, we notice that the feature maps of the shared network have a very large receptive field so that it can be hardly matched to small objects. The semantic information in the high-layer is significant for feature representation [21]. Based on these two considerations, the transfer connection block is presented to combine high semantic features from higher layers with fine details from lower layers, which is shown in Figure 3. To match the dimensions between them, the de-convolution operation is used to enlarge the high-level feature maps and sum them in the element-wise way. To be specific, the modified backbone network produces feature maps through three TCBs, starting from the last layer of the backbone network, which has high semantic information. Then the feature maps of the last layer are transmitted back to combine bottom-up feature maps at middle layers by top-down and skip connections. The TCP is sequentially embedded into the last three stages of the backbone network. By default, ResNet_50 is used to be the backbone network [22].

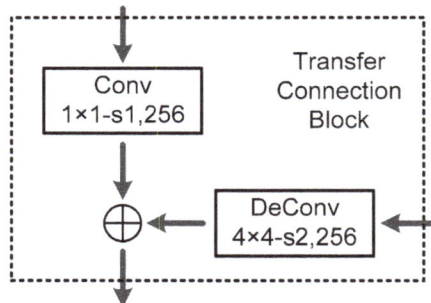

Figure 3. The overview of the transfer connection block.

2.3. Random Covering

Occlusion caused by fog or cloud is a critical influencing factor on the generalization ability of CNNs in optical remote sensing images. It is desirable to achieve invariance to various levels of occlusion. When some parts of an object are occluded, a strong detection model should recognize its category and locate it from the overall object structure. However, the collected training samples usually reveal limited variance in occlusion. In an extreme case when no occlusion happens in all the training objects, the learned CNN model will work well on the testing images without occlusion. But it may fail to recognize objects with partial occlusion because of the limited generalization ability of the CNN model. While we can manually augment occluded images to the training data, this process is costly and the levels of occlusion can be limited.

To address the occlusion problem and improve the generalization ability of CNNs, Random Covering is introduced as a new data augmentation approach. This idea is inspired by another data augmentation approach named Random Erasing [23]. In the training phase, Random Covering happens with a certain probability. For an image I, within a mini-batch in the training phase, it is randomly chosen to undergo either Random Covering with probability p, or kept unchanged with probability $1 - p$. Random Covering randomly selects a rectangle region I_{rc} in the image and adds random values on these selected pixels. Assume the size of the image is $W \times H$ and its area is $S = W \times H$. We randomly initialize the area of the covering rectangle region to S_{rc}, where S_{rc}/S is in the range specified by minimum s_l and maximum s_h. The aspect ratio r_{rc} of covering rectangle region is randomly initialized between r_1 and r_2. Then the size of covering region I_{rc} is $H_{rc} = \sqrt{S_{rc} \times r_{rc}}$ and $W_{rc} = \sqrt{S_{rc}/r_{rc}}$. A point $p = (x_{rc}, y_{rc})$ in the image I is randomly initialized as the center of the covering region I_{rc}, where the left-top location p_{lu} and the right-bottom location p_{rb} are $\left(max\left(1, x_{rc} - \frac{W_{rc}}{2}\right), max\left(1, y_{rc} - \frac{H_{rc}}{2}\right)\right)$ and $\left(min\left(W, x_{rc} + \frac{W_{rc}}{2}\right), min\left(H, y_{rc} + \frac{H_{rc}}{2}\right)\right)$. After selecting the covering region I_{rc}, each pixel in I_{rc} is assigned to the weighted summation of the original pixel and a random value. The weight coefficient λ is randomly initialized in a range specified by minimum λ_1 and maximum λ_2. The Random Covering procedure is shown in Algorithm 1. In the case of object detection, we select covering region in the bounding box of each object. If there are multiple objects in the image, Random Covering is applied on each object separately.

Algorithm 1: Random Covering Procedure

Input: Input image I;
        Area of image $S = W \times H$;
        Covering probability p;
        Area ratio range s_l and s_h;
        Aspect ratio range r_1 and r_2;
        Weight coefficient range λ_1 and λ_2;
Output: Covering image I^*.
Initialization: $p_1 \leftarrow Rand(0, 1)$.
if $p_1 \geq p$ **then**
    $I^* \leftarrow I$;
    return I^*.
else
    $S_{rc} \leftarrow Rand(s_l, s_h) \times S$;
    $r_{rc} \leftarrow Rand(r_1, r_2)$;
    $\lambda \leftarrow Rand(\lambda_1, \lambda_2)$;
    $H_{rc} \leftarrow \sqrt{S_{rc} \times r_{rc}}$, $W_{rc} \leftarrow \sqrt{S_{rc}/r_{rc}}$;
    $x_{rc} \leftarrow Rand(1, W)$, $y_{rc} \leftarrow Rand(1, H)$;
    $p_{lu} \leftarrow \left(max\left(1, x_{rc} - \frac{W_{rc}}{2}\right), max\left(1, y_{rc} - \frac{H_{rc}}{2}\right)\right)$;
    $p_{rb} \leftarrow \left(min\left(W, x_{rc} + \frac{W_{rc}}{2}\right), min\left(H, y_{rc} + \frac{H_{rc}}{2}\right)\right)$;
    $I_{rc} \leftarrow (p_{lu}, p_{rb})$;
    $I(I_{rc}) \leftarrow \lambda \cdot Rand(0, 1) + (1 - \lambda) \cdot I(I^*)$;
    $I^* \leftarrow I$;
    return I^*.
end

---

3. Dataset and Experimental Settings

To evaluate and validate the effectiveness of deformable Faster R-CNN on the optical remote sensing images, the datasets, experimental settings, and the corresponding evaluation metrics of the experimental results are described in this section.

3.1. Evaluation Metrics

Here, we explain two universally agreed and widely applied standard measures for evaluating the object detection methods, namely the Precision–Recall Curve (PRC) and Average Precision (AP). The first evaluation metric is based on the overlapping area between detections and ground truth. The Precision measures the fraction of detections that are true positives and the Recall measures the fraction of positives that are correctly identified. Let TP, FP, and FN denote the number of true positives, the number of false positives, and the number of false negatives, respectively. The Precision and Recall can be formulated as:

$$Precision = \frac{TP}{TP + FP} \qquad (5)$$

$$Recall = \frac{TP}{TP + FN} \qquad (6)$$

In an object-level evaluation, detections are recognized as TP if the area overlap ratio α between detections and ground truth object exceeds a predefined threshold λ by the formula

$$\alpha = \frac{Area(detection \cap ground\_truth)}{Area(detection \cup ground\_truth)} > \lambda \qquad (7)$$

where $Area(detection \cap ground\_truth)$ denotes the intersection of the detection and ground truth and $Area(detection \cup ground\_truth)$ denotes their union. Otherwise they are considered as FP. In addition, if several detections overlap with the same ground truth object, only one is considered as the true positive and the others are considered as false positives.

The second evaluation metric called AP is based on the area under the PRC. The AP computes the average value of Precision over the interval from $Recall = 0$ to $Recall = 1$. Mean AP (mAP) computes the average value of AP over all object categories. AP and mAP are used as the quantitative indicators in object detection. Typically, the higher the AP and mAP is, the better the detection performance, and vice versa.

3.2. Dataset and Implementation Details

To evaluate the performance of deformable Faster R-CNN, we conduct experiments on various optical remote sensing datasets. We chose three datasets, including the *NWPU VHR-10* [24], *SORSI* [25], and *HRRS* [26] datasets. The *NWPU VHR-10* dataset is a 10-class geospatial object detection dataset that contains a total of 650 annotated optical remote sensing images in the manner of *VOC 2007*. The ratios of training, validation and testing dataset are set to 20%, 20%, and 60%, respectively. Then, we randomly selected 130, 130, and 390 images to fill these three subsets, respectively. To make the model more robust to various input object sizes and shapes, each training image is sampled by the following options: (1) Using the original/flipped input image; and (2) rotating the input image by an angle step of 18°. The *SORSI* dataset contains only two categories: Ship and plane which includes 5922 optical remote sensing images—5216 images for ship and 706 images for plane. The numbers of this dataset in different classes are highly imbalanced, which poses great challenges for model training. To make a fair comparison, the *SORSI* dataset is randomly split into 80% for training, and 20% for testing as well. Some samples of these three datasets are shown in Figure 4. Besides, a more challenging occlusion dataset is collected by Qiu et al., which is available on https://github.com/QiuWhu/Data. This dataset includes 47 images with total 184 airplanes, 105 airplanes of which are partially occluded by cloud or hangar or truncated by image border.

(a) Some examples in *NWPU VHR-10* dataset

(b) Some examples in *SORSI* dataset

(c) Some examples in *HRRS* dataset

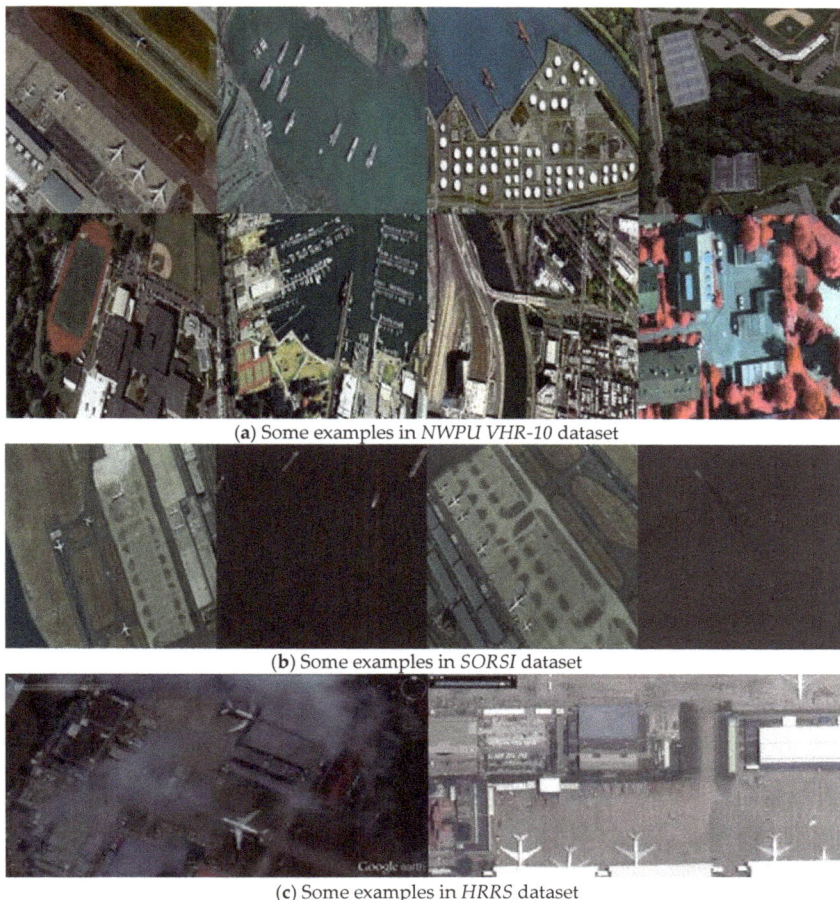

Figure 4. Example images of *NWPU VHR-10/SORSI/HRRS* datasets.

Adopting the alternating training strategy in this paper, we trained and tested both RPN and Fast R-CNN on images of a single scale based on Caffe [27] in all of the experiments. The images were resized such that their shorter side is 608 pixels under the premise of ensuring the longer side less than 1024 pixels. We used the pre-training model ResNet-50 to initialize the network. The deformable Faster R-CNN is constructed by substituting the standard convolution layer with a deformable convolution layer in the last three-network stage. For other newly added layers, we initialized the parameters by drawing weights from a zero-mean Gaussian distribution with standard deviation of 0.01. Furthermore, it is easy for our method to adopt Online Hard Example Mining (OHEM) [28] during training. Assuming N proposals per image generated by RPN, in the forward pass, we evaluate the loss of all N proposals. Then we sort all RoIs (positive and negative) by loss and select B RoIs that have the highest loss. Back-propagation [29] is performed based on the selected proposals.

For the *NWPU VHR-10* dataset, we trained a total of 80 K iterations, with a learning rate of 10^{-3} for the first 60 K iterations, 10^{-4} for the next 20 K iterations. The iteration was halved for the *SORSI* datasets. Weight decay and momentum were 0.0005 and 0.9, respectively. For anchors, we adopted three scales with box areas of 16^2, 40^2, and 100^2 pixels, and an aspect ratio of 1:1, which were adjusted for better coverage of the size distribution of our optical remote sensing dataset. At the RPN stage, we sampled a total of 256 anchors as a mini-batch for training (128 proposals for the Fast RCNN stage), where the ratio of positive to negative samples was 1:1. The evaluation metric is AP of each object and mAP with the Interception-of-Union (IoU) threshold set to 0.5. Non-Maximum Suppression (NMS) is adopted to reduce redundancy on the proposal regions based on their box-classification scores. The IoU threshold is fixed for NMS at 0.7. All experiments were performed on Intel i7-6700K CPU and NVIDIA GTX1080 GPU.

4. Experimental Results and Discussion

4.1. Quantitative Evaluation of NWPU VHR-10 Dataset

To evaluate the proposed deformable Faster RCNN with TCB quantitatively, we compared it with the AP values with four state-of-the-art CNN-based methods: (1) A rotation-invariant CNN (RICNN) model which considers rotation-invariant information with a rotation-invariant layer and other fine-tuned layers; (2) the SSD model with an input image size of 512 × 512 pixels; (3) the R-P-Faster RCNN [30] object detection framework; and (4) deformable R-FCN with the aspect ratio constrained NMS. The results of these methods all come out of the previous papers [31].

As shown in Table 1, the proposed deformable Faster RCNN with TCB, which is fine-tuned on the ResNet-50 ImageNet pre-trained model, obtains the best mean AP value of 84.4% among all the object detection methods. It also indicates that our deformable faster RCNN with TCB achieves the best AP values for most classes, except baseball diamond, harbor, and bridge. In particular, the AP values of small objects like vehicle increase more than other objects, which illustrate the good performance of our methods for small object detection. This will be further verified through the results on the SORSI dataset in the next subsection. Compared with the second best method of deformable R-FCN with arcNMS, the AP values of seven objects are increased, including airplane (0.873 to 0.907), ship (0.814 to 0.871), storage tank (0.636 to 0.705), tennis court (0.816 to 0.893), basketball court (0.741 to 0.873), Ground track field (0.903 to 0.972), and Vehicle (0.755 to 0.888). Figure 5 plots the PRCs of our method over ten testing classes, respectively. The recall ratio evaluates the ability of detecting more targets, while the precision evaluates the quality of detecting correct objects rather than containing many false alarms. Obviously, the ground track field obtains the best performance, in comparison to other objects adopting the proposed method.

Table 1. The AP values of the object detection methods on the *NWPU VHR-10* dataset.

Method	RICNN	SSD	R-P-Faster R-CNN	Deformable R-FCN (ResNet-101) with arcNMS	Deformable Faster RCNN (ResNet-50) with TCB
Airplane	0.884	0.957	0.904	0.873	0.907
Ship	0.773	0.829	0.75	0.814	0.871
Storage tank	0.853	0.856	0.444	0.636	0.705
Baseball diamond	0.881	0.966	0.899	0.904	0.895
Tennis court	0.408	0.821	0.79	0.816	0.893
Basketball court	0.585	0.856	0.776	0.741	0.873
Ground track field	0.867	0.582	0.877	0.903	0.972
Harbor	0.686	0.548	0.791	0.753	0.735
Bridge	0.615	0.419	0.682	0.714	0.699
Vehicle	0.711	0.756	0.732	0.755	0.888
mean AP	0.726	0.759	0.765	0.791	0.844

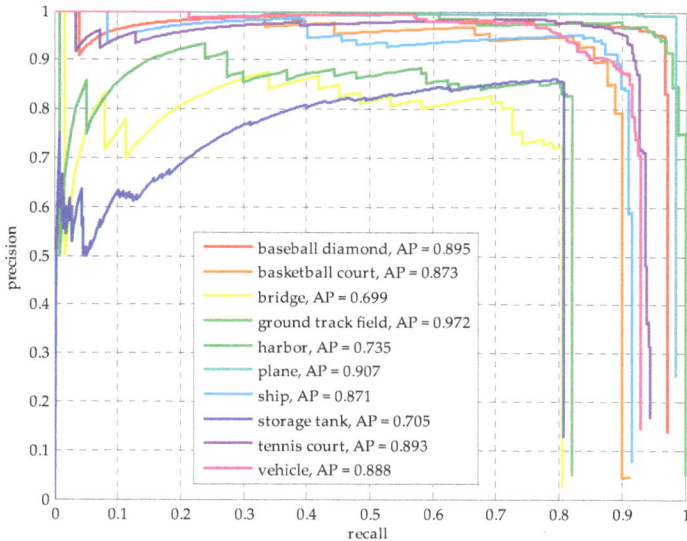

Figure 5. Precision versus recall curve for the proposed method over the *NWPU VHR-10* dataset.

4.2. Quantitative Evaluation of SORSI Dataset

To verify the performance on detecting small objects in optical remote sensing images, we conduct experiments on the SORSI dataset, only including two categories: Plane and ship. Besides, the areas of bounding boxes falling in the ship category dominate from 10^2 to 50^2 pixels while those in the plane category possess from 50^2 to 100^2 pixels. In other words, the ship has smaller scale than the plane, which indicates that detecting ships is considerably more challenging. The results of the baseline come from [25]. From Table 2, it can be seen that the AP value for ship grows by five percentage points while adopting the TCB module, which manifests the TCB module, which is significant to detect smaller object. Besides, AP values for ship and plane steadily improves by one percentage point when deformable convolution layers are used. In addition, the final AP values for all objects have a big improvement while adding the OHEM mechanism in the training phase, especially for the ship category. This demonstrates that the TCB module works well with the OHEM mechanism for detecting small objects.

Table 2. The results of modified Faster R-CNN on SORSI dataset.

Method	Baseline	Faster RCNN with TCB	Deformable Faster RCNN with TCB	Deformable Faster RCNN with TCB (+OHEM)
plane	0.729	0.778	0.792	0.862
Ship	0.850	0.826	0.831	0.903
mean AP	0.789	0.802	0.812	0.883

4.3. Quantitative Evaluation of HRRS Dataset

To verify the effectiveness of the proposed Random Covering on the partial occlusion problem, experiments are conducted on the HRRS dataset. This dataset only includes one category: Airplane. This dataset includes 47 images with total 184 airplanes, 105 airplanes of which are partially occluded by cloud or hangar or truncated by image border. Therefore, we only randomly cover the images, which contain one airplane at least. First, we conduct an experiment on the SORSI dataset. It is surprising that the AP value for plane gets improvement by 0.4 percentage points while the AP value for ship remains unchanged. This shows that the proposed Random Covering can work well on an un-occluded dataset and improve the generalization ability of our model. Second, all the images of the HRRS dataset are tested by the previous model. Figure 6 shows a comparison of PRC while the model trains with or without Random Covering. In addition, we count up the number of true positives for the partially occluded objects, as illustrated in the Table 3. The results indicate that both the AP value and the *TP* increase by a large margin while adopting the Random Covering in the training phase.

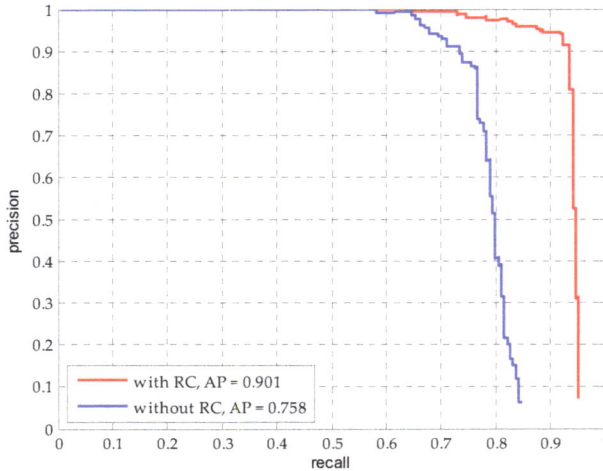

Figure 6. Precision versus recall curve for the HRRS dataset with/without RC.

Table 3. The AP and #*TP* on the HRRS dataset with or without RC.

Method	with RC	without RC
AP/#*TP*	0.901/96	0.758/77

5. Conclusions

In this paper, a unified deformable Faster R-CNN is introduced for modeling geometric variations in optical remote sensing images. Besides, we presented a transfer connection block aggregating multi-layer features to produce a single high-level feature map of a fine resolution, which is significant for detecting small objects. To improve the generalization ability of the CNN model and address the occlusion problem, we proposed a simple data augmentation approach named Random Covering, which was used in the training phase. Experiments conducted on three datasets show the effectiveness of our method. In the future work, we will focus on the balance between the TCB module and the average running time per image, and the effect of deformable convolution in the feature extraction network.

Author Contributions: Y.R. provided the original idea for the study; C.Z. and S.X. contributed to the discussion of the design; Y.R. conceived and designed the experiments; C.Z. supervised the research and contributed to the article's organization; and Y.R. drafted the manuscript, which was revised by all authors. All authors read and approved the final manuscript.

Funding: This research received no external funding.

Acknowledgments: The authors would like to thank all the colleagues who generously provided their image dataset with the ground truth. The authors would also like to thank the anonymous reviewers for their very competent comments and helpful suggestions.

Conflicts of Interest: The authors declare no conflict of interest.

References

1. Lecun, Y.; Bengio, Y. Convolutional networks for images, speech, and time series. In *Handbook of Brain Theory and Neural Networks*; MIT Press: Cambridge, MA, USA, 1995.
2. Krizhevsky, A.; Sutskever, I.; Hinton, G.E. Imagenet classification with deep convolutional neural networks. In Proceedings of the Advances in Neural Information Processing Systems 2012, Lake Tahoe, NV, USA, 3–8 December 2012; pp. 1097–1105.
3. Long, J.; Shelhamer, E.; Darrell, T. Fully convolutional networks for semantic segmentation. In Proceedings of the IEEE Conference on Computer Vision and Pattern Recognition, Boston, MA, USA, 7–12 June 2015.
4. Girshick, R.; Donahue, J.; Darrell, T.; Malik, J. Rich feature hierarchies for accurate object detection and semantic segmentation. In Proceedings of the IEEE Conference on Computer Vision and Pattern Recognition, Columbus, OH, USA, 24–27 June 2014.
5. Stankov, K.; He, D.C. Detection of buildings in multispectral very high spatial resolution images using the percentage occupancy hit-or-miss transform. *IEEE J. Sel. Top. Appl. Earth Obs. Remote Sens.* **2014**, *7*, 4069–4080. [CrossRef]
6. Ok, A.O.; Başeski, E. Circular oil tank detection from panchromatic satellite images: A new automated approach. *IEEE Geosci. Remote Sens. Lett.* **2015**, *12*, 1347–1351. [CrossRef]
7. Wen, X.; Shao, L.; Fang, W.; Xue, Y. Efficient Feature Selection and Classification for Vehicle Detection. *IEEE Trans. Circuits Syst. Video Technol.* **2015**, *25*, 508–517.
8. An, Z.; Shi, Z.; Teng, X.; Yu, X.; Tang, W. An automated airplane detection system for large panchromatic image with high spatial resolution. *Optik-Int. J. Light Electron Opt.* **2014**, *125*, 2768–2775. [CrossRef]
9. Jain, A.K.; Ratha, N.K.; Lakshmanan, S. Object detection using Gabor filters. *Pattern Recognit.* **1997**, *30*, 295–309. [CrossRef]
10. Leninisha, S.; Vani, K. Water flow based geometric active deformable model for road network. *ISPRS J. Photogramm. Remote Sens.* **2015**, *102*, 140–147. [CrossRef]
11. Ok, A.O. Automated detection of buildings from single VHR multispectral images using shadow information and graph cuts. *ISPRS J. Photogramm. Remote Sens.* **2013**, *86*, 21–40. [CrossRef]
12. Ok, A.O.; Senaras, C.; Yuksel, B. Automated detection of arbitrarily shaped buildings in complex environments from monocular VHR optical satellite imagery. *IEEE Trans. Geosci. Remote Sens.* **2013**, *51*, 1701–1717. [CrossRef]
13. Blaschke, T.; Hay, G.J.; Kelly, M.; Lang, S.; Hofmann, P.; Addink, E.; Feitosa, R.Q.; Van der Meer, F.; Van der Werff, H.; Van Coillie, F.; et al. Geographic object-based image analysis—Towards a new paradigm. *ISPRS J. Photogramm. Remote Sens.* **2014**, *87*, 180–191. [CrossRef] [PubMed]
14. He, K.; Zhang, X.; Ren, S.; Sun, J. Spatial pyramid pooling in deep convolutional networks for visual recognition. In *European Conference on Computer Vision*; Springer: Cham, Switzerland, 2014.
15. Girshick, R. Fast R-CNN. In Proceedings of the 2015 IEEE International Conference on Computer Vision, Santiago, Chile, 7–13 December 2015.
16. Ren, S.; He, K.; Girshick, R.; Sun, J. Faster R-CNN: Towards real-time object detection with region proposal networks. In Proceedings of the Advances in Neural Information Processing Systems, Montreal, QC, Canada, 7–12 December 2015.
17. Redmon, J.; Divvala, S.; Girshick, R.; Farhadi, A. You only look once: Unified, real-time object detection. In Proceedings of the IEEE Conference on Computer Vision and Pattern Recognition, Las Vegas, NV, USA, 27–30 June 2016.
18. Liu, W.; Anguelov, D.; Erhan, D.; Szegedy, C.; Reed, S.; Fu, C.Y.; Berg, A.C. SSD: Single shot multibox detector. In *European Conference on Computer Vision*; Springer: Cham, Switzerland, 2016.
19. Cheng, G.; Zhou, P.; Han, J. Learning rotation-invariant convolutional neural networks for object detection in VHR optical remote sensing images. *IEEE Trans. Geosci. Remote Sens.* **2016**, *54*, 7405–7415. [CrossRef]
20. Dai, J.; Qi, H.; Xiong, Y.; Li, Y.; Zhang, G.; Hu, H.; Wei, Y. Deformable Convolutional Networks. 2017. Available online: http://openaccess.thecvf.com/content_ICCV_2017/papers/Dai_Deformable_Convolutional_Networks_ICCV_2017_paper.pdf (accessed on 22 August 2018).

21. Lin, T.Y.; Dollár, P.; Girshick, R.B.; He, K.; Hariharan, B.; Belongie, S.J. Feature Pyramid Networks for Object Detection. 2017. Available online: http://openaccess.thecvf.com/content_cvpr_2017/papers/Lin_Feature_ Pyramid_Networks_CVPR_2017_paper.pdf (accessed on 22 August 2018).
22. Ren, Y.; Zhu, C.; Xiao, S. Object Detection Based on Fast/Faster RCNN Employing Fully Convolutional Architectures. *Math. Probl. Eng.* **2018**, *2018*, 3598316. [CrossRef]
23. Zhong, Z.; Zheng, L.; Kang, G.; Li, S.; Yang, Y. Random erasing data augmentation. *arXiv*, 2017; arXiv:1708.04896.
24. Cheng, G.; Han, J.A. A survey on object detection in optical remote sensing images. *ISPRS J. Photogramm. Remote Sens.* **2016**, *117*, 11–28. [CrossRef]
25. Ren, Y.; Zhu, C.; Xiao, S. Small Object Detection in Optical Remote Sensing Images via Modified Faster R-CNN. *Appl. Sci.* **2018**, *8*, 813. [CrossRef]
26. Qiu, S.; Wen, G.; Fan, Y. Occluded object detection in high-resolution remote sensing images using partial configuration object model. *IEEE J. Sel. Top. Appl. Earth Obs. Remote Sens.* **2017**, *10*, 1909–1925. [CrossRef]
27. Jia, Y.; Shelhamer, E.; Donahue, J.; Karayev, S.; Long, J.; Girshick, R.; Guadarrama, S.; Darrell, T. Caffe: Convolutional architecture for fast feature embedding. In Proceedings of the 22nd ACM International Conference on Multimedia, Orlando, FL, USA, 3–7 November 2014.
28. Shrivastava, A.; Gupta, A.; Girshick, R. Training region-based object detectors with online hard example mining. In Proceedings of the IEEE Conference on Computer Vision and Pattern Recognition, Las Vegas, NV, USA, 27–30 June 2016.
29. LeCun, Y.; Boser, B.; Denker, J.S.; Henderson, D.; Howard, R.E.; Hubbard, W.; Jackel, L.D. Backpropagation applied to handwritten zip code recognition. *Neural Comput.* **1989**, *1*, 541–551. [CrossRef]
30. Han, X.; Zhong, Y.; Zhang, L. An efficient and robust integrated geospatial object detection framework for high spatial resolution remote sensing imagery. *Remote Sens.* **2017**, *9*, 666. [CrossRef]
31. Xu, Z.; Xu, X.; Wang, L.; Yang, R.; Pu, F. Deformable ConvNet with Aspect Ratio Constrained NMS for Object Detection in Remote Sensing Imagery. *Remote Sens.* **2017**, *9*, 1312. [CrossRef]

remote sensing

MDPI

Article

Evaluating the Effects of Image Texture Analysis on Plastic Greenhouse Segments via Recognition of the OSI-USI-ETA-CEI Pattern

Yao Yao [1,2] and **Shixin Wang** [1,*]

1   Institute of Remote Sensing and Digital Earth, Chinese Academy of Sciences, No.20 Datun Road, Chaoyang District, Beijing 100101, China; yaoyao2016@radi.ac.cn
2   University of Chinese Academy of Sciences, No.19(A) Yuquan Road, Shijingshan District, Beijing 100049, China
*   Correspondence: wangsx@radi.ac.cn; Tel.: +86-10-6487-9460

Received: 11 January 2019; Accepted: 18 January 2019; Published: 23 January 2019

Abstract: Compared to multispectral or panchromatic bands, fusion imagery contains both the spectral content of the former and the spatial resolution of the latter. Even though the Estimation of Scale Parameter (ESP), the ESP 2 tool, and some segmentation evaluation methods have been introduced to simplify the choice of scale parameter (SP), shape, and compactness, many challenges remain, including obtaining the natural border of plastic greenhouses (PGs) from a GaoFen-2 (GF-2) fusion imagery, accelerating the progress of follow-up texture analysis, and accurately evaluating over-segmentation and under-segmentation of PG segments in geographic object-based image analysis. Considering the features of high-resolution images, the heterogeneity of fusion imagery was compressed using texture analysis before calculating the optimal scale parameter in ESP 2 in this study. As a result, we quantified the effects of image texture analysis, including increasing averaging operator size (AOS) and decreasing greyscale quantization level (GQL) on PG segments via recognition of a proposed Over-Segmentation Index (OSI)-Under-Segmentation Index (USI)-Error Index of Total Area (ETA)-Composite Error Index (CEI) pattern. The proposed pattern can be used to reasonably evaluate the quality of PG segments obtained from GF-2 fusion imagery and its derivative images, showing that appropriate texture analysis can effectively change the heterogeneity of a fusion image for better segmentation. The optimum setup of GQL and AOS are determined by comparing CEI and visual analysis.

Keywords: texture analysis; multi-resolution segmentation (MRS); greenhouse extraction; over-segmentation index (OSI); under-segmentation index (USI); error index of total area (ETA); composite error index (CEI); GaoFen-2 (GF-2)

1. Introduction

Extracting plastic greenhouse (PG) segments from well-segmented high-resolution imagery is a basic goal for many applications, such as area monitoring, production forecast, and the accurate inversion of land surface temperature; and it is more effective than traditional manual drawing when many samples have to be selected as the reference polygons in large-scale research.

Segmentation, its evaluation, and texture analysis are crucial steps in geographic object-based image analysis (GEOBIA). According to 254 case studies in Ma et al. [1], 80.9% used eCognition (Trimble, Munich, Germany) for segmentation, whereas the remaining segmentation software mainly includes ENVI (Harris Geospatial Solutions, Inc., Broomfield, USA), SPRING (National Institute for Space Research, São José dos Campos, Brazil) and ERDAS (Hexagon Geospatial, Madison, USA). Generally, objects can be obtained via chessboard, quadtree-based, contrast split, contrast filter, multi-threshold superpixel [2–5], watershed [6,7], and multi-resolution segmentation (MRS) [8,9] in eCognition software [10], or the active contours (snakes) [11–13] method in MATLAB (MathWorks, Natick, USA). MRS is most widely and successfully employed method under the context of remote sensing GEOBIA applications [14–18]. Even though thematic vector data can improve the quality of the segmentation [19], the decision of the optimal value of scale parameter (SP), shape, and compactness in MRS is not easy, since the conventional try-and-evaluate method [19,20] is too complicated, time consuming, and provides incomplete results. Therefore, Estimation of Scale Parameter (ESP) and ESP 2 are methods that have been introduced to calculate variance among segmentation results that are produced by the given shape, compactness, and step-changing scale levels. ESP estimates the SP for MRS on single-layer image data or other continuous data (e.g., digital surface models) semi-automatically [21], and ESP 2 can automatically obtain optimal scale parameter (OSP) on multiple layers [22]. As an updated version, ESP 2 has been adopted to find the specific scale levels for specific target objects [23], and is also employed to determine the optimal parameters for extracting greenhouses from WorldView-2 and WorldView-3 multispectral imagery [17,18]. The segmentation results of GaoFen-2 (GF-2) multispectral and panchromatic fusion imagery based on the ESP 2 tool still do not meet the requirements for the degree of over-segmentation and struggle to delineate the natural boundary of PGs, which is an obstacle to fully using the panchromatic band. Namely, over-segmentation and under-segmentation [14,24] are still two critical issues for PG segments, which is called problem I.

The pixels of one class display some texture features that differ from other categories in satellite imagery. To illustrate, textural information can be used as an additional band to improve the object-oriented classification of urban areas in Quickbird imagery [25]; however, a similar pixel-based maximum likelihood PG classification in Agüera et al.'s research [26] showed that the inclusion of a band with texture information did not significantly improve the overwhelming majority quality index values compared to those found when only multi-spectral bands were considered. Another object-based work conducted by Hasituya et al. [27] showed that adding textural features from medium-resolution imagery provides only limited improvement in accuracy, whereas spectral features more significantly contribute to monitoring plastic-mulched farmland. Some researchers treated the grey-level co-occurrence matrix (GLCM) [28] parameter values as available features of separated objects for sample training [20,29]. However, these schemes were executed in a so-called "black box" without a practical physical mechanism, so they are not easily reproducible for another similar task. The recognition and use of texture information in eCognition is another formidable time-consuming task [10], even if optimal SP, shape, and compactness are derived from ESP or ESP 2 based on initial fusion imagery. As an ancillary feature for mapping greenhouses, texture should be further studied both in pixel-based and object-based extraction, which can be called problem II.

Purposive preprocessing operations based on pixel-level imagery are important prior to MRS. Apart from frequently-used orthorectification, radiometric and atmospheric correction [18], and pan sharpening, texture analysis of these images can also generate derivative input data, and then influence the results of MRS. Thus, our first process involved compressing the heterogeneity of the fusion image by different texture analysis to produce some derivative images, and then exploring what effects image texture analysis would exert on PG segments. This led to our second idea that, in order to compare the accuracy of different PG segments, a reliable evaluation system is indispensable, which can be called problem III.

Many evaluation methods have been proposed. Depending on whether a human evaluator examines the segmented image visually or not, Zhang et al. [30] introduced a hierarchy of segmentation evaluation methods and a survey of unsupervised methods. Zhang et al. [31], Gao et al. [32], and Wang et al. [33] each proposed novel unsupervised methods respectively to evaluate the segmentation quality; however, these methods still need supervised evaluation for verification. Supervised evaluation [34–36], also known as relative evaluation [37], is a method used for comparing the resulting segments against a manually-delineated reference polygons. For instance, Lucieer et al. [38] quantified the uncertainty of segments by those with the largest overlapping area with corresponding reference polygons. Möller et al. [39] and Clinton et al. [40] used the area of each overlapping polygon partitioned by segments and reference polygons. Persello et al. [41] and

Marpu et al. [24] used the largest area of overlapping polygons. Clinton et al. [40] also summarized goodness, area-based, location-based, and combined measures that facilitate the identification of optimal segmentation results relative to a training set. Marpu et al. [24] provided a detailed view of the segmentation quality in respect to over- and under-segmentation compared with reference polygons, which proved that MRS performs well under a reasonable SP. Liu et al. [42] proposed three discrepancy indices—Potential Segmentation Error (PSE), Number-of-Segments Ratio (NSR), and Euclidean Distance 2 (ED2)—to measure the discrepancy between the reference polygons and the corresponding image segments. PSE, NSR and ED2 were used [43] and adopted by Aguilar et al. [17] and Gao et al. [44] to evaluate the effects of different segmentation parameters on MRS, and modified by Novelli et al. [45] to the evaluation of object-based greenhouse detection. Cai et al. [46] presented four kinds of supervised measurement methods based on area, object number, feature similarity, and distance to study the influence of different object characteristics on extraction accuracy. With defined variables, the pros and cons of these supervised evaluation methods are discussed in Section 3.4. of this paper, and more detailed reviews of accuracy assessment for object-based image analysis can be found in Ye et al. [47] and Chen et al. [48].

The three main contributions of this study were: (1) to improve the PG segments derived from eCognition, we tried a two texture analysis method that involved increasing AOS or decreasing GQL prior to MRS; (2) to evaluate the quality of PG segments generated from different derivative images, we designed a supervised evaluation pattern named the Over-Segmentation Index (OSI)-Under-Segmentation Index (USI)-Error Index of Total Area (ETA)-Composite Error Index (CEI) based on pixel level and independent from the number of manual delineated reference polygons; and (3) to prove the availability of the proposed pattern, we compared it with several supervised evaluation methods theoretically, and contrasted it with the PSE-NSR-ED2 method by numerical and visualized analysis.

The remainder of this paper is organized as follows: Section 2 introduces the study area and data source, Section 3 explains the methodologies applied in the analysis, Section 4 outlines the effects of image texture analysis on PG segments via recognition of the OSI-USI-ETA-CEI pattern and explains our hypothesis, Section 5 discusses several key points and provides a comparison of our method with some related methods, and Section 6 summarizes the conclusions.

2. Study Area and Data Sets

2.1. Study Area

This study was conducted in Shouguang City, Shandong Province, P.R.China, which is an agricultural region called the "hometown of Chinese vegetable greenhouses" (Figure 1).

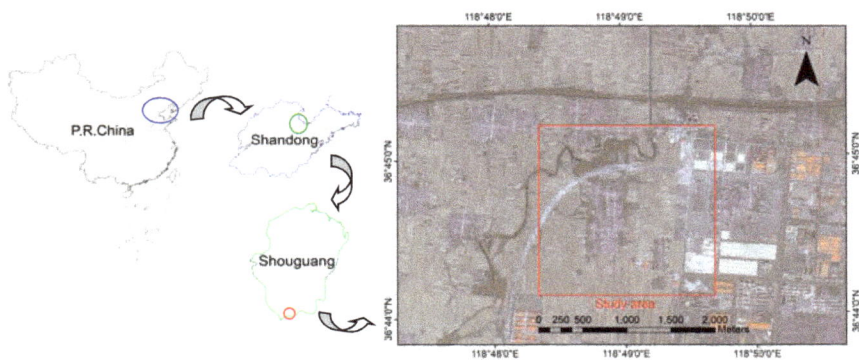

Figure 1. Location of the study area on a Red-Green-Blue GaoFen-2 image taken on 25 April 2016. Coordinate system: WGS_1984_UTM_Zone_50N.

The study area (36°44′40″N and 118°49′0″E) was chosen for these reasons: (a) greenhouses are the main local production mode and are developing rapidly in Shouguang City; (b) even though the greenhouses account for nearly half the area in the selected region, they are adjacent to various land cover types such as water, trees, buildings with high reflectance, residences, and barren land, which form a representative common image; and (c) both continuous and scattered greenhouse can be found in the selected region.

2.2. GF-2 Data and Pretreatment

As shown in Figure 1, the GF-2 imagery selected in this study was acquired on April 25, 2016, which is a high-yield period for greenhouse crops [49].

GF-2 is equipped with two high-resolution scanners with 1 m panchromatic and 4 m multispectral, and was launched on August 19, 2014. GF-2 started imaging and transmitting data on August 21, 2014. Table 1 introduces the payload parameters of the GF-2 satellite [50].

Table 1. Payload parameters of GF-2 satellite

Camera	Band No.	Spectral Range (μm)	Spatial Resolution (m)	Swath Width (km)	Side-Looking Ability	Repetition Cycle (Days)
Panchromatic	1	0.45–0.90	1			
Multispectral	2	0.45–0.52	4	45 (2 Camera Stitching with)	±35°	5
	3	0.52–0.59				
	4	0.63–0.69				
	5	0.77–0.89				

To take full advantage of both panchromatic and multispectral bands, the first pretreatment step is Rational Polynomial Coefficients (RPC) orthorectification, and then image fusion. The Gram-Schmidt Pan Sharpening method in ENVI 5.3 was adopted in this study, and the depth of the resulting fusion image is 16 bits; thus, the greyscale quantization level (GQL) of the GF-2 fusion imagery is 65,536 (2^{16}). The computer employed in the experiments had the following specification:

(1) Processor: Intel® Core™ i7-8700K CPU @ 3.70GHz (12 CPUs);
(2) Graphics adapter: NVIDIA® GeForce® GTX™ 1080 Ti, 11 GB;
(3) Memory: SAMSUNG® DDR4 2400MHz, 2 × 8 GB and SAMSUNG® DDR4 2400MHz, 2 × 16 GB;
(4) Hard disk: SAMSUNG® MZVLW256HEHP-000H1, 256 GB and Seagate® ST2000DM001-1ER164, 2 TB;
(5) Operating system: Microsoft® Windows® 10 Professional, 64-bit.

2.3. Reference Polygons and Field Validation

To verify extraction results, reference polygons were first manually delineated from the GF-2 fusion image. Polygons that were hard to judge whether or not they represent greenhouses from the image were validated or amended by field investigation. To illustrate, four verification points are demonstrated in Figure 2. Three statistical parameters of the reference polygons were obtained in ArcGIS 10.3 (Esri, Redlands, CA, USA): the number of reference polygons was 151, and the summation area was 1,659,078 m², and the total area of study area was 4,000,000 m².

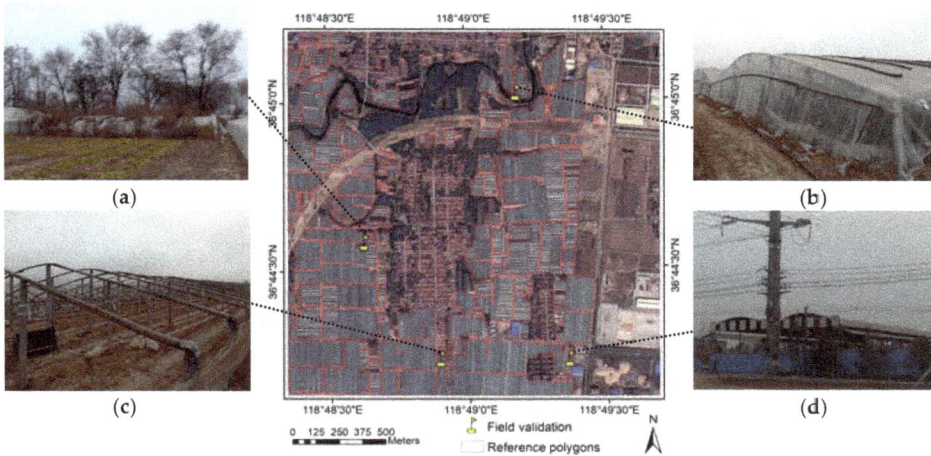

Figure 2. Sketch map of field validation, reference polygons, and (**a**) abandoned greenhouse covered with weeds and shrubs, (**b**) greenhouses with some pixels with high reflectance, (**c**) unsheathed greenhouses, and (**d**) another shed used for storage, which is much taller than greenhouses.

3. Methodology

A flowchart of experiment design, methods, variables, and indicator system for the evaluation of the effects of texture analysis on PG segments is shown in Figure 3.

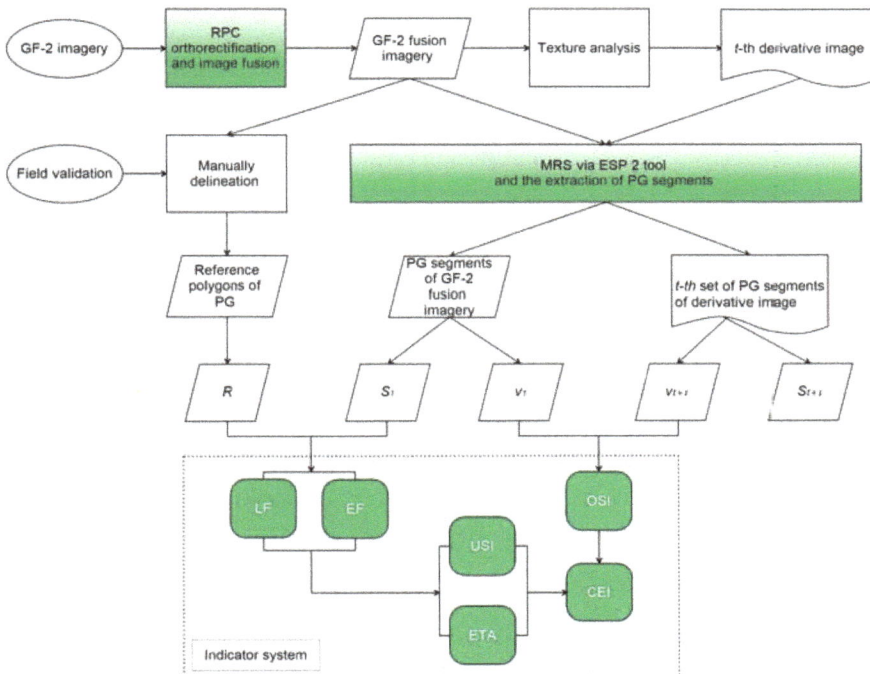

Figure 3. Flowchart of experiment design, methods, variables, and indicator system.

3.1. Texture Analysis

Texture is the visual effect caused by spatial variation in tonal quantity over relatively small areas [51], among which the homogeneity and heterogeneity are a pair of coupled features. Even though homogeneity is more frequently employed in texture analysis, we choses the concept of heterogeneity to explain our method and enable understanding. The definition of heterogeneity refers to the distinctly nonuniformity in composition or character (i.e. color, shape, size, texture, etc.)

PG can be more discernible in very high resolution satellite imageries such as Quickbird, Worldview, and GF-2, whereas the heterogeneity is a nonnegligible obstacle when segmenting these images based on GEOBIA. If the heterogeneity of a PG surface can be compressed, a better segmentation result might be derived from the processed image. Considering the nature of heterogeneity in a digital number image, image preprocessing that increases the average operator size (AOS) or decrease the greyscale quantization level (GQL) is the method used to produce derivative images with different heterogeneities in this study.

For an averaging operator [11], the template weighting functions are unity (such as $1/9$ in AOS 3×3). The goal of averaging is to reduce noise, which is its foundation for compressing the heterogeneity. Averaging is a low-pass filter, since it allows low spatial frequencies to be retained and to suppress high frequency components. The size of an averaging operator is then equivalent to the reciprocal of the bandwidth of the low-pass filter it implements. A larger template, say 11×11 or 13×13, will remove more noise (high frequencies) but reduce the level of detail.

The GQL size is dependent on the maximum quantization level in a monochromatic image or a single channel of a multichannel image. It can be decreased according to the assigned maximum quantization level and a particular weighted combination of frequencies, which is a redistribution of the greyscale value at each pixel, so that the values can be clustered in a certain range if they are spread over a broad range. As long as GQL decreases, the heterogeneity of each band is compressed.

By increasing AOS or decreasing GQL, information of the pixel's neighborhood can be effectively used, preceding the MRS. To evaluate the effects of AOS and GQL on MRS segmentation, four increased AOSs (3×3, 5×5, 7×7, and 9×9) and three decreased GQLs (128, 64, and 32) were adopted to produce another 19 images based on the initial fusion imagery (GQL initial). Hence, there were 20 input data that were used for segmentation, rather than merely evaluating the segmentation results from the sole data source.

The 19 derivative images were also produced in ENVI 5.3, in which the co-occurrence measures tool can simultaneously change the AOS and GQL of multi-bands among GQL initial, 64, and 32. The derivative images of GQL 128 were produced using the stretch tool, and averaging operations on GQL 128 were conducted using low pass convolution filters, since the co-occurrence measures tool does not support the conversion between GQL initial and GQL 128.

3.2. MRS via ESP 2 Tool

MRS in eCognition is based on the Fractal Net Evolution Approach (FNEA) principle and is widely used for segmentation. It is a region-growing process, and the optimization procedure starts with single-image objects of one pixel and repeatedly merges them in pairs to larger units until an upper threshold is not exceeded locally [8,17,18]. For this purpose, a scale parameter (SP) is proposed to adjust the threshold calculation. Higher values of the scale parameter would result in larger image objects, and smaller values result in smaller image objects. The basic goal of an optimization procedure is to minimize the incorporated heterogeneity at each single merge [8]. If the resulting increase in heterogeneity when fusing two adjacent objects exceeds a threshold determined by the SP, then no further fusion occurs, and the segmentation stops [33]. The SP criteria are defined as a combination of shape and color criteria (color = $1 -$ shape), whereas shape is interiorly divided as compactness and smoothness criteria; thus, the three parameter values that must be set are SP, shape, and compactness.

ESP 2 is a generic tool for eCognition software that employs local variance (*LV*) to measure the difference in the MRS under increment scales [22]. When the *LV* value at a given level (LV_e) is equal to or lower than the value recorded at the previous level (LV_{e-1}). The level $e - 1$ is then selected as the OSP for segmentation. Based on this concept, ESP 2 can help derive the dependent SP, whereas shape and compactness can be deduced from the try-and-error experiment within different assessment systems [17,18], which recommend obtaining the SP parameter by fixing the compactness at 0.5 and the testing shape values around 0.3.

Since this study focused on the effect of texture analysis on MRS, the uniform shape and compactness were set to 0.3 and 0.5, respectively. Thus, the OSP was automatically calculated by the ESP 2 tool with the algorithm parameters set as shown in Table 2. The Level 1 and its segments in the exported results were adopted for the next step of analysis.

Table 2. Algorithm parameters and settings in the ESP 2 tool.

Parameter	Value	Parameter	Value
Use of Hierarchy (0 = no; 1 = yes)	1	Starting scale_Level 3	10
Hierarchy: TopDown = 0 or BottomUp = 1?	1	Step size_Level 3	100
Starting scale_Level 1	10	Shape (between 0.1 and 0.9)	0.3
Step size_Level 1	1	Compactness (between 0.1 and 0.9)	0.5
Starting scale_Level 2	10	Produce LV Graph (0 = no; 1 = yes)	1
Step size_Level 2	10	Number of Loops	200

3.3. Extraction of PG Segments

As different derivative images required different samples, features, parameters, or threshold values in automatic extraction, and ensuring good quality is difficult, the greenhouse objects in this study were manually selected by artificial visual interpretation using the single select button on the manual editing toolbar in eCognition 9.0, so that each segmentation object can be as precisely evaluated as possible. Theoretically, manual extraction would have a maximum precision on the criterion of geometric accuracy, but this is only credible for the criterion of the area, since, in other methods, the commission area also has a probability of offsetting the omissions while the geometric error can only be accumulated.

The principle used to assign an object as a greenhouse is when the proportion of greenhouse area is more than 60% [24] and the feature of other categories is negligible from visual analysis. Otherwise, the object's feature is deemed to be unusable to extract the greenhouse contained within, which would be evaluated as omission error in follow-up work.

After exporting from eCognition 9.0, two statistical parameters of the extracted PG segments were obtained in ArcGIS 10.3: number of PG segments and summation area.

3.4. Establishment of OSI-USI-ETA-CEI Pattern

3.4.1. Case Study and Variable Definition

To better understand the problems in PG segmentation, the definitions of variables, and the establishment of OSI-USI-ETA-CEI pattern, five cases of PG segments that were extracted from initial GF-2 fusion imagery and four derivative images are demonstrated in Figure 4, and all images were segmented under their optimal scale parameter provided by the ESP 2 tool. Notably, these cases cannot represent the segmentation quality of a whole image.

Without decreasing the GQL, the degree of over-segmentation of the PG segments that were extracted from the initial GF-2 fusion imagery (Figure 4a) or images derived from the treatment of AOS 3 × 3 (Figure 4b) are much worse than those extracted from other derivative images (Figure 4c–e), since the dark and the sunny sides of the PG in the two images are segmented as different parts, making it hard to delineate the PGs' boundaries, which is not convenient for subsequent feature recognition and extraction.

Apart from the number of PG segments, the number and area of the fragments (the smaller polygons that are partitioned jointly by reference polygons and PG segments) are also need to be explored in depth.

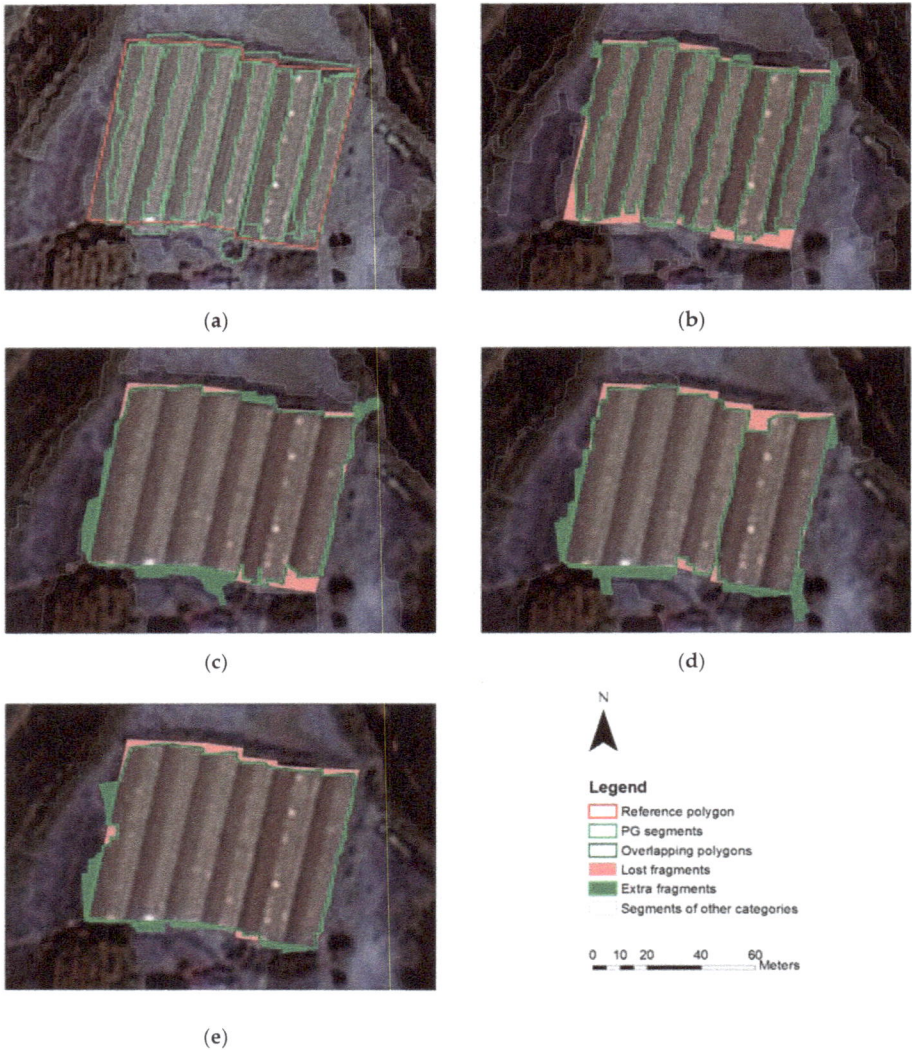

Figure 4. (**a**) Reference polygon and PG segments resulting from initial GF-2 fusion imagery and overlapping polygons, lost fragments, extra fragments, and derivative images resulting from the treatment of (**b**) GQL initial and AOS 3 × 3, (**c**) GQL 128 and AOS 3 × 3, (**d**) GQL 64 and AOS 3 × 3, and (**e**) GQL 32 and AOS 3 × 3.

To parameterize the relationship between the reference polygons and PG segments, four quantity-based variables, seven area-based variables and their assemblies were defined in Table 3:

Table 3. Four quantity-based variables, seven area-based variables and their assemblies.

Variable	Definition
m	total number of reference polygons
v	total number of PG segments
n	number of reference polygons that have no PG segments overlapping with them, $n \leq m$
u_x	x-th number of corresponding segments found for one single reference geometry, $x \in [0, m-n]$ [45]
r_i	i-th reference polygon of assembly R; R indicates the real area of PG, $i \in (0, m]$
s_j	j-th extracted PG segment of assembly S; S indicates the extraction area of PG, $j \in (0, v]$
o_k	k-th polygon of assembly O; O indicates the overlapping area between R and S, $O = R \cap S$
bs_h	h-th element of Biggest Segments (BS); BS is the assembly of PG segments representing the biggest overlapping polygon within its corresponding reference polygon, indicating the total area of biggest segments, $BS \subseteq S$
bo_h	h-th element of Biggest Overlaps (BO); BO is the assembly of overlapping polygons where each is partitioned by every bs_h and its corresponding reference polygon, indicating the total area of biggest overlaps, $BO \subseteq O$
lf_p	p-th element of Lost Fragments (LF); LF is the assembly of fragments where each is part of R and also part of a segment that cannot represent PG (fragments in R but outside of O, which are shown in coral red in Figure 4), LF indicates the total area of lost fragments [24]
ef_q	q-th element of Extra Fragments (EF); EF is the assembly of fragments where each is part of S but not part of R itself (fragments in S but outside O, shown in dark green in Figure 4), EF indicates the total area of extra fragments [24]

It is generally thought that a high-quality image segmentation should result in a minimum amount of over- and under-segmentation, and different area-based or number-based indicators have been designed based on selected samples and their corresponding reference polygons [14,24,39,41–43,45,52], which we rewrote using variables defined above for comparison, as shown in Table 4.

Table 4. Different area-based or number-based indicators of over- and under-segmentation.

Year	Reference	Over-Segmentation	Under-Segmentation
2002	Lucieer et al. [38]	$\frac{r_i - bs_h}{r_i} > 0$	$\frac{r_i - bs_h}{r_i} < 0$
2007	Möller et al. [39]	$\frac{o_k}{r_i}$	$\frac{o_k}{s_j}$
2010	Clinton et al. [40]	$1 - \frac{o_k}{r_i}$	$1 - \frac{o_k}{s_j}$
2010	Persello et al. [41]	$1 - \frac{bo_h}{r_i}$	$1 - \frac{bo_h}{s_j}$
2010	Marpu et al. [24]	$\frac{bo_h}{r_i}$	$\frac{lf_p}{r_i}, \frac{ef_q}{r_i}$
2012	Liu et al. [42]	$NSR = \frac{|m-v|}{m}$	$PSE = \frac{EF}{R}$
2016	Novelli et al. [45]	$NSR_{new} = \frac{|m-v-n \times \max(u_x)|}{m-n}$	$PSE_{new} = \frac{EF + n \times \max(ef_q)}{\sum_{i=1}^{m-n} r_i}$

Some feature similarity-based, location-based or distance-based [41,46] methods are available for measuring the quality of segmentation, but these methods only work when segments have an approximately one-to-one relationship with the reference polygons, whereas the segmentation results of continuous greenhouses always have the relationship with the reference of poly-to-one.

The OSI-USI-ETA-CEI pattern is based on the reference polygons that were manually delineated in Section 2.3. and the PG segments extracted in Section 3.3. The method is designed for evaluating the segmentation quality of PG segments from images with different heterogeneities. In short, OSI denotes the extent to which the number of PG segments may affect the USI and ETA, USI indicates the absolute geometric error of PG segments, ETA indicates the discrepancy in the total area between PG segments and reference polygons, and CEI indicates the composite error.

3.4.2. Over-Segmentation Index (OSI)

Segmentation results that are over-segmented are more likely to cause omission and commission errors in follow-up classification because the number and some feature values (like mean value) of both the interested and non-interested objects that are over-segmented would range widely compared with those that are not over-segmented. An extreme example of this is when an image is segmented on the pixel level. Even though it is much easier to compare the number rather than other feature values for two assemblies of polygons, indicators that compare the number of reference polygons with segments [17,18,42–45] and that compare the area of a reference polygon with its biggest corresponding segment [24,38,53] were both designed or applied in over-segmentation analysis. However, these criteria are designed for pursuing perfect segmentation that is similar to manually delineated reference polygons, which is not suitable for evaluating the segments of continuous PGs since drawing those reference polygons is different from segmenting an image. Reference polygons prefer to draw the outline of a single or continuous greenhouse rather than divide pixels with different grey levels, whereas segmentation prefers the latter, especially when some pixels' material or Bidirectional Reflectance Distribution Function (BRDF) varies significantly from their surrounding pixels. To manually draw the outline of continuous greenhouses is so subjective that it is hard to determine the size of a reference polygon as well as the total number of reference polygons, i.e., no wonderful polygons can be used to define whether a segmentation is over-segmented or not. A similar view was reported by Zhang et al. [30]. Thus, continuous greenhouse extraction in high-resolution imagery does not require a similar segment number compared to the manual reference polygons, nor accordant outlines or even skeletons (Figure 4). However, we should consider counting the segment numbers in the initial fusion imagery as an effective reference to assess over-segmentation instead, since the high heterogeneity among greenhouse pixels in an initial fusion image tends to lead to the worst over-segmenting result compared with derivative images with lower heterogeneity. Therefore, the segment numbers of the initial fusion imagery under its OSP using ESP 2 can be regarded as ancillary data of reference polygons in Section 2.3. The ancillary data provides a numerical reference and the manually delineated polygons provides a geometric reference.

Synthesizing the situation stated above, over-segmentation of PG segments is indicated by a new OSI in this study, which is a relative value calculated by Equation (1):

$$OSI = \frac{v}{v_1} \tag{1}$$

where v denotes the number of extracted PG segments when the corresponding image is under the optimal segmentation using ESP 2 tool, v_1 denotes the number of PG segments extracted from the initial GF-2 fusion image, and v_{t+1} denotes the number of PG segments extracted from the t-th derivative image. A higher OSI indicates a larger error of over-segmentation.

3.4.3. Under-Segmentation Index (USI)

When an image is over-segmented, it is still possible to construct the object, but when an image is under-segmented, the object may not be recovered [24]. Under-segmentation is more worthwhile to be exactly evaluated.

From Figure 4, both lost and extra fragments are shown to have many members with very tiny areas, and the boundaries of the reference polygons usually have fewer polylines than that of PG segments. The number of lost and extra fragments are not only caused by geometric errors but also changed by how we draw the outline of single or continuous greenhouses, so it is not appropriate to neither count the number nor calculate the mean value of those fragments [24] when evaluating the geometric errors of continuous greenhouses.

In general, the area of extra-segments are parts of under-segmentation error according to some studies (Table 4), as the PG segments should slice those pixels that are not representing a PG. However, lost fragments can be considered as a result of under-segmentation of those segments that do not contain enough PG pixels, i.e., the lost fragments can be regarded as the extra fragments of another category (Figure 4). Therefore, it is necessary to adequately evaluate both the *LF* and *EF* error rather than to consider only one and then combine the two errors into a single index (USI) to indicate the intension of under-segmentation of PG segments. The theoretical value should range between zero and one. The index can be calculated as:

$$\text{USI} = \frac{LF + EF}{R} \tag{2}$$

where *LF*, *EF*, and *R* are the total area of lost fragments, extra fragments, and the real area of PG, respectively. A higher USI indicates a larger under-segmentation error.

3.4.4. Error Index of Total Area (ETA)

Lost and extra fragments have an opposite influence on the final area of extraction even though both directly contribute to the under-segmentation. Although area-based measures were discussed in Clinton et al. [40] and some new measures based on area were designed after that [24,42,46], these sample-based studies only focused on the proportion of the omission or commission area in a segmentation, but the percentage of the difference in the total area between segmentation results and corresponding reference polygons seems to be ignored, which should be fully considered when evaluating the precision of the total area of extraction and the consequence of under-segmentation. Thus, the Error Index of Total Area (ETA) can be used to indicate the discrepancy, which can be calculated by:

$$\text{ETA} = \frac{|S - R|}{R} = \frac{|EF - LF|}{R} \tag{3}$$

where S denotes the summation area of extracted PG segments when corresponding image is under the optimal segmentation using the ESP 2 tool, S_1 denotes the summation area of PG segments extracted from the initial GF-2 fusion image, and S_{t+1} denotes the summation area of PG segments extracted from the t-th derivative image. A higher ETA value indicates a larger total area error.

3.4.5. Composite Error Index (CEI)

In general, the more the PG segments are over-segmented, the larger the omission and commission error produced indirectly in automatic classification or extraction. Under-segmentation causes geometric errors and directly leads to an area difference from the reference. Given this consideration, a new CEI is presented in Equation (4) to consider the composite error of segmentation results when comparing to a set of reference polygons:

$$\text{CEI} = \lambda \times \text{OSI} + \text{USI} + \text{ETA} \tag{4}$$

where λ is a possible weight used to rescale the value of quantity-based OSI so that the indicator will not overwhelm the value of area-based USI and ETA; thus, the OSI multiplied by λ denotes the indirect influence of the number of PG segments on extraction in CEI.

When omission or commission segments in an extraction are generated due to over-segmenting, their geometric error and area difference from the real value (indicated by USI and ETA, respectively) couple on the extraction. Thus, the value of λ in this study is defined as the sum of USI and ETA as:

$$\lambda = \text{USI} + \text{ETA} \tag{5}$$

4. Results

4.1. Derivative Images with Different Heterogeneities

We provide two images (Figure 5) as examples to show the visual disparity of different heterogeneities. Even though there is no significant difference at first sight, some subtle distinctions can be found from the white roof (in red frame), and the texture in image that derived from the treatment of GQL 64 and AOS 3 × 3 is more distinct than in initial GF-2 fusion imagery. Details emerge in the images as long as they are segmented (Figure 6).

(a)   (b)

Figure 5. (a) Initial GF-2 fusion imagery; (b) image derived from the treatment of GQL 64 and AOS 3 × 3.

4.2. PG Segments in Images with Different Heterogeneities

Seven sets of PG segments are shown in Figure 6 as examples to demonstrate the visual disparity. The number of PG segments extracted from initial fusion imagery is outdistancing the other situations as well as reference polygons. Another significant difference is the number of segments of the white roof (in red frame). In Figure 6a–c, the roof's boundary is hard to distinguish from the thumbnails, while the other examples are much better. The boundaries of the PG segments in Figure 6c,d are more orderly both horizontally and vertically, which conforms to the outlines of the greenhouses, whereas the segmentation results in Figure 6e–g irregularly delineate the greenhouses.

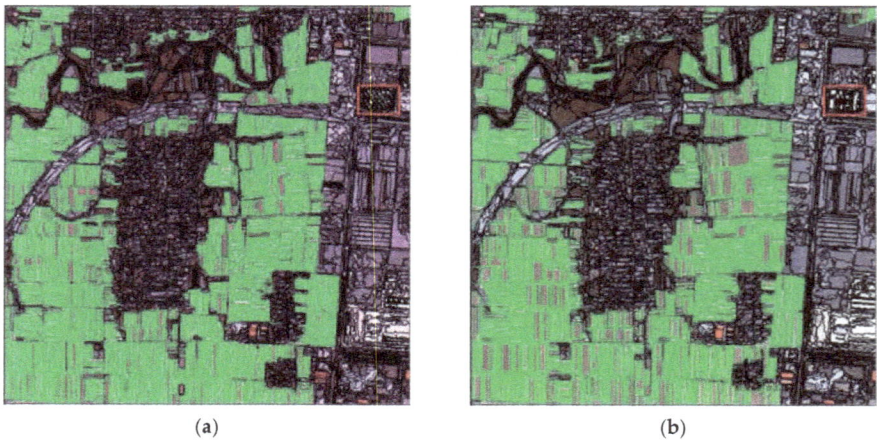

(a)   (b)

Figure 6. *Cont.*

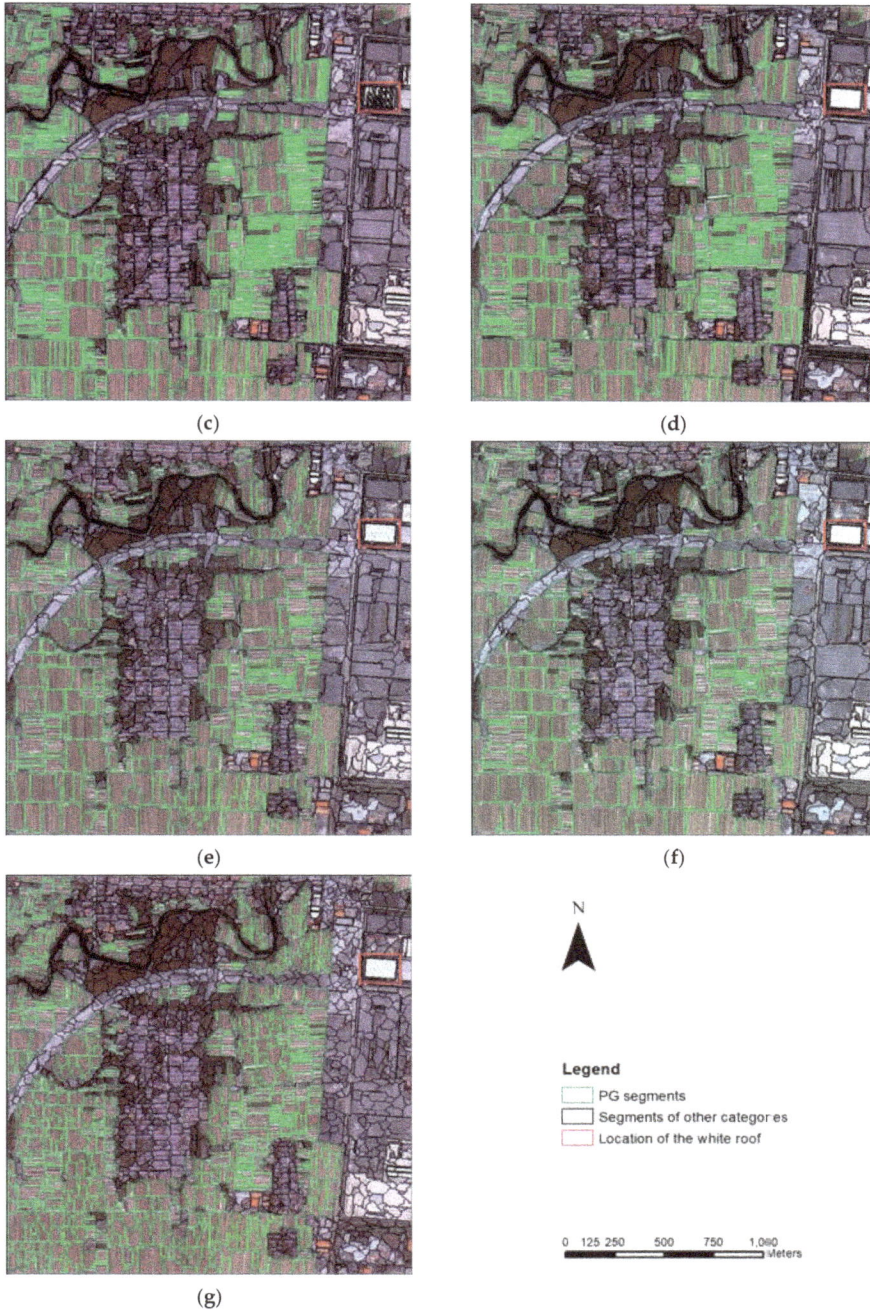

Figure 6. (a) PG segments based on initial GF-2 fusion imagery; and derivative images as well as PG segments resulting from the treatments of (**b**) GQL initial and AOS 3 × 3, (**c**) GQL 128 without an averaging operator, (**d**) GQL 128 and AOS 3 × 3, (**e**) GQL 64 without an averaging operator, (**f**) GQL 64 and AOS 3 × 3, and (**g**) GQL 32 without an averaging operator.

Apart from the examples in Figure 6, the indicator values of each set of PG segments that were extracted from different images are shown in Table 5, while the summation area of reference polygons (*R*) was 1,659,078 m², and OSP is the optimal scale parameter calculated by the ESP 2 tool with fixed shape of 0.3 and compactness of 0.5. We use DIF denotes the value of CEI of each set of PG segments minus that of PG segments of the initial fusion imagery.

Table 5. Indicator values of each set of PG segments that were extracted from different images.

GQL	AOS	OSP	v	S (m²)	LF (m²)	EF (m²)	OSI	USI	ETA	CEI	DIF
initial	none	81	3007	1,690,287	86,211	117,420	1.000	0.123	0.019	0.283	0.000
	3 × 3	103	1737	1,659,104	109,546	109,572	0.578	0.132	0.000	0.208	−0.075
	5 × 5	161	711	1,733,075	89,339	163,336	0.236	0.152	0.045	0.243	−0.040
	7 × 7	197	446	1,710,681	114,154	165,757	0.148	0.169	0.031	0.229	−0.054
	9 × 9	151	507	1,693,593	122,930	157,805	0.169	0.169	0.021	0.222	−0.061
128	none	82	808	1,700,510	102,356	143,788	0.269	0.148	0.025	0.220	−0.063
	3 × 3	94	561	1,660,182	128,221	129,325	0.187	0.155	0.001	0.185	−0.098
	5 × 5	109	380	1,679,966	133,606	154,494	0.126	0.174	0.013	0.210	−0.073
	7 × 7	100	400	1,689,608	123,859	154,389	0.133	0.168	0.018	0.211	−0.072
	9 × 9	115	267	1,703,666	117,728	162,316	0.089	0.169	0.027	0.213	−0.070
64	none	50	457	1,781,987	87,601	210,510	0.152	0.180	0.074	0.292	0.009
	3 × 3	56	352	1,796,653	90,983	228,558	0.117	0.193	0.083	0.308	0.025
	5 × 5	64	255	1,827,083	83,023	251,028	0.085	0.201	0.101	0.328	0.045
	7 × 7	67	250	1,870,392	72,295	283,609	0.083	0.215	0.127	0.370	0.087
	9 × 9	82	171	1,903,843	81,584	326,349	0.057	0.246	0.148	0.416	0.133
32	none	31	821	1,738,822	103,795	183,539	0.273	0.173	0.048	0.282	−0.001
	3 × 3	39	527	1,793,144	96,658	230,724	0.175	0.197	0.081	0.327	0.044
	5 × 5	46	355	1,773,550	130,505	244,977	0.118	0.226	0.069	0.330	0.047
	7 × 7	49	361	1,792,180	119,801	252,903	0.120	0.225	0.080	0.341	0.058
	9 × 9	46	379	1,802,023	110,171	253,116	0.126	0.219	0.086	0.344	0.061

The experiment was designed to find the optimal set of GQL and AOS, which could result in optimum PG segments for the extraction. PG segments in derivative image with the treatment of GQL 128 and AOS 3 × 3 has the lowest CEI, which is consistent with visual analysis in Section 3.

4.3. Effects of Image Texture Analysis on PG Segments

4.3.1. Effects of Increasing AOS on PG Segments

Images under four kinds of GQLs can be employed to process by four AOSs, so we could evaluate the effects of increasing AOS on PG Segments as Figures 7 and 8 shown.

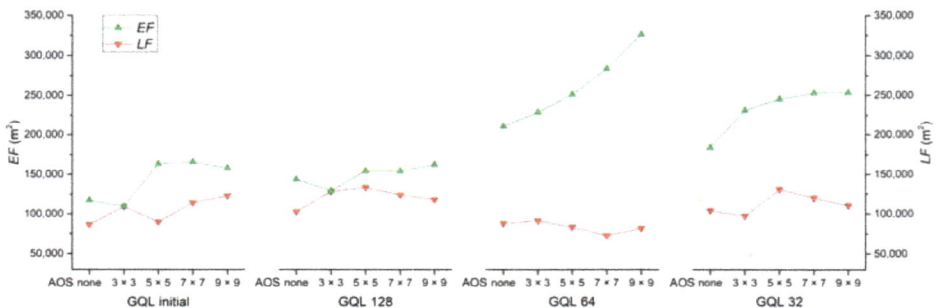

Figure 7. Effects of increasing AOS on PG segments using the area of lost and extra fragments.

With the increase in AOS (Figure 7), both the sum of *LF* and *EF* (related to USI) and the distances between them (related to ETA) have relatively low values under the treatments of GQL initial and

GQL 128 than that of GQL 64 and GQL 32. Variation of *LF* and *EF* is the foundation to understand the value of USI and ETA, while the sum of USI and ETA is main source of CEI (Figure 8).

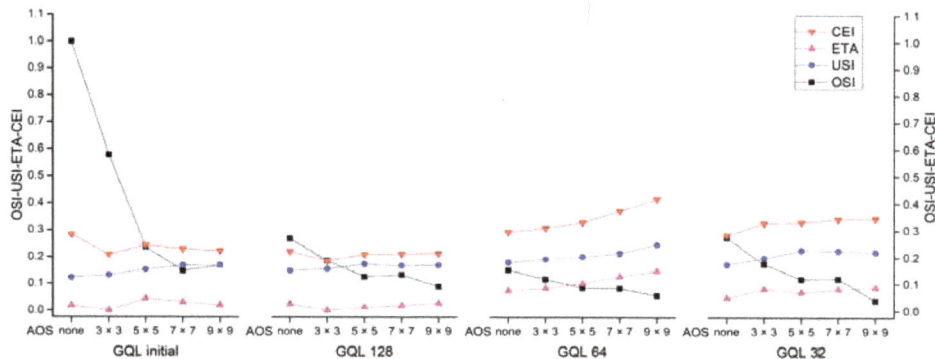

Figure 8. Effects of increasing AOS on PG segments using the OSI-USI-ETA-CEI pattern.

For GQL initial and GQL 128, the AOS 3 × 3 setup can let the values of ETA and CEI reach their minimum simultaneously, and significantly decrease the value of OSI, whereas USI increases somewhat.

For GQL 64 and GQL 32, the increase in AOS lead to the increase in CEI, which was not expected.

For each kind of GQL, the increase in AOS lead to opposite change trends of USI and OSI, whereas CEI had the same with ETA. The curves under GQL 128 are smoother and steadier than other GQLs with increasing AOS.

4.3.2. Effects of Decreasing GQL on PG Segments

Five AOS setups were used to evaluate the effects of decreasing GQL on PG Segments as shown in Figures 9 and 10.

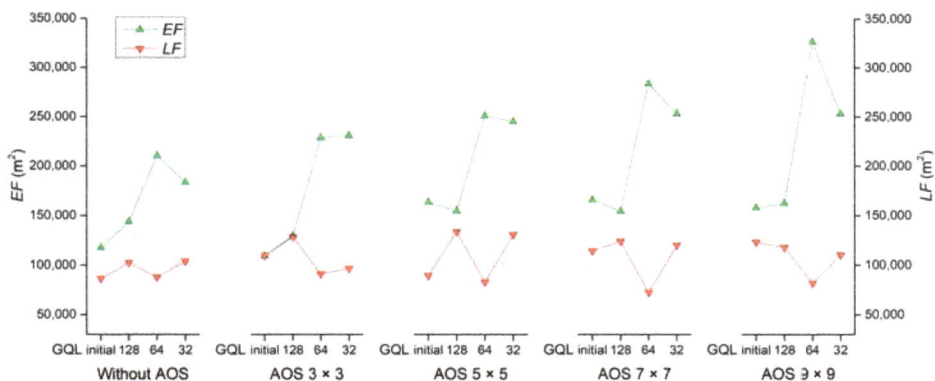

Figure 9. Effects of decreasing GQL on PG segments using the area of lost and extra fragments.

Similar to Figure 7, Figure 9 demonstrates the superiority of GQL initial and GQL 128 on both the sum of *LF* and *EF* (related to USI) and the distances between them (related to ETA), which shows the lower omission and commission errors of PG segments than GQL 64 and GQL 32.

From Figure 10, the treatment of GQL 128 has the lowest CEI values under each setup of AOS compared to other GQLs, among of which AOS 3 × 3 produced the minimum value.

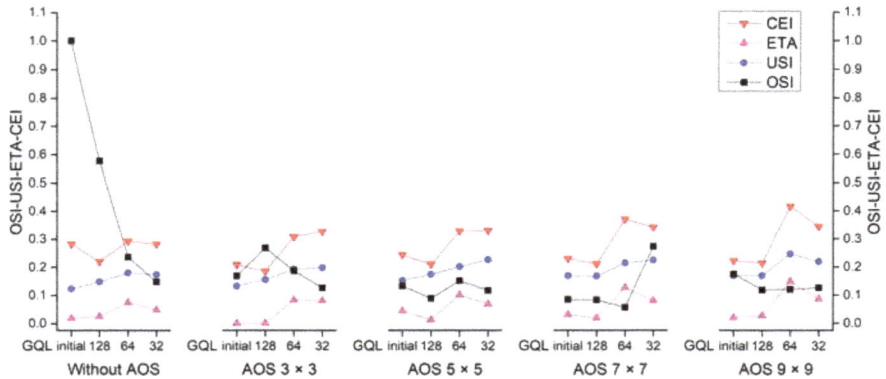

Figure 10. Effects of decreasing GQL on PG segments using the OSI-USI-ETA-CEI pattern.

4.3.3. Combined Effects of AOS and GQL on PG Segments

Compared with the initial fusion imagery, the increase in AOS and the decrease in GQL can reduce the influence of over-segmentation but can also increase the error of both under-segmentation and the extraction area. An interaction of the effects occurs among OSI, USI, and ETA, all of which synthetically decide the CEI.

After the treatment of different AOSs on GQL initial and GQL 128, the CEI was reduced by 4.0 to 9.8% compared to the initial fusion imagery, whereas the treatment of different AOSs on GQL 64 and GQL 32 increase the CEI by 0.9 to 13.0%, except that of GQL 32 without an averaging operator, which was very close to the initial fusion imagery with only a 0.1% reduction.

Thus, the optimum texture analysis setup for GF-2 fusion imagery is GQL 128 and AOS 3 × 3, since the PG segments of the derivate image can reduce the CEI by 9.8% than PG segments of the GF-2 fusion imagery.

5. Discussion

Since the segmentation results from the initial fusion image are highly fragmented compared with the boundary of real-world entities, compressing the heterogeneity of adjacent pixels before segment was notable. To improve the PG segments derived from eCognition, the innovation of our method is that we evaluate the effects of texture analysis on PG segments using OSI-USI-ETA-CEI pattern, based on the nature of segmentation and heterogeneity in a digital number image.

5.1. Evaluation Problems

Several problems need to be considered when measurement methods are used for the evaluation of PG segments.

First, the confusion matrix is a common method for evaluation or verification, but in object-based extraction, a problem occurs for each segmented object: omission and commission errors may coexist compared to the reference polygon. Thus, the extraction result cannot be accurately evaluated by choosing those segmented objects with an inaccurate boundary as the true values [19,20]. Each greenhouse object's omission and commission errors should be considered on the pixel-based rather than the object-based level.

If the specified proportion of a greenhouse that is adopted to define a segment as a PG increases or reduces, the percentage of *LF* and *EF* would change either [24].

As a non-negligible indicator, the difference in the area between the extraction results and the reference polygons must also be evaluated quantitatively.

5.2. Comparison with Related Evaluation Research

Several pointed discussions in Section 3.4. theoretically explained why the OSI-USI-ETA-CEI method differs from existing indexes, which is summarized in Table 6. Notably, the existing indexes are designed based on scattered reference polygons and the corresponding overlapped PG segments, whereas the OSI-USI-ETA-CEI method in this study was built on the whole image.

Table 6. Comparing the OSI-USI-ETA-CEI pattern with related evaluation research.

Proposed Pattern	Comparison with Related Evaluation Research
OSI	Calculated by the ratio of the number of PG segments of derivate image to the number of PG segments of initial fusion imagery, instead of ignoring the impact of the number, or assuming the delineated polygons have a dependable quantity
USI	Calculated by the proportion of area of lost and extra fragments together as a consequence of under-segmentation, instead of calculating only one of them or separately
ETA	Considers the difference in area between extraction results and reference polygons
CEI	Rescale the OSI by geometry and area discrepancy first and then simply sum the rescaled OSI with USI and ETA up, instead of calculating the Euclidean Distance of indicator values

Even though the effect of different shape and compactness values on PG segments are not discussed in this study, we selected them on the basis of Aguilar et al. [17,18], which considered ESP 2 to be an effective tool on PG segmentation. Aguilar et al. [17] evaluated the effects of shape and compactness on the quality of PG segments by the ED2 in Equation (6), and Aguilar et al. [18] compared the PG segments derived from multispectral, panchromatic, and atmospheric correctional multispectral orthoimages under different shape and compactness values by the modified ED2 in Equation (7):

$$ED2 = \sqrt{PSE^2 + NSR^2} \tag{6}$$

$$modified\ ED2 = \sqrt{PSE_{new}^2 + NSR_{new}^2} \tag{7}$$

where PSE, NSR, PES$_{new}$, and NSR$_{new}$ were defined in Section 3.4.1.

The ED2, modified ED2 and CEI are all evaluation methods based on both numeral and areal indicators. Since the ED2 has a similar principle to the modified ED2 and is more computable than that, we use the PSE-NSR-ED2 method contrasted with the OSI-USI-ETA-CEI pattern by numerical and visualized analysis to support our availability.

The best texture analysis setup for GF-2 fusion imagery according to ED2 value is GQL 64 and AOS 9×9, which was judged as the worst one by USI, ETA, CEI and even PSE. Furthermore, the ED2 almost determined by the NSR (Table 7 and Figure 11) shows that ED2 is not suit to evaluate the quality of segmentation results in this study.

Table 7. Values of PSE, NSR, and ED2 of each set of PG segments under different GQL and AOS.

GQL	AOS	PSE	NSR	ED2	GQL	AOS	PSE	NSR	ED2
Initial	none	0.071	18.914	18.914	64	none	0.127	2.026	2.030
	3×3	0.066	10.503	10.504		3×3	0.138	1.331	1.338
	5×5	0.098	3.709	3.710		5×5	0.151	0.689	0.705
	7×7	0.100	1.954	1.956		7×7	0.171	0.656	0.678
	9×9	0.095	2.358	2.360		9×9	0.197	0.132	0.237
128	none	0.087	4.351	4.352	32	none	0.111	4.437	4.438
	3×3	0.078	2.715	2.716		3×3	0.139	2.490	2.494
	5×5	0.093	1.517	1.519		5×5	0.148	1.351	1.359
	7×7	0.093	1.649	1.652		7×7	0.152	1.391	1.399
	9×9	0.098	0.768	0.774		9×9	0.153	1.510	1.518

Therefore, the drawbacks of the ED2 [42] are: (1) equated the attribute of numerical indicator with that of areal indicator; (2) gave the larger indicator a bigger weight than the smaller one in calculation. The modified ED2 does this as well, which is not expected to happen when compositing the numerical and areal indicators.

Although the PSE-NSR-ED2 method loses its efficacy in this study, it did work in several studies [17,42–44]. A possible reason is the heterogeneity of the images of their study areas was not as high as that of the GF-2 fusion data in this study [17].

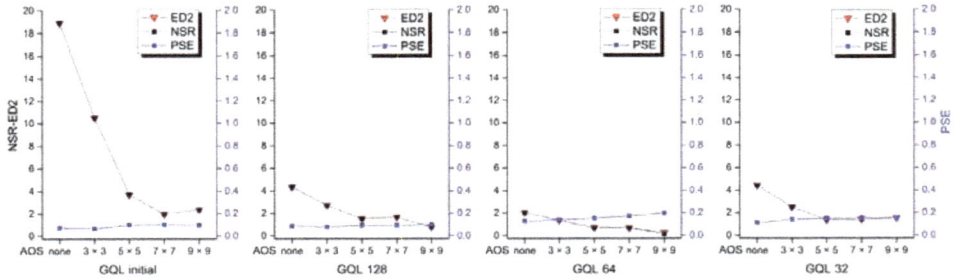

Figure 11. Effects of increasing AOS on PG segments using the PSE-NSR-ED2 pattern.

5.3. Next Steps

Although the experiments are all based on MRS in eCognition, the proposed methods are presented in a general sense and may helpful for practitioners who suffer from segmentation issues in GEOBIA.

The drawback of the proposed method is the number of PG segments relies on the validity of the ESP 2 tool and the PG segments must be obtained by manual selection. Thus, other effective segmentation schemes or automatic extraction methods are needed for experimentation with the method in the future.

We only analyzed the effects of two preprocessing operations (increasing AOS and decreasing GQL) on PG segments, whereas the influence of atmospheric correction on PG segments was evaluated by Aguilar et al. [18], but different methods or tools to change the heterogeneity of input imagery are available, such as median filter, Gaussian averaging operator, Region-Scalable Fitting (RSF) model, and the Laplacian of Gaussian (LoG) operator [11,13].

Since segmentation results are highly scene-dependent, the investigation should also be applied to other scenes and data sources in future studies.

6. Conclusions

This study was designed to examine the ability of extracting greenhouses from GF-2 imagery. To complete the process, compressing the heterogeneity of the initial fusion image was designed to effectively use the texture analysis and improve the MRS, and a new OSI-USI-ETA-CEI pattern was proposed to evaluate the effects of texture analysis on PG segments.

Although this work should be only considered as an initial approach, the following conclusions are drawn:

(1) Appropriate texture analysis applied to a fusion image can change its heterogeneity effectively for better segmentation.
(2) When shape and compactness are fixed at 0.3 and 0.5 respectively, the optimum treatment of GF-2 fusion imagery prior to segmenting the plastic greenhouses using ESP 2 tool is compresses the GQL to 128 and uses the AOS 3 × 3 setup, which reduces the CEI by 9.8% compared with the initial fusion imagery in this study.

(3)   The proposed OSI-USI-ETA-CEI pattern can be applied to evaluate the effects of image processing on the quality of PG segments, which is more accurate but requires a higher workload than the PSE-NSR-ED2 method.

Author Contributions: Y.Y. conceived the idea, designed the research, performed the experiments, analyzed the data and wrote the paper; S.W. contributed materials tools and gave constructive suggestions on review and editing.

Funding: This research was funded by the National Key R&D Program of China, grant number 2017YFB0503805.

Acknowledgments: We thank the China Center for Resources Satellite Date and Application for providing the GF-2 imagery. We thank Wenliang Liu for his funding support on this research. We thank Hongjie Wang and Yumin Gu for their assistance with the field validation. We would like to express our gratitude to the reviewers for helping us better evaluate our ideas in this article. We also thank the Remote Sensing Editorial Office for their helpful work and the MDPI English Editing Service for improving our expressions.

Conflicts of Interest: The authors declare no conflict of interest.

References

1.   Ma, L.; Li, M.C.; Ma, X.X.; Cheng, L.; Du, P.J.; Liu, Y.X. A review of supervised object-based land-cover image classification. *ISPRS J. Photogramm. Remote Sens.* **2017**, *130*, 277–293. [CrossRef]

2.   Levinshtein, A.; Stere, A.; Kutulakos, K.N.; Fleet, D.J.; Dickinson, S.J.; Siddiqi, K. TurboPixels: Fast superpixels using geometric flows. *IEEE Trans. Pattern Anal. Mach. Intell.* **2009**, *31*, 2290–2297. [CrossRef] [PubMed]

3.   Schick, A.; Fischer, M.; Stiefelhagen, R. An evaluation of the compactness of superpixels. *Pattern Recognit. Lett.* **2014**, *43*, 71–80. [CrossRef]

4.   Tian, X.; Jiao, L.; Yi, L.; Guo, K.; Zhang, X. The image segmentation based on optimized spatial feature of superpixel. *J. Vis. Commun. Image Represent.* **2015**, *26*, 146–160. [CrossRef]

5.   Stutz, D.; Hermans, A.; Leibe, B. Superpixels: An evaluation of the state-of-the-art. *Comput. Vis. Image Underst.* **2018**, *166*, 1–27. [CrossRef]

6.   Cousty, J.; Bertrand, G.; Najman, L.; Couprie, M. Watershed cuts: Thinnings, shortest path forests, and topological watersheds. *IEEE Trans. Pattern Anal. Mach. Intell.* **2010**, *32*, 925–939. [CrossRef]

7.   Ciecholewski, M. Automated coronal hole segmentation from Solar EUV Images using the watershed transform. *J. Vis. Commun. Image Represent.* **2015**, *33*, 203–218. [CrossRef]

8.   Baatz, M.; Schäpe, A. Multiresolution segmentation: An optimization approach for high quality multi-scale image segmentation. In *Angewandte Geographische Informationsverarbeitung*; Herbert Wichmann Verlag: Heidelberg, Germany, 2000; pp. 12–23.

9.   Tian, J.; Chen, D.M. Optimization in multi-scale segmentation of high-resolution satellite images for artificial feature recognition. *Int. J. Remote Sens.* **2007**, *28*, 4625–4644. [CrossRef]

10.   Trimble. *eCognition Developer 9.3 Reference Book*; Trimble Germany GmbH: Munich, Germany, 2017.

11.   Nixon, M.S.; Aguado, A.S. *Feature Extraction & Image Processing for Computer Vision*, 3rd ed.; Elservier and Pte Ltd.: Singapore, 2012.

12.   Balla-Arabé, S.; Gao, X. Geometric active curve for selective entropy optimization. *Neurocomputing* **2014**, *139*, 65–76. [CrossRef]

13.   Ding, K.; Xiao, L.; Weng, G. Active contours driven by region-scalable fitting and optimized Laplacian of Gaussian energy for image segmentation. *Signal Process.* **2017**, *134*, 224–233. [CrossRef]

14.   Neubert, M.; Herold, H.; Meinel, G. *Assessing Image Segmentation Quality—Concepts, Methods and Application*; Springer: Berlin, Germany, 2008.

15.   Neubert, M.; Herold, H. Assessment of remote sensing image segmentation quality. In Proceedings of the GEOBIA, Calgary, AB, Canada, 6–7 August 2008; p. 5.

16.   Blaschke, T. Object based image analysis for remote sensing. *ISPRS J. Photogramm. Remote Sens.* **2010**, *65*, 2–16. [CrossRef]

17.   Aguilar, M.A.; Aguilar, F.J.; García Lorca, A.; Guirado, E.; Betlej, M.; Cichon, P.; Nemmaoui, A.; Vallario, A.; Parente, C. Assessment of Multiresolution Segmentation for Extracting Greenhouses from Worldview-2 Imagery. *ISPRS Int. Arch. Photogramm. Remote Sens. Spat. Inf. Sci.* **2016**, *XLI-B7*, 145–152. [CrossRef]

18. Aguilar, M.A.; Novelli, A.; Nemamoui, A.; Aguilar, F.J.; García Lorca, A.; González-Yebra, Ó. Optimizing Multiresolution Segmentation for Extracting Plastic Greenhouses from WorldView-3 Imagery. In Proceedings of the Intelligent Interactive Multimedia Systems and Services, Gold Coast, Australia, 20–22 May 2017; pp. 31–40.

19. Coslu, M.; Sonmez, N.K.; Koc-San, D. Object-Based Greenhouse Classification from High Resolution Satellite Imagery: A Case Study Antalya-Turkey. *ISPRS Int. Arch. Photogramm. Remote Sens. Spat. Inf. Sci.* **2016**, *XLI-B7*, 183–187. [CrossRef]

20. Wu, C.; Deng, J.; Wang, K.; Ma, L.; Shah, T.A.R. Object-based classification approach for greenhouse mapping using Landsat-8 imagery. *Int. J. Agric. Biol. Eng.* **2016**, *9*, 79–88.

21. Drăguţ, L.; Tiede, D.; Levick, S.R. ESP: A tool to estimate scale parameter for multiresolution image segmentation of remotely sensed data. *Int. J. Geogr. Inf. Sci.* **2010**, *24*, 859–871. [CrossRef]

22. Dragut, L.; Csillik, O.; Eisank, C.; Tiede, D. Automated parameterisation for multi-scale image segmentation on multiple layers. *ISPRS J Photogramm Remote Sens* **2014**, *88*, 119–127. [CrossRef]

23. d'Oleire-Oltmanns, S.; Tiede, D. Specific target objects—Specific scale levels? Application of the estimation of scale parameter 2 (ESP 2) tool for the identification of scale levels for distinct target objects. *South-East. Eur. J. Earth Obs. Geomat.* **2014**, *3*, 580–583.

24. Marpu, P.R.; Neubert, M.; Herold, H.; Niemeyer, I. Enhanced evaluation of image segmentation results. *J. Spat. Sci.* **2010**, *55*, 55–68. [CrossRef]

25. Su, W.; Li, J.; Chen, Y.; Liu, Z.; Zhang, J.; Low, T.M.; Suppiah, I.; Hashim, S.A.M. Textural and local spatial statistics for the object-oriented classification of urban areas using high resolution imagery. *Int. J. Remote Sens.* **2008**, *29*, 3105–3117. [CrossRef]

26. Agüera, F.; Aguilar, F.J.; Aguilar, M.A. Using texture analysis to improve per-pixel classification of very high resolution images for mapping plastic greenhouses. *Isprs J. Photogramm. Remote Sens.* **2008**, *63*, 635–646. [CrossRef]

27. Hasituya; Chen, Z.; Wang, L.; Wu, W.; Jiang, Z.; Li, H. Monitoring Plastic-Mulched Farmland by Landsat-8 OLI Imagery Using Spectral and Textural Features. *Remote Sens.* **2016**, *8*, 353. [CrossRef]

28. Haralick, R.M.; Shanmugam, K.; Dinstein, I. Texture features for image classifications. *IEEE Trans. Syst. Man Cybern.* **1973**, *3*, 610–621. [CrossRef]

29. Aguilar, M.A.; Nemmaoui, A.; Novelli, A.; Aguilar, F.J.; García Lorca, A. Object-Based Greenhouse Mapping Using Very High Resolution Satellite Data and Landsat 8 Time Series. *Remote Sens.* **2016**, *8*, 513. [CrossRef]

30. Zhang, H.; Fritts, J.E.; Goldman, S.A. Image segmentation evaluation: A survey of unsupervised methods. *Comput. Vis. Image Underst.* **2008**, *110*, 260–280. [CrossRef]

31. Zhang, X.; Xiao, P.; Feng, X. An Unsupervised Evaluation Method for Remotely Sensed Imagery Segmentation. *IEEE Geosci. Remote Sens. Lett.* **2012**, *9*, 156–160. [CrossRef]

32. Gao, H.; Tang, Y.; Jing, L.; Li, H.; Ding, H. A Novel Unsupervised Segmentation Quality Evaluation Method for Remote Sensing Images. *Sensor* **2017**, *17*, 2427. [CrossRef]

33. Wang, Y.; Qi, Q.; Liu, Y. Unsupervised Segmentation Evaluation Using Area-Weighted Variance and Jeffries-Matusita Distance for Remote Sensing Images. *Remote Sens.* **2018**, *10*, 1193. [CrossRef]

34. Yang, L.; Albregtsen, F.; Lønnestad, T.; Grøttum, P. A supervised approach to the evaluation of image segmentation methods. *Comput. Anal. Images Patterns* **1995**, 759–765.

35. Zhang, Y.J. A survey on evaluation methods for image segmentation. *Pattern Recognit.* **1996**, *29*, 1335–1346. [CrossRef]

36. Chabrier, S.; Laurent, H.; Emile, B.; Rosenberger, C.; Marche, P. A comparative study of supervised evaluation criteria for image segmentation. In Proceedings of the EUSIPCO, Vienna, Austria, 6–10 September 2004; pp. 1143–1146.

37. Correia, P.; Pereira, F. Objective evaluation of relative segmentation. In Proceedings of the International Conference on Image Processing, Vancouver, BC, Canada, 10–13 September 2000; pp. 308–311.

38. Lucieer, A.; Stein, A. Existential uncertainty of spatial objects segmented from satellite sensor imagery. *IEEE Trans. Geosci. Remote Sens.* **2002**, *40*, 2518–2521. [CrossRef]

39. Möller, M.; Lymburner, L.; Volk, M. The comparison index: A tool for assessing the accuracy of image segmentation. *Int. J. Appl. Earth Obs. Geoinf.* **2007**, *9*, 311–321. [CrossRef]

40. Clinton, N.; Holt, A.; Scarborough, J.; Yan, L.; Gong, P. Accuracy Assessment Measures for Object-based Image Segmentation Goodness. *Photogramm. Eng. Remote Sens.* **2010**, *76*, 289–299. [CrossRef]

41. Persello, C.; Bruzzone, L. A Novel Protocol for Accuracy Assessment in Classification of Very High Resolution Images. *IEEE Trans. Geosci. Remote Sens.* **2010**, *48*, 1232–1244. [CrossRef]

42. Liu, Y.; Bian, L.; Meng, Y.; Wang, H.; Zhang, S.; Yang, Y.; Shao, X.; Wang, B. Discrepancy measures for selecting optimal combination of parameter values in object-based image analysis. *ISPRS J. Photogramm. Remote Sens.* **2012**, *68*, 144–156. [CrossRef]

43. Liu, Y.; Zhang, Y.D.; Huang, Z.; Wang, M.M.; Yang, D.; Ma, H.M.; Zhang, Y.X.; Li, Y.F.; Li, H.W.; Hu, X.G. Segmentation optimization via recognition of the PSE-NSR-ED2 patterns along with the scale parameter in object-based image analysis. In Proceedings of the GEOBIA 2016: Solutions and Synergies, Enschede, The Netherlands, 14–16 September 2016.

44. Gao, M.; Qunou, J.; Yiyang, Z.; Wentao, Y.; Mingchang, S. Comparison of plastic greenhouse extraction method based on GF-2 remote-sensing imagery. *J. China Agric. Univ.* **2018**, *23*, 125–134.

45. Novelli, A.; Aguilar, M.A.; Nemmaoui, A.; Aguilar, F.J.; Tarantino, E. Performance evaluation of object based greenhouse detection from Sentinel-2 MSI and Landsat 8 OLI data: A case study from Almería (Spain). *Int. J. Appl. Earth Obs. Geoinf.* **2016**, *52*, 403–411. [CrossRef]

46. Cai, L.; Shi, W.; Miao, Z.; Hao, M. Accuracy Assessment Measures for Object Extraction from Remote Sensing Images. *Remote Sens.* **2018**, *10*, 303. [CrossRef]

47. Ye, S.; Pontius, R.G.; Rakshit, R. A review of accuracy assessment for object-based image analysis: From per-pixel to per-polygon approaches. *ISPRS J. Photogramm. Remote Sens.* **2018**, *141*, 137–147 [CrossRef]

48. Chen, Y.; Ming, D.; Zhao, L.; Lv, B.; Zhou, K.; Qing, Y. Review on High Spatial Resolution Remote Sensing Image Segmentation Evaluation. *Photogramm. Eng. Remote Sens.* **2018**, *84*, 25–42. [CrossRef]

49. Yang, D.; Chen, J.; Zhou, Y.; Chen, X.; Chen, X.; Cao, X. Mapping plastic greenhouse with medium spatial resolution satellite data: Development of a new spectral index. *ISPRS J. Photogramm. Remote Sens.* **2017**, *128*, 47–60. [CrossRef]

50. China Centre For Resources Satellite Data and Application. GF-2. Available online: http://www.cresda.com/EN/satellite/7157.shtml (accessed on 5 November 2015).

51. Anys, H.; He, D.-c. Evaluation of textural and multipolarization radar features for crop classification. *IEEE Trans. Geoscience Remote Sens.* **1995**, *33*, 1170–1181. [CrossRef]

52. Kim, M.; Madden, M.; Warner, T. *Estimation of Optimal Image Object Size for the Segmentation of Forest Stands with Multispectral IKONOS Imagery*; Springer: Berlin/Heidelberg, Germany, 2008.

53. Lucieer, A. *Uncertainties in Segmentation and Their Visualisation*; International Institute for Geo-Information Science and Earth Observation: Enschede, The Netherlands, 2004.

remote sensing

MDPI

Article

Remote Sensing Image Scene Classification Using CNN-CapsNet

Wei Zhang [1,2] ⬤, **Ping Tang [1] and Lijun Zhao [1,]***

[1] National Engineering Laboratory for Applied Technology of Remote Sensing Satellites,
    Institute of Remote Sensing and Digital Earth of Chinese Academy of Sciences, Beijing 100101, China;
    zhangwei@aircas.ac.cn (W.Z.); tangping@aircas.ac.cn (P.T.)
[2] University of Chinese Academy of Sciences, Beijing 100049, China
*   Correspondence: zhaolj201934@aircas.ac.cn; Tel.: +86-010-64855178

Received: 7 January 2019; Accepted: 22 February 2019; Published: 28 February 2019

Abstract: Remote sensing image scene classification is one of the most challenging problems in understanding high-resolution remote sensing images. Deep learning techniques, especially the convolutional neural network (CNN), have improved the performance of remote sensing image scene classification due to the powerful perspective of feature learning and reasoning. However, several fully connected layers are always added to the end of CNN models, which is not efficient in capturing the hierarchical structure of the entities in the images and does not fully consider the spatial information that is important to classification. Fortunately, capsule network (CapsNet), which is a novel network architecture that uses a group of neurons as a capsule or vector to replace the neuron in the traditional neural network and can encode the properties and spatial information of features in an image to achieve equivariance, has become an active area in the classification field in the past two years. Motivated by this idea, this paper proposes an effective remote sensing image scene classification architecture named CNN-CapsNet to make full use of the merits of these two models: CNN and CapsNet. First, a CNN without fully connected layers is used as an initial feature maps extractor. In detail, a pretrained deep CNN model that was fully trained on the ImageNet dataset is selected as a feature extractor in this paper. Then, the initial feature maps are fed into a newly designed CapsNet to obtain the final classification result. The proposed architecture is extensively evaluated on three public challenging benchmark remote sensing image datasets: the UC Merced Land-Use dataset with 21 scene categories, AID dataset with 30 scene categories, and the NWPU-RESISC45 dataset with 45 challenging scene categories. The experimental results demonstrate that the proposed method can lead to a competitive classification performance compared with the state-of-the-art methods.

Keywords: remote sensing; scene classification; CNN; capsule; PrimaryCaps; CapsNet

1. Introduction

With the development of Earth observation technology, many different types (e.g., multi/hyperspectral [1] and synthetic aperture radar [2]) of high-resolution images of the Earth's surface are readily available. Therefore, it is particularly important to effectively understand their semantic content, and more intelligent identification and classification methods of land use and land cover (LULC) are definitely demanded. Remote sensing image scene classification, which aims to automatically assign a specific semantic label to each remote sensing image scene patch according to its contents, has become an active research topic in the field of remote sensing image interpretation because of its vital applications in LULC, urban planning, land resource management, disaster monitoring, and traffic control [3–6].

During the last decades, several methods have been developed for remote sensing image scene classification. The early methods for scene classification were mainly based on low-level features or hand-crafted features, which focus on designing various human-engineering features locally or globally, such as color, texture, shape, and spatial information. Representative features, including the scale invariant feature transform (SIFT), color histogram (CH), local binary pattern (LBP), Gabor filters, grey level cooccurrence matrix (GLCM), and the histogram of oriented gradients (HOG) or their combinations, are usually used for scene classification [7–12]. It is worth noting that methods relying on these low-level features perform well on some images with uniform texture or spatial arrangements, but they are still limited for distinguishing images with more challenging and complex scenes, which is because the involvement of humans in feature design significantly influences the effectiveness of the representation capacity of scene images. In contrast to low-level feature-based methods, the mid-level feature approaches attempt to compute a holistic image representation formed by local visual features such as SIFT, color histogram, or LBP of local image patches. The general pipeline of building mid-level features is to extract local attributes of image patches first and then to encode them to obtain the mid-level representation of remote sensing images. The well-known bag-of-visual-words (BoVW) model is the most popular mid-level approach and has been widely adopted for remote sensing image scene classification because of its simplicity and effectiveness [13–18]. The methods based on the BoVW have improved the classification performance, but due to the limitation of representation capability of the BOVW model, no further breakthroughs have been achieved for remote sensing image scene classification.

Recently, with the prevalence of deep learning methods, which have achieved impressive performance on many applications including image classification [19], object recognition [20], and semantic segmentation [21], the feature representation of images has stepped into a new era. Unlike low-level and mid-level features, deep learning models can learn more powerful, abstract and discriminative features via deep-architecture neural networks without a considerable amount of engineering skill and domain expertise. All of these deep learning models, especially the convolutional neural network (CNN), are more applicable for remote sensing image scene classification and have achieved state-of-the-art results [22–34]. Although the CNN-based methods have dramatically improved classification accuracy, some scene classes are still easily mis-classified. Taking the AID dataset as an example, the class-specific classification accuracy of 'school' is only 49% [35], which is usually confused with 'dense residential'. As shown in Figure 1, two images labelled 'school' and two images labelled 'dense residential' have been selected from the AID dataset. We can see that the contexts among these four images have similar image distribution and all contain many buildings and trees. However, different from the arrangement irregularity of buildings in 'school', the buildings in 'dense residential' are arranged closely and orderly. This spatial layout difference between them is very helpful in distinguishing the two classes and should be given more consideration in the phase of classification. However, the use of the fully connected layer at the end of the CNN model compresses the two-dimensional feature map into a one-dimensional feature map and cannot fully consider the spatial relationship, which makes it difficult to distinguish the two classes.

school   dense residential

Figure 1. Sample images labelled school and dense residential in the AID dataset.

Recently, the advent of the capsule network (CapsNet) [36], which is a novel architecture to encode the properties and spatial relationship of the features in an image and is a more effective image recognition algorithm, shows encouraging results on image classification. Although the CapsNet is still in its infancy [37], it has been successfully applied in many fields [38–49] in recent years, such as brain tumor classification, sound event detection, object segmentation, and hyperspectral image classification. The CapsNet uses a group of neurons as a capsule to replace a neuron in the traditional neural network. In addition, the capsule is a vector to represent internal properties that can be used to learn part–whole relationships between various entities, such as objects or object parts, to achieve equivariance [36] and can solve the problem of traditional neural networks using fully connected layers cannot efficiently capture the hierarchical structure of the entities in images to preserve the spatial information [50].

To further improve the accuracy of the remote sensing image scene classification and motivated by the powerful ability of feature learning of deep CNN and the property of equivariance of CapsNet, a new architecture named CNN-CapsNet is proposed to deal with the task of remote sensing image scene classification in this paper. The proposed architecture is composed of two parts. First, a pretrained deep CNN, such as VGG-16 [51], is fully trained on the ImageNet [52] dataset, and its intermediate convolutional layer is used as an initial feature maps extractor. Then, the initial feature maps are fed into a newly designed CapsNet to label the remote sensing image scenes. Experimental results on three challenging benchmark datasets show that the proposed architecture achieves a more competitive accuracy compared with state-of-the-art methods. In summary, the major contributions of this paper are as follows:

- To further improve classification accuracy, especially classes that have high homogeneity in the image content, a new novel architecture named CNN-CapsNet is proposed to deal with the remote sensing image scene classification problem, which can discriminate scene classes effectively.
- By combining the CNN and the CapsNet, the proposed method can obtain a superior result compared with the state-of-the-art methods on three challenging datasets without any data-augmentation operation.
- This paper also analyzes the influence of different factors in the proposed architecture on the classification result, including the routing number in the training phase, the dimension of capsules in the CapsNet and different pretrained CNN models, which can provide valuable guidance for subsequent research on the remote sensing image scene classification using CapsNet.

The remainder of this paper is organized as follows. In Section 2, the materials are illustrated. Section 3 introduces the theory of CNN and CapsNet first, and then describes the proposed method in detail. Section 4 analyzes the influence of different factors, and discusses the experimental results of the proposed method. Finally, conclusions are drawn in Section 5.

2. Materials

Three popular remote sensing datasets (UC Merced Land-Use [14], AID [35], and NWPU-RESISC45 [53]) with different visual properties are chosen to better demonstrate the robustness and effectiveness of the proposed method. In addition, details about the datasets are described in Sections 2.1–2.3.

2.1. UC Merced Land-Use Dataset

The UC Merced Land-Use dataset is composed of 2100 aerial scene images divided into 21 land use scene classes, as shown in Figure 2. Each class contains 100 images with size of 256 × 256 pixels with a pixel spatial resolution of 0.3 m in the red green blue (RGB) color space. These images were selected from aerial orthoimagery downloaded from the United States Geological Survey (USGS) National Map of the following US regions: Birmingham, Boston, Buffalo, Columbus, Dallas, Harrisburg, Houston, Jacksonville, Las Vegas, Los Angeles, Miami, Napa, New York, Reno, San Diego, Santa Barbara, Seattle,

Tampa, Tucson, and Ventura. It is not only the diversity of land-use categories contained in the dataset that makes it challenging. Some highly overlapped classes such as dense residential, medium residential and sparse residential are included in this dataset, which are mainly different in the density of structures and makes the dataset more difficult to classify. This dataset has been widely used for the task of remote sensing image scene classification [18,23–25,27,28,30,32,54–58].

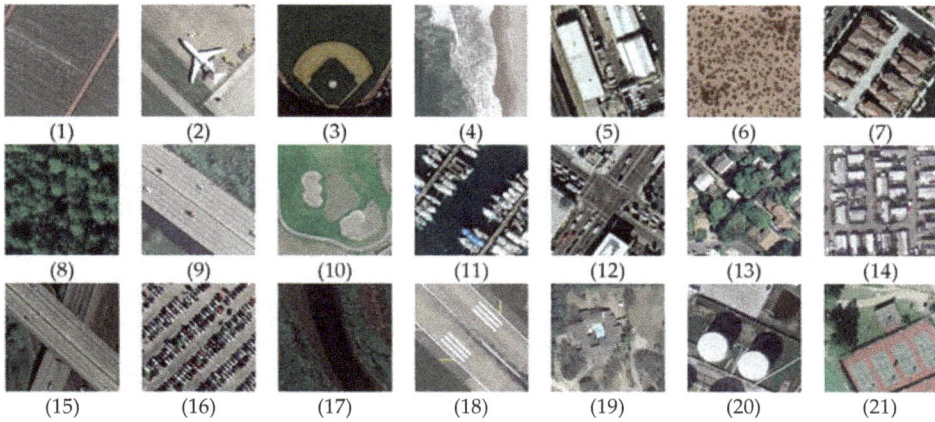

Figure 2. Example images of UC Merced Land-Use dataset: (1) agriculture; (2) airplane; (3) baseball diamond; (4) beach; (5) buildings; (6) chaparral; (7) dense residential; (8) forest; (9) freeway; (10) golf course; (11) harbor; (12) intersection; (13) medium residential; (14) mobile home park; (15) overpass; (16) parking lot; (17) river; (18) runway; (19) sparse residential; (20) storage tanks; and (21) tennis court.

2.2. AID Dataset

AID is large-scale aerial image dataset, which was collected from Google Earth imagery and is a more challenging dataset compared with the UC Merced Land-Use dataset because of the following reasons. First, the AID dataset contains more scene types and images. In detail, it has 10,000 images with a fixed size of 600×600 pixels within 30 classes as shown in Figure 3. Some similar classes make the interclass dissimilarity smaller, and the number of images of different scene types differs from 220 to 420. Moreover, AID images were chosen under different times and seasons and different imaging conditions, and from different countries and regions around the world, including China, the United States, England, France, Italy, Japan, and Germany, which definitely increases the intraclass diversities. Finally, AID images have the property of multiresolution, changing from approximately 8 m to about half a meter.

Figure 3. *Cont.*

(21)　　(22)　　(23)　　(24)　　(25)　　(26)　　(27)　　(28)　　(29)　　(30)

Figure 3. Example images of AID dataset: (1) airport; (2) bare land; (3) baseball field; (4) beach; (5) bridge; (6) centre; (7) church; (8) commercial; (9) dense residential; (10) desert; (11) farmland; (12) forest; (13) industrial; (14) meadow; (15) medium residential; (16) mountain; (17) park; (18) parking; (19) playground; (20) pond; (21) port; (22) railway station; (23) resort; (24) river; (25) school; (26) sparse residential; (27) square; (28) stadium; (29) storage tanks; and (30) viaduct.

2.3. NWPU-RESISC45 Dataset

NWPU-RESISC45 dataset is more complex than UC Merced Land-Use and AID datasets and consists of a total of 31,500 remote sensing images divided into 45 scene classes as shown in Figure 4. Each class includes 700 images with a size of 256×256 pixels in the RGB color space. This dataset was extracted from Google Earth by the experts in the field of remote sensing image interpretation. The spatial resolution varies from approximately 30 to 0.2 m per pixel. This dataset covers more than 100 countries and regions all over the world with developing, transitional, and highly developed economies.

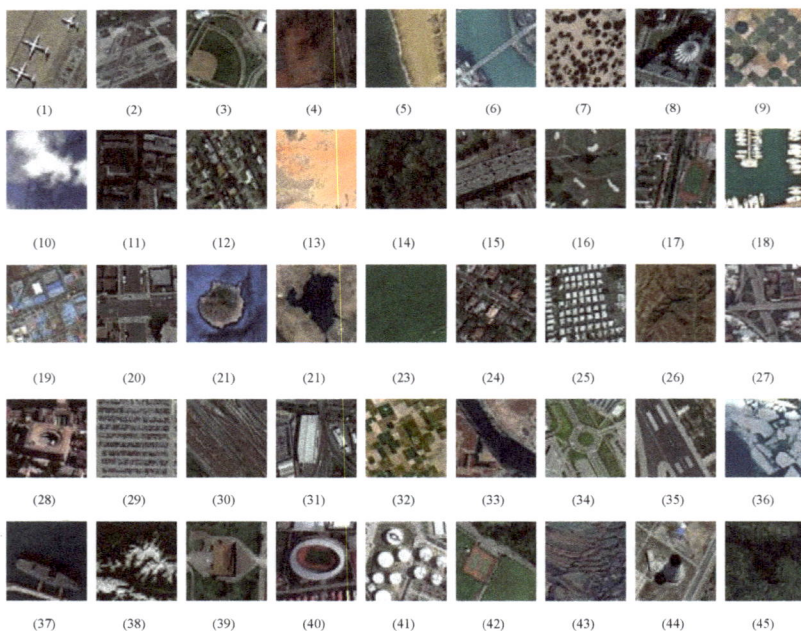

Figure 4. Example images of NWPU-RESISC45 dataset: (1) airplane; (2) airport; (3) baseball diamond; (4) basketball court; (5) beach; (6) bridge; (7) chaparral; (8) church; (9) circular farmland; (10) cloud; (11) commercial area; (12) dense residential; (13) desert; (14) forest; (15) freeway; (16) golf course; (17) ground track field; (18) harbor; (19) industrial area; (20) intersection; (21) island; (22) lake; (23) meadow; (24) medium residential; (25) mobile home park; (26) mountain; (27) overpass; (28) palace; (29) parking lot; (30) railway; (31) railway station; (32) rectangular farmland; (33) river; (34) roundabout; (35) runway; (36) sea ice; (37) ship; (38) snow berg; (39) sparse residential; (40) stadium; (41) storage tank; (42) tennis court; (43) terrace; (44) thermal power station; and (45) wetland.

3. Method

In this section, a brief introduction about CNN and CapsNet will be made first and then the proposed architecture will be detailed.

3.1. CNN

The convolutional neural network is a type of feed-forward artificial neural network, which is biologically inspired by the organization of the animal visual cortex. They have wide applications in image and video recognition, recommender systems and natural language processing. As shown in Figure 5, CNN is generally made up of two main parts: convolutional layers and pooling layers. The convolutional layer is the core building block of a CNN, which outputs feature maps by computing a dot product between the local region in the input feature maps and a filter. Each of the feature maps is followed by a nonlinear function for approximating arbitrarily complex functions and squashing the output of the neural network to be within certain bounds, such as the rectified linear unit (ReLU) nonlinearity, which is commonly used because of its computational efficiency. The pooling layer performs a downsampling operation to feature maps by computing the maximum or average value on a sub-region. Usually, the fully connected layers follow several stacked convolutional and pooling layers and the last fully connected layer is the softmax layer computing the scores for each class.

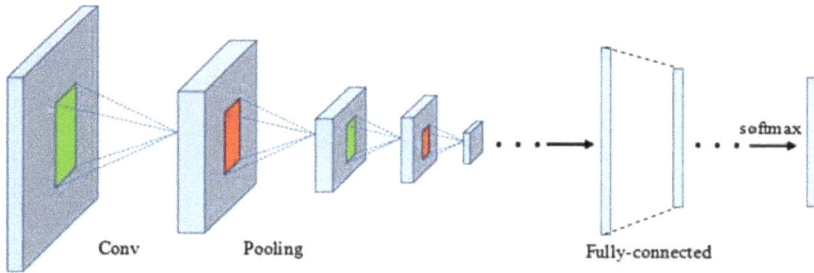

Figure 5. The convolutional neural network (CNN) architecture.

3.2. CapsNet

CapsNet is a completely novel deep learning architecture, which is robust to affine transformation [41]. In CapsNet, a capsule is defined as a vector that consists of a group of neurons, whose parameters can represent various properties of a specific type of entity that is presented in an image, such as position, size, and orientation. The length of each activity vector provides the existence probability of the specific object, and its orientation indicates its properties. Figure 6 illustrates the way that CapsNet routes the information from one layer to another layer by a dynamic routing mechanism [36], which means capsules in lower levels predict the outcome of capsules in higher levels and higher level capsules are activated only if these predictions agree.

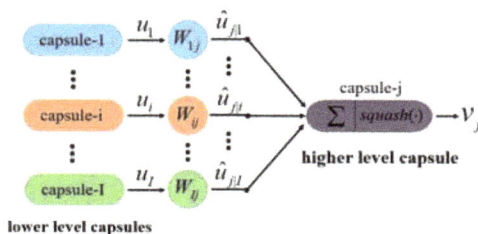

Figure 6. Connections between the lower level and higher level capsules.

Considering u_i as the output of lower-level capsule i, its prediction for higher level capsule j is computed as:

$$\hat{u}_{j|i} = W_{ij}u_i \tag{1}$$

where W_{ij} is the weighting matrix that can be learned by back-propagation. Each capsule tries to predict the output of higher level capsules, and if this prediction conforms to the actual output of higher level capsules, the coupling coefficient between these two capsules increases. Based on the degree of conformation, coupling coefficients are calculated using the following softmax function:

$$c_{ij} = \frac{\exp(b_{ij})}{\sum_k \exp(b_{ik})} \tag{2}$$

where b_{ij} is set to 0 initially at the beginning of routing by an agreement process and is the log probability of whether lower-level capsule i should be coupled with higher level capsule j. Then, the input vector to the higher level capsule j can be calculated as follows:

$$s_j = \sum_i c_{ij}\hat{u}_{j|i} \tag{3}$$

Because the length of the output vector represents the probability of existence, the following nonlinear squash function, which is an activation function to ensure that short vectors are decreased to almost zero, and the long vectors are close to one, is used on the output vector computed in Equation (3) to prevent the output vectors of capsules from exceeding one.

$$v_j = \frac{\|s_j\|^2}{1 + \|s_j\|^2} \frac{s_j}{\|s_j\|} \tag{4}$$

where s_j and v_j represent the input vector and output vector, respectively, of capsule j. In addition, the log probabilities b_{ij} is updated in the routing process based on the agreement between v_j and $\hat{u}_{j|i}$ according to the rule that if the two vectors agree, they will have a large inner product. Therefore, agreement a_{ij} for updating log probabilities b_{ij} and coupling coefficients c_{ij} is calculated as follows:

$$a_{ij} = \hat{u}_{j|i}v_j \tag{5}$$

As mentioned above, Equations (2)–(5) make up one whole routing procedure for computing v_j. The routing algorithm consists of several iterations of the routing procedure [36], and the number of iterations can be described as the routing number. Take the 'school' scene type detection as an example for a clearer explanation. Lengths of the outputs of the lower-level capsules ($u_1, u_2, \ldots, u_I$) encode the existence probability of their corresponding entities (e.g., building, tree, road, and playground). Directions of the vectors encode various properties of these entities, such as size, orientation, and position. In training, the network gradually encodes the corresponding part-whole relationship by a routing algorithm to obtain a higher-level capsule (v_j), which encodes the whole scene contexts that the 'school' represents. Thus, the capsule can learn the spatial relationship between entities within an image.

Each capsule k in the last layer is associated with a loss function l_k, which can be computed as follows:

$$l_k = T_k\max(0, m^+ - \|v_k\|)^2 + \lambda(1 - T_k)\max(0, \|v_k\| - m^-)^2 \tag{6}$$

where T_k is 1 when class k is actually present, m^+, m^- and λ are hyper-parameters that should be indicated while training. The total loss is simply the sum of the loss of all output capsules of the last layer.

A typical CapsNet is shown in Figure 7 and contains three layers: one convolutional layer (Conv1), the PrimaryCaps layer and the FinalCaps layer. The Conv1 converts the input image (raw pixels) to initial feature maps, whose size can be described as H × W × L. Then, by two reshape functions and one squash operation, the PrimaryCaps can be computed, which contains H × W × L/S1 capsules (each capsule in the PrimaryCaps is an S1 dimension vector and is denoted as the S1-D vector in Figure 7). The FinalCaps has T (number of total predict classes) capsules (each capsule in the FinalCaps is an S2 dimension vector and is denoted as the S2-D vector in Figure 7), and each of these capsules receives input from all the capsules in the PrimaryCaps layer. The detail of FinalCaps is illustrated in Figure 8. At the end of the CapsNet, the length of each capsule in FinalCaps is computed by an L_2 norm function, the corresponding scene category represented by the maximum value is the final classification result.

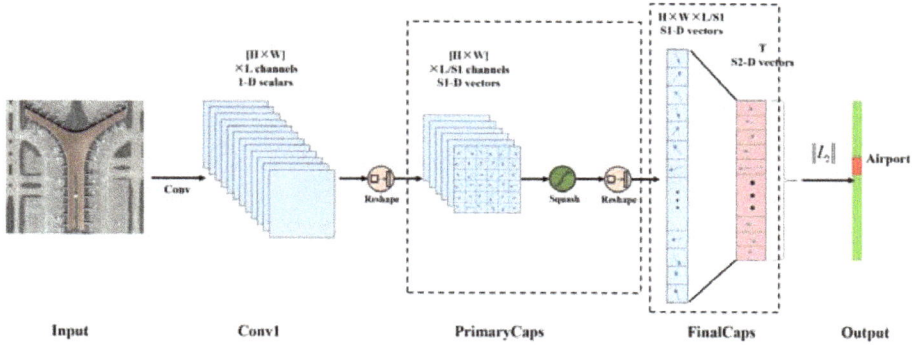

Figure 7. A typical CapsNet architecture.

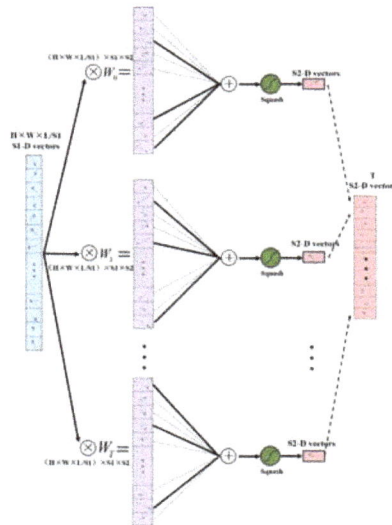

Figure 8. The detail of FinalCaps.

3.3. Proposed Method

As illustrated in Figure 9, the proposed architecture CNN-CapsNet can be divided into two parts: CNN and CapsNet. First, a remote sensing image is fed into a CNN model, and the initial feature maps are extracted from the convolutional layers. Then, the initial feature maps are fed into CapsNet to obtain the final classification result.

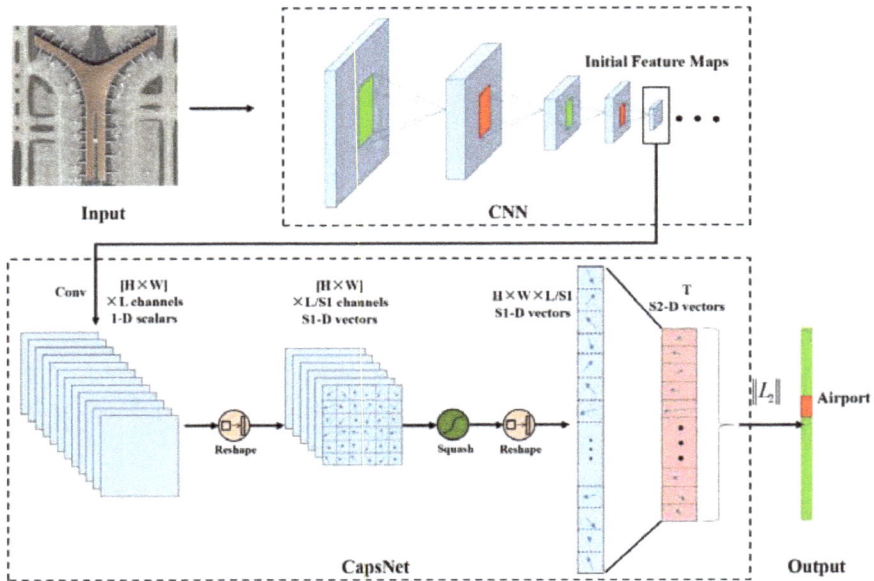

Figure 9. The architecture of the proposed classification method.

As for CNN, two representative CNN models (VGG-16 and Inception-V3) fully trained on the ImageNet dataset are used as initial feature map extractors, considering their popularity in the remote sensing field [25,27,28,35,53,56,59]. The "block4_pool" layer of VGG-16 and the "mixd7" of Inception-V3 are selected as the layer of initial feature maps, whose sizes are 16 × 16 × 512 and 14 × 14 × 768, respectively, if the input image size is 256 × 256 pixels. The influence of the two pretrained CNN models on the classification results is discussed in Section 4.2. In addition, a brief introduction about them follows.

- VGG-16: Simonyan et al. [51] presented the very deep CNN models that secured the first and the second places in the localization and classification tracks, respectively, on ILSVRC2014. The two best-performing deep models, named VGG-16 (containing 13 convolutional layers and 3 fully connected layers) and VGG-19 (containing 16 convolutional layers and 3 fully connected layers) are the basis of their team's submission, which demonstrates the important aspect of the model's depth. Rather than using relatively large receptive fields in the convolutional layers, such as 11 × 11 with stride 4 in the first convolutional layer of AlexNet [60], VGGNet uses very small 3 × 3 receptive fields through the whole network. VGG-16 is the most representative sequence-like CNN architecture as shown in Figure 5 (consisting of a simple chain of blocks such as the convolution layer and pooling layer), which has achieved great success in the field of remote sensing image scene classification.

- Inception-v3: Unlike the sequence-like CNN architecture such as VGG-16, which only increases the depth of the convolution layers, the Inception-like CNN architecture attempts to increase the width of a single convolution layer, which means different sizes of kernels are used on the single convolution layer and can extract different scales of features. As shown in Figure 10, it is the core component of GoogLeNet [61] named Inception-v1. Inception-v3 [62] is an improved version of Inception-v1 and is designed on the following four principles: to avoid representation bottlenecks, especially early in the network; higher dimensional representations are easier to process locally within a network; spatial aggregation can be done over lower dimensional embedding without much or any loss in representation; to balance the width and depth of the network. The Inception-v3 reached 21.2% top-1 and 5.6% top-5 error on the ILSVR 2012 classification.

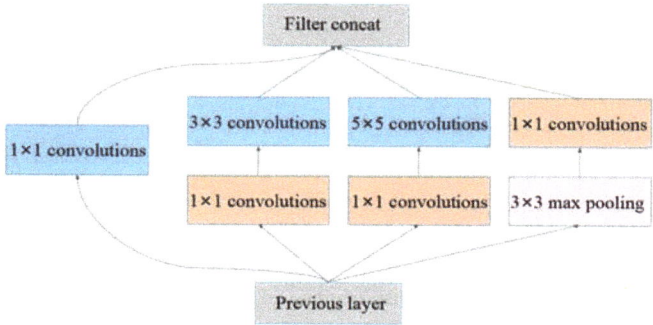

Figure 10. Inception-v1 module.

For CapsNet, a CapsNet with an analogical architecture as shown in Figure 7 is designed, including three layers: one convolutional layer, one PrimaryCaps layer and one FinalCaps layer. A 5 × 5 convolution kernel with a stride of 2, and a ReLU activation function is used in the convolution layer. The number of output feature maps (the variable L) is set as 512. The dimension of the capsules in the PrimaryCaps and FinalCaps layers (the variables S1 and S2) are the vital parameters of the CapsNet and their influence on the classification result is discussed in Section 4.2. The variable T is determined by the remote sensing datasets and is set as 21, 30, and 45 for the UC Merced Land-Use dataset, AID dataset and NWPU-RESISC45 dataset, respectively. In addition, 50% dropout was used between the PrimaryCaps layer and the FinalCaps layer to prevent overfitting.

As shown in Figure 11, the proposed method includes two training phases. In the first training phase, the parameters in the pretrained CNN model are frozen, and weights in the CapsNet are initialized by Gaussian distribution with zero mean and unit variance. Then, they are trained with a learning rate of lr1 to minimize the sum of the margin losses in Equation (6). When the CapsNet is fully trained, the second training phase begins with a lower learning rate lr2 to fine-tune the whole architecture until convergence. The parameters between the adjacent capsule layers except for the coupling coefficient can be updated by a gradient descent algorithm, while the coupling coefficients are determined by the iterative dynamic routing algorithm [36]. The optimal routing number in the iterative dynamic routing algorithm is discussed in Section 3.2. When the training finishes, the testing images are fed into the fully trained CNN-CapsNet architecture to evaluate the classification result.

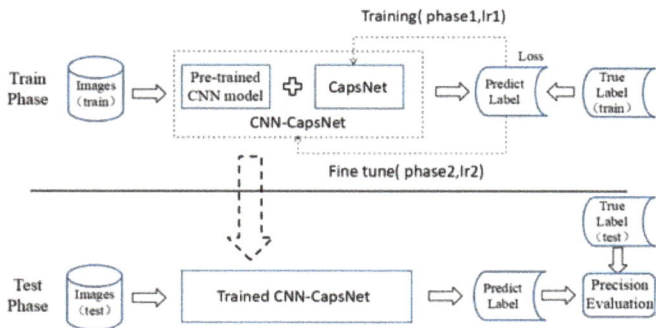

Figure 11. The flowchart of the proposed method.

4. Results and Analysis

4.1. Experimental Setup

4.1.1. Implementation Details

In this work, the Keras framework was used to implement the proposed method. The hyperparameters used in the training stage were set by trial and error as follows. For the Adam optimization algorithm, the batch-size was set as 64 and 50 to cater to the computer memory (due to the different volume of training parameters of the model in two training phases); the learning rates lr1 and lr2 were set as 0.001 and 0.0002 separately for two training phases. The sum of all classes' margin losses in Equation (6) was used for the loss function, and m^+, m^-, and λ were set as 0.9, 0.1 and 0.5. All models were trained until the training loss converged. At the same time, for a fair comparison, the same ratios were applied in the following experiments according to the experimental settings in works [23–25,27,28,30,35,53–57,63–68]. For the UC Merced Land-Use dataset, the 80% and 50% training ratio were set separately. For the AID dataset, 50% and 20% of the images were randomly selected as the training samples, and the rest were left for testing. In addition, a 20% and 10% training ratio were used for the NWPU-RESISC45 dataset. Here, two training ratios were considered for each of the three datasets to comprehensively evaluate the proposed method. Moreover, different ratios were used for different datasets because the numbers of images for the three datasets are different. A small ratio can usually satisfy the full training requirement of the models when a dataset has a large amount of data. Note that all images in the AID dataset were resized to 256 × 256 pixel from the original 600 × 600 pixel because of memory overflow in the training phase. All the implementations were evaluated on an Ubuntu 16.04 operating system with one 3.6 GHz 8-core i7-4790CPU and 32GB memory. Additionally, a NVIDIA GTX 1070 graphics processing unit (GPU) was used to accelerate computing.

4.1.2. Evaluation Protocol

The overall accuracy (OA) and confusion matrix were computed to evaluate experimental results and to compare with the state-of-the-art methods. The OA was defined as the number of correctly classified images divided by the total number of test images, which is a valuable measure to reveal the classification method performance on the whole test images. The value of OA is in the range of 0 to 1, and a higher value indicates a better classification performance. The confusion matrix is an informative table that can allow direct visualization of the performance on each class and can be used for easily analyzing the errors and confusion between different classes, in which the column represents the instances in a predicted class and the row represents the instances in an actual class. Thus, each item x_{ij} in the matrix is the proportion of images that are predicted to be the i-th class while truly belonging to the j-th class.

To compute the overall accuracy, the dataset was randomly divided into training and testing sets according to the ratios in Section 4.1.1 and repeated ten times to reduce the influence of the randomness for a reliable result. The mean and standard deviation of overall accuracies on the testing sets from each individual run were reported. Additionally, the confusion matrix was obtained from the best classification results by fixing the ratios of the training sets of the UC Merced Land-Use dataset, AID dataset and NWPU-RESISC45 dataset to be 50%, 20%, and 20%, respectively.

4.2. Experimental Results and Analysis

4.2.1. Analysis of Experimental Parameters

In this section, three parameters including the routing number, the dimension of the capsule in the CapsNet, and different pretrained CNN models, were tested to analyze how these parameters affect the classification result. In addition, the optimal parameters used in the experiments of Sections 4.2.2

and 4.2.3. Training rations of 80%, 50%, 20% were selected for the UC Merced Land-Use dataset, AID dataset and NWPU-RESISC45 dataset, respectively, in this section's experiments.

1. The routing number

In the dynamic routing algorithm, the routing number is a vital parameter for determining whether the CapsNet can obtain the best coupling coefficients. Therefore, it is necessary to select an optimal routing number. Thus, the routing number was set to (1, 2, 3, 4) while other parameters in the proposed architecture were kept the same. The pretrained VGG-16 model was selected as the primary feature extractor, and the dimension of the capsule in the PrimaryCaps and FinalCaps layers were set to 8 and 16, respectively. As shown in Figure 12, the OAs first increased and then decreased with the increase in the routing number for all three datasets and all reached their peaks at the routing number of 2. A smaller value may generate inadequate training, and a larger value will lead to missing the optimal fitting. In addition, the bigger the value is, the longer the required training time. Comprehensively, the routing number 2 was chosen as the optimal number, considering the training time and was applied in remaining experiments.

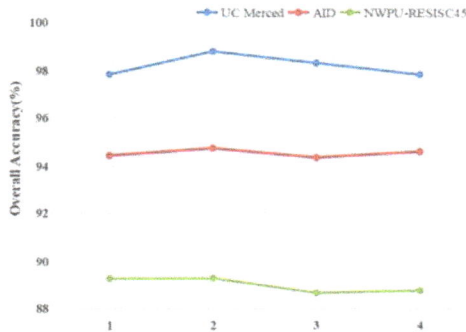

Figure 12. The influence of the routing number on the classification accuracy.

2. The dimension of the capsule

The capsule is the core component of CapsNet and consists of many neurons, and their activities within a capsule represent the various properties of a remote sensing scene image. The primary capsules in the PrimaryCaps are the lower-level capsules that are learned from the primary feature maps extracted from the pretrained CNN models, and they can represent some small entities in the remote sensing image. The capsules with a higher dimension in the FinalCaps are in a higher level and represent more complex entities such as the scene class that the image presents. Thus, the dimension of the capsule in the CapsNet should be considered for its importance in the final classification result. When the dimension of the capsule is low, the representation ability of the capsule is weak, which leads to confusion between two scene classes with high similarity in image context. In contrast, the capsule with a high dimension may contain redundant information or noise, e.g., two neurons may represent very similar properties. Both of them will have a negative influence on the classification result. Thus, a set of values ((6,12), (8,16), (10,20), (12,24)) were set to evaluate the capsule's influence. Additionally, other parameters were fixed with the pretrained VGG-16 model as the primary feature extractor, and the routing number was set to 2. The experimental results are shown in Figure 13. As expected, in all three datasets, the curves of OAs had their single peaks. The value (8, 16) obtained the best performance, and thus it was used in the next experiments.

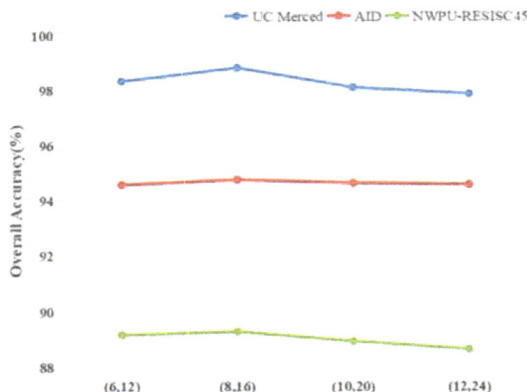

Figure 13. The influence of the dimension of the capsule on the classification accuracy.

3. Different pretrained CNN models

As described in Section 3.3, two representative CNN architectures (VGG-16 and Inception-v3) were selected as feature extractors to evaluate the effectiveness of convolutional features on classification. The "block4_pool" layer of VGG-16 and the "mixd7" of Inception-V3 were selected as the layer of initial feature maps. Other parameters remained unchanged in the experiment. As shown in Figure 14, the Inception-v3 model achieved the highest classification accuracy on all three datasets. This can be explained by the fact that the Inception-v3 consists of the inception modules, which can extract multiscale features and have a stronger ability to extract effective features than VGG-16; however, compared with the Inception-v3, the VGG-16 may lose considerable information due to the consistent existence of pooling layers. Moreover, the OA differences between VGG-16 and Inception-v3 on the AID and NWPU-RESISC45 datasets were more conspicuous than those on the UC Merced dataset.

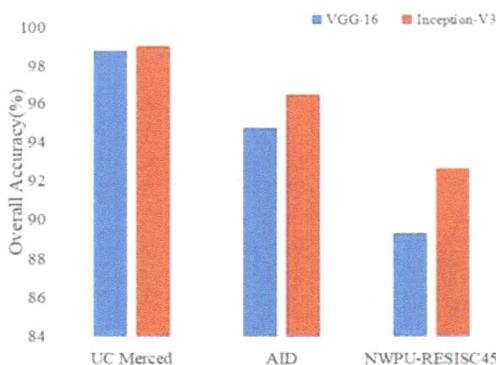

Figure 14. The influence of pretrained CNN models on the classification accuracy.

Compared with the UC Merced dataset, the other two datasets have more classes, higher intraclass variations and smaller interclass dissimilarity. Since the Inception-v3 shows its effectiveness in extracting features with more complex datasets, it was chosen as the final feature extractor on the evaluation of the proposed method.

4.2.2. Experimental Results

1. Classification of the UC Merced Land-Use dataset

To evaluate the classification performance of the proposed method, a comparative evaluation against several state-of-the-art classification methods on the UC Merced Land-Use dataset is shown in Table 1. As seen from Table 1, the proposed architecture CNN-CapsNet using pretrained Inception-v3 as the initial feature maps extractor (denoted as Inception-v3-CapsNet) achieved the highest OA of 99.05% and 97.59% for 80% and 50% training ratio, respectively, among all methods. The CNN-CapsNet using pretrained VGG-16 as the initial feature maps extractor (denoted as VGG-16-CapsNet) also outperformed most methods. This demonstrates that the CNN-CapsNet architecture can learn a higher level representation of scene images by combining CNN and CapsNet.

Table 1. Overall accuracy (%) and standard deviations of the proposed method and the comparison methods under the training ratios of 80% and 50% on the UC-Merced dataset.

Method	80% Training Ratio	50% Training Ratio
CaffeNet [35]	95.02 ± 0.81	93.98 ± 0.67
GoogLeNet [35]	94.31 ± 0.89	92.70 ± 0.60
VGG-16 [35]	95.21 ± 1.20	94.14 ± 0.69
SRSCNN [24]	95.57	/
CNN-ELM [65]	95.62	/
salM^{3}LBP-CLM [63]	95.75 ± 0.80	94.21 ± 0.75
TEX-Net-LF [64]	96.62 ± 0.49	95.89 ± 0.37
LGFBOVW [18]	96.88 ± 1.32	/
Fine-tuned GoogLeNet [25]	97.10	/
Fusion by addition [28]	97.42 ± 1.79	/
CCP-net [66]	97.52 ± 0.97	/
Two-Stream Fusion [30]	98.02 ± 1.03	96.97 ± 0.75
DSFATN [54]	98.25	/
Deep CNN Transfer [27]	98.49	/
GCFs+LOFs [56]	99 ± 0.35	97.37 ± 0.44
VGG-16-CapsNet (ours)	98.81 ± 0.22	95.33 ± 0.18
Inception-v3-CapsNet (ours)	**99.05 ± 0.24**	**97.59 ± 0.16**

Figure 15 shows the confusion matrix generated from the best classification result by Inception-v3-CapsNet with the training ratio of 50%. As shown in the confusion matrix, 20 categories achieved accuracies greater than 94%, half of which achieved an accuracy of 100%. In addition, only the class of 'dense residential', which were easily confused with 'medium residential', achieved an accuracy of 80%. This may have resulted from the fact that the two classes have similar image distributions, such as the building structure and density which cannot be well utilized to distinguish each other.

2. Classification of AID dataset

The AID dataset was also tested to demonstrate the effectiveness of the proposed method, compared with other state-of-the-art methods on the same dataset. The results are shown in Table 2. It can be seen that the proposed method of the Inception-v3-CapsNet model generated the best performance with OAs of 96.32% and 93.79% by using 50% and 20% samples, respectively, for training, except for approximately 0.53% lower performance than the method of GCFs + LOFs in a 50% training ratio. This can be explained that the process of downsampling from 600 × 600 to 256 × 256 for the AID dataset in the preprocessing causes some loss of important information and has a negative effect on the classification result. However, in the 20% training ratio, the proposed method outperforms GCFs + LOFs by approximately 1.31%. In addition, data augmentation was used in GCFs + LOFs. Thus, overall, the proposed method yields the state-of-the-art result on AID dataset comprehensively.

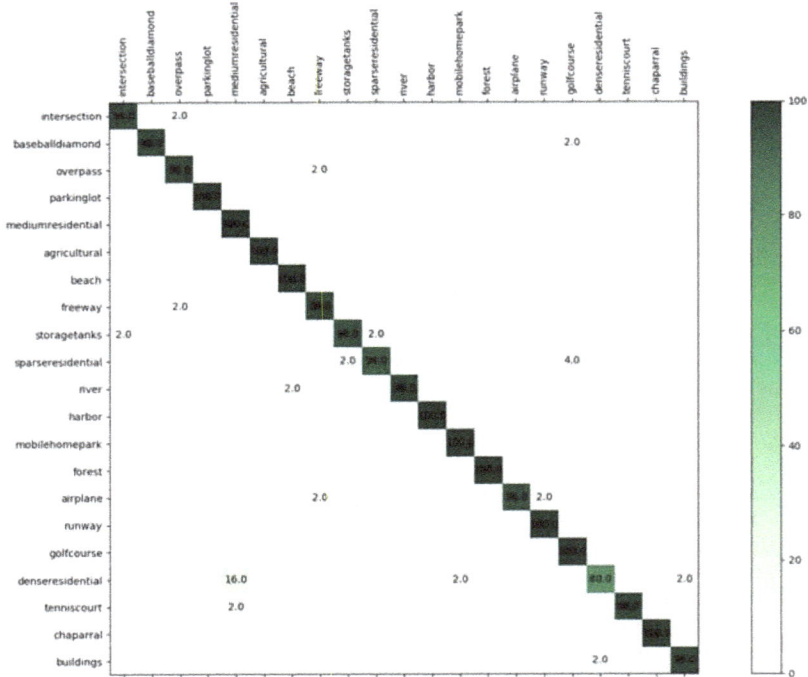

Figure 15. Confusion matrix of the proposed method on UC Merced Land-Use dataset by fixing the training ratio to 50%.

Table 2. Overall accuracy (%) and standard deviations of the proposed method and the comparison methods under the training ratios of 50% and 20% on the AID dataset.

Method	50% Training Ratio	20% Training Ratio
CaffeNet [35]	89.53 ± 0.31	86.86 ± 0.47
GoogLeNet [35]	86.39 ± 0.55	83.44 ± 0.40
VGG-16 [35]	89.64 ± 0.36	86.59 ± 0.29
salM³LBP-CLM [63]	89.76 ± 0.45	86.92 ± 0.35
TEX-Net-LF [64]	92.96 ± 0.18	90.87 ± 0.11
Fusion by addition [28]	91.87 ± 0.36	/
Two-Stream Fusion [30]	94.58 ± 0.25	92.32 ± 0.41
GCFs+LOFs [56]	**96.85 ± 0.23**	92.48 ± 0.38
VGG-16-CapsNet (ours)	94.74 ± 0.17	91.63 ± 0.19
Inception-v3-CapsNet (ours)	96.32 ± 0.12	**93.79 ± 0.13**

As for the analysis of the confusion matrix, shown in Figure 16, 80% of all 30 categories achieved classification accuracies greater than 90% where the mountain class achieved the 100% accuracy. Some categories with small interclass dissimilarity, such as 'sparse residential', 'medium residential', and 'dense residential' were also classified accurately with 99.17%, 94.83% and 95.73%, respectively. The classes of 'school' and 'resort' had relatively low classification accuracies with 67.92% and 72.84. In detail, the 'school' class was easily confused with 'commercial' because they had the same image distribution. In addition, the resort class was usually misclassified as 'park' due to the existence of some analogous objects such as green belts and ponds. Even so, great improvements were achieved by the proposed method, compared with the classification accuracy of 49% and 60% in [35]. This means that the CNN-CapsNet could learn the differences of spatial information between these scene classes with the same image distribution and distinguish them effectively.

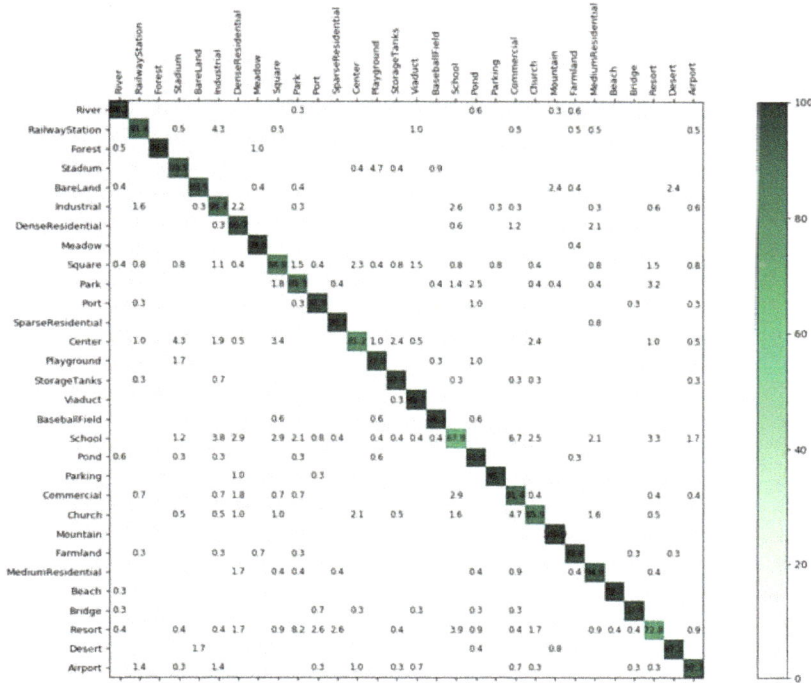

Figure 16. Confusion matrix of the proposed method on the AID dataset by fixing the training ratio as 20%.

3. Classification of NWPU-RESISC45 dataset

Table 3 shows the classification performance comparison of the proposed architecture and the existing state-of-the-art methods using the most challenging NWPU-RESISC45 dataset. It can be observed that the Inception-v3-CapsNet model also achieved remarkable classification results, with OA improvements of 0.27% and 1.88% over the second best model using 20% and 10% training ratios, respectively. The good performance of the proposed method further verifies the effectiveness of combining the pretrained CNN model and CapsNet.

Table 3. Overall accuracy (%) and standard deviations of the proposed method and the comparison methods under the training ratios of 20% and 10% on NWPU-RESISC45 dataset.

Method	20% Training Ratio	10% Training Ratio
GoogLeNet [53]	78.48 ± 0.26	76.19 ± 0.38
VGG-16 [53]	79.79 ± 0.15	76.47 ± 0.18
AlexNet [53]	79.85 ± 0.13	76.69 ± 0.21
Two-Stream Fusion [30]	83.16 ± 0.18	80.22 ± 0.22
BoCF [67]	84.32 ± 0.17	82.65 ± 0.31
Fine-tuned AlexNet [53]	85.16 ± 0.18	81.22 ± 0.19
Fine-tuned GoogLeNet [53]	86.02 ± 0.18	82.57 ± 0.12
Fine-tuned VGG-16 [53]	90.36 ± 0.18	87.15 ± 0.45
Triple networks [68]	92.33 ± 0.20	/
VGG-16-CapsNet (ours)	89.18 ± 0.14	85.08 ± 0.13
Inception-v3-CapsNet (ours)	**92.6 ± 0.11**	**89.03 ± 0.21**

Figure 17 gives the confusion matrix generated from the best classification result by Inception-v3-CapsNet with the training ratio of 20%. From the confusion matrix, 36 categories among all 45 categories achieved classification accuracies greater than 90%. The major confusion was in 'palace' and 'church' because both of them have similar styles of buildings. In spite of that, substantial improvements were still achieved with 79.3% and 68% compared with 75% and 64% in [53], respectively.

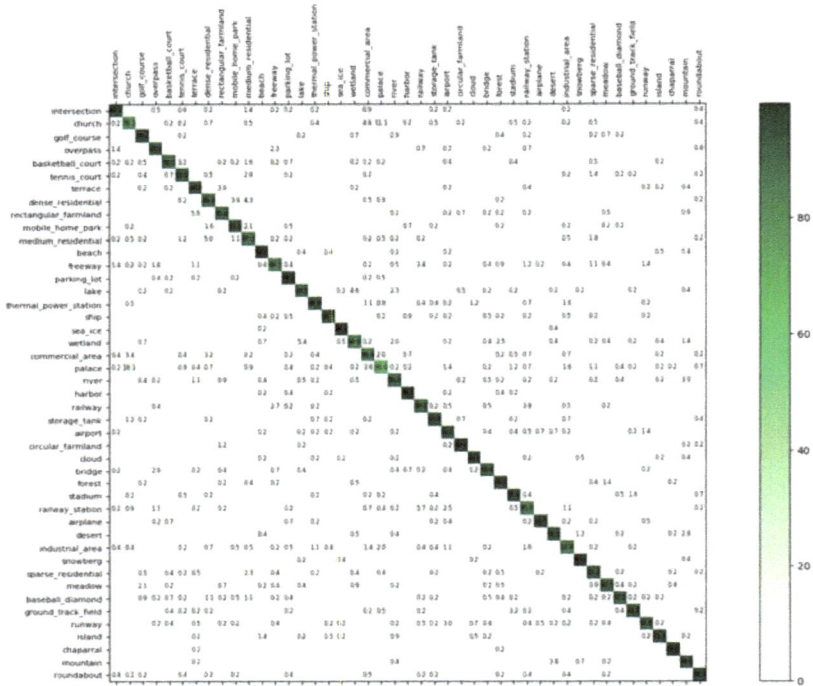

Figure 17. Confusion matrix of the proposed method on the NWPU-RESISC45 dataset by fixing the training ratio as 20%.

4.2.3. Further Explanation

In Section 4.2.2, it was found that the proposed method obtains state-of-the-art classification results. This mainly benefits from the following three factors: fine-tuning, capsules and the pretrained CNN model. In this section, further analysis will be performed on how they accomplish the significant performance for classification. In addition, the training ratios of the three datasets were the same as those in Section 4.2.1.

1. Effectiveness of fine-tuning

First, the strategy of fine-tuning was added to train the proposed architecture. To evaluate the effectiveness of fine-tuning in the proposed method, a comparison was made between the classification results with and without fine-tuning. As shown in Figure 18, the methods with fine-tuning obtained a significant improvement compared with no fine-tuning operation. The reason is that the features extracted from the pretrained CNN models have a strong relationship with the original task. Fine-tuning can adjust the parameters of the pretrained CNN model to cater to the current training datasets for an accuracy improvement.

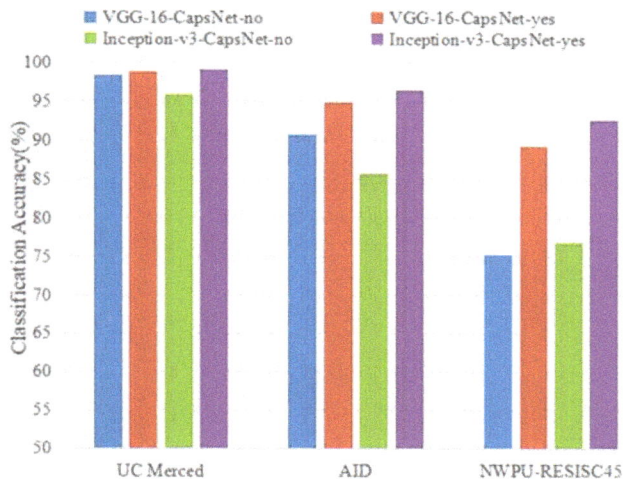

Figure 18. Overall accuracy (%) of the proposed method with and without fine-tuning on three datasets.

2. Effectiveness of capsules

In the design of the proposed architecture, the CapsNet is used as the classifier to label the remote sensing image, which uses the capsule to replace the neuron in traditional neural networks. To prove the validity of the positive impact on classification results with this replacement, a comparative experiment was conducted. In detail, a new CNN architecture was designed as the classifier, which consists of one convolutional layer and two fully connected layers. In addition, the only difference between the new CNN architecture and the CapsNet described in Section 3.3 was using the neuron to replace the capsule while other parameters including the training hyperparameters were all kept the same. The experimental results are shown in Figure 19 (the VGG-16-CNN and Inception-v3-CNN in Figure 19 mean that using pretrained VGG-16 and Inception-v3 as feature extractors, respectively, and using the newly designed CNN architecture as the classifier). For three datasets, the models using capsules all achieved better performance than those using traditional neurons. This further demonstrates that the CapsNet can learn more representative information of scene images.

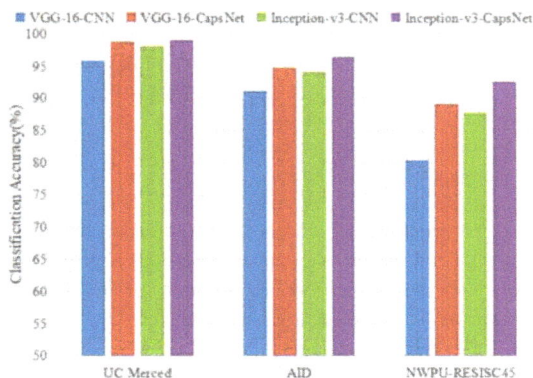

Figure 19. Overall accuracy (%) of the proposed method with neuron and capsule on three datasets.

3. Effectiveness of the pretrained CNN model

The pretrained CNN model was selected as the initial feature maps extractor instead of designing a new CNN architecture. This is also a great factor for the success of the proposed architecture. For comparison, a CNN architecture (Self-CNN) was designed, which only contained four consecutive convolutional layers and the size of its output feature maps was $16 \times 16 \times 512$, the same as that of the pretrained VGG-16 used in this paper. The parameters of the CapsNet were the same. The new Self-CNN-CapsNet architecture was trained from scratch. The classification results are presented in Figure 20. From the Figure, the classification accuracy of CNN-CapsNet with the pretrained CNN model as the feature extractor was much higher than that with self-CNN. This is because the existing datasets cannot fully train the model and further proves the effectiveness of using pretrained CNN models as feature extractors.

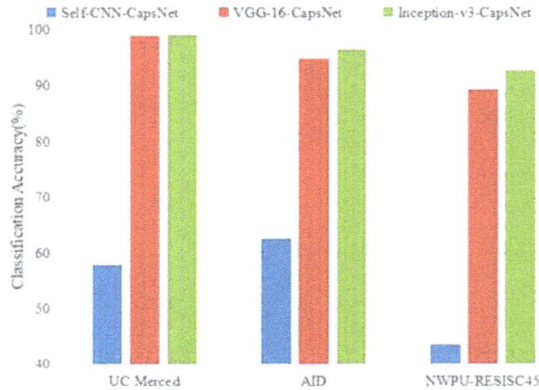

Figure 20. Overall accuracy (%) of the proposed method with the pretrained model and self-model on three datasets.

5. Conclusions

In recent years, the prevalence of deep learning methods especially the CNN has made the performance of remote sensing scene classification state-of-the-art. However, the scene classes with the same image distribution are still not distinguished effectively. This is mainly because some fully connected layers are added to the end of the CNN, which gives less consideration to the spatial relationship that is vital to classification. To preserve the spatial information, the new architecture CapsNet is proposed, which uses the capsule to replace the neuron in the traditional neural network. In addition, the capsule is a vector to represent internal properties that can be used to learn part-whole relationships within an image. In this paper, to further improve the classification accuracy of remote sensing image scene classification and inspired by the CapsNet, a novel architecture named CNN-CapsNet is proposed for remote sensing image scene classification. The proposed architecture consists of two parts: CNN and CapsNet. The CNN part is transferring the original remote sensing images to the original feature maps. In addition, the CapsNet part converts the original feature maps into various levels of capsules and to obtain the final classification result. Experiments were performed on three public challenging datasets, and the experimental results demonstrate the effectiveness of the proposed CNN-CapsNet and show that the proposed method outperforms the current state-of-the-art methods. In future work, different from using feature maps from only one CNN model, in this paper, feature maps from different pretrained CNN models will be merged for remote sensing image scene classification.

Author Contributions: W.Z. conceived and designed the whole framework and the experiments, as well as wrote the manuscript. L.Z. contributed to the discussion of the experimental design. L.Z. and P.T. all helped to organize the paper and performed the experimental analysis. P.T. helped to revise the manuscript, and all authors read and approved the submitted manuscript.

Remote Sens. **2019**, *11*, 494

Acknowledgments: This research was funded by the National Natural Science Foundation of China under grant 41701397; and the Major Project of High-Resolution Earth Observation System of China under Grant 03-Y20A04-9001-17/18.

Conflicts of Interest: The authors declare no conflict of interest.

References

1. Plaza, A.; Plaza, J.; Paz, A.; Sanchez, S. Parallel hyperspectral image and signal processing. *IEEE Signal Process. Mag.* **2011**, *28*, 119–126. [CrossRef]
2. Hubert, M.J.; Carole, E. Airborne SAR-efficient signal processing for very high resolution. *Proc. IEEE* **2013**, *101*, 784–797. [CrossRef]
3. Cheriyadat, A.M. Unsupervised feature learning for aerial scene classification. *IEEE Trans. Geosci. Remote Sens.* **2014**, *52*, 439–451. [CrossRef]
4. Shao, W.; Yang, W.; Xia, G.S. Extreme value theory-based calibration for the fusion of multiple features in high-resolution satellite scene classification. *Int. J. Remote Sens.* **2013**, *34*, 8588–8602. [CrossRef]
5. Estoque, R.C.; Murayama, Y.; Akiyama, C.M. Pixel-based and object-based classifications using high- and medium-spatial-resolution imageries in the urban and suburban landscapes. *Geocarto Int.* **2015**, *30*, 1113–1129. [CrossRef]
6. Zhang, X.; Wang, Q.; Chen, G.; Dai, F.; Zhu, K.; Gong, Y.; Xie, Y. An object-based supervised classification framework for very-high-resolution remote sensing images using convolutional neural networks. *Remote Sens. Lett.* **2018**, *9*, 373–382. [CrossRef]
7. Yang, Y.; Newsam, S. Comparing SIFT descriptors and Gabor texture features for classification of remote sensed imagery. In Proceedings of the 15th IEEE International Conference on Image Processing (ICIP), San Diego, CA, USA, 12–15 October 2008; pp. 1852–1855.
8. Dos Santos, J.A.; Penatti, O.A.B.; da Silva Torres, R. Evaluating the Potential of Texture and Color Descriptors for Remote Sensing Image Retrieval and Classification. In Proceedings of the VISAPP, Angers, France, 17–21 May 2010; pp. 203–208.
9. Chen, C.; Zhang, B.; Su, H.; Li, W.; Wang, L. Land-use scene classification using multi-scale completed local binary patterns. *Signal Image Video Process.* **2016**, *10*, 745–752. [CrossRef]
10. Li, H.T.; Gu, H.Y.; Han, Y.S.; Yang, J.H. Object oriented classification of high-resolution remote sensing imagery based on an improved colour structure code and a support vector machine. *Int. J. Remote Sens.* **2010**, *31*, 1453–1470. [CrossRef]
11. Penatti, O.A.; Nogueira, K.; dos Santos, J.A. Do deep features generalize from everyday objects to remote sensing and aerial scenes domains? In Proceedings of the IEEE Conference on Computer Vision and Pattern Recognition Workshops, Boston, MA, USA, 7–12 June 2015; pp. 44–51.
12. Luo, B.; Jiang, S.J.; Zhang, L.P. Indexing of remote sensing images with different resolutions by multiple features. *IEEE J. Sel. Top. Appl. Earth Obs. Remote Sens.* **2013**, *6*, 1899–1912. [CrossRef]
13. Zhou, L.; Zhou, Z.; Hu, D. Scene classification using a multi-resolution bag-of-features model. *Pattern Recognit.* **2013**, *46*, 424–433. [CrossRef]
14. Yang, Y.; Newsam, S. Bag-of-visual-words and spatial extensions for land-use classification. In Proceedings of the 18th SIGSPATIAL International Conference on Advances in Geographic Information Systems, San Jose, CA, USA, 3–5 November 2010; pp. 270–279.
15. Zhao, L.; Tang, P.; Huo, L. A 2-D wavelet decomposition-based bag-of-visual-words model for land-use scene classification. *Int. J. Remote Sens.* **2014**, *35*, 2296–2310. [CrossRef]
16. Zhao, L.; Tang, P.; Huo, L. Land-use scene classification using a concentric circle-structured multiscale bag-of-visual-words model. *IEEE J. Sel. Top. Appl. Earth Obs. Remote Sens.* **2014**, *7*, 4620–4631. [CrossRef]
17. Sridharan, H.; Cheriyadat, A. Bag of lines (bol) for improved aerial scene representation. *IEEE Geosci. Remote Sens. Lett.* **2015**, *12*, 676–680. [CrossRef]
18. Zhu, Q.; Zhong, Y.; Zhao, B.; Xia, G.; Zhang, L. Bag-of-visual-words scene classifier with local and global features for high spatial resolution remote sensing imagery. *IEEE Geosci. Remote Sens. Lett.* **2016**, *13*, 747–751. [CrossRef]
19. He, K.M.; Zhang, X.Y.; Ren, S.Q.; Sun, J. Deep residual learning for image recognition. In Proceedings of the 2016 IEEE Conference on Computer Vision and Pattern Recognition (CVPR), Las Vegas, NV, USA, 26–30 June 2016; pp. 770–778. [CrossRef]

20. Girshick, R.; Donahue, J.; Darrell, T.; Malik, J. Rich feature hierarchies for accurate object detection and semantic segmentation. In Proceedings of the 2014 IEEE Conference on Computer Vision and Pattern Recognition (CVPR), Ohio, CO, USA, 24–27 June 2014; pp. 580–587. [CrossRef]

21. Long, J.; Shelhamer, E.; Darrell, T. Fully convolutional networks for semantic segmentation. In Proceedings of the 2015 IEEE Conference on Computer Vision and Pattern Recognition (CVPR), Boston, MA, USA, 7–12 June 2015; pp. 3431–3440. [CrossRef]

22. Zhong, Y.; Fei, F.; Zhang, L. Large patch convolutional neural networks for the scene classification of high spatial resolution imagery. *J. Appl. Remote Sens.* **2016**, *10*, 025006. [CrossRef]

23. Zhang, F.; Du, B.; Zhang, L. Scene classification via a gradient boosting random convolutional network ramework. *IEEE Trans. Geosci. Remote Sens.* **2016**, *54*, 1793–1802. [CrossRef]

24. Liu, Y.; Zhong, Y.; Fei, F.; Zhu, Q.; Qin, Q. Scene Classification Based on a Deep Random-Scale Stretched Convolutional Neural Network. *Remote Sens.* **2018**, *10*, 444. [CrossRef]

25. Castelluccio, M.; Poggi, G.; Sansone, C.; Verdoliva, L. Land use classification in remote sensing images by convolutional neural networks. *arXiv* **2015**, arXiv:1508.00092.

26. Nogueira, K.; Penatti, O.A.B.; dos Santos, J.A. Towards better exploiting convolutional neural networks for remote sensing scene classification. *Pattern Recognit.* **2017**, *61*, 539–556. [CrossRef]

27. Hu, F.; Xia, G.S.; Hu, J.; Zhang, L. Transferring deep convolutional neural networks for the scene classification of high-resolution remote sensing imagery. *Remote Sens.* **2015**, *7*, 14680–14707. [CrossRef]

28. Chaib, S.; Liu, H.; Gu, Y.; Yao, H. Deep feature fusion for VHR remote sensing scene classification. *IEEE Trans. Geosci. Remote Sens.* **2017**, *55*, 4775–4784. [CrossRef]

29. Zou, Q.; Ni, L.; Zhang, T.; Wang, Q. Deep learning based feature selection for remote sensing scene classification. *IEEE Geosci. Remote. Sens. Lett.* **2015**, *12*, 2321–2325. [CrossRef]

30. Yu, Y.; Liu, F. A two-stream deep fusion framework for high-resolution aerial scene classification. *Comput. Intell. Neurosci.* **2018**, *2018*, 8639367. [CrossRef] [PubMed]

31. Othman, E.; Bazi, Y.; Alajlan, N.; Alhichri, H.; Melgani, H. Using convolutional features and a sparse autoencoder for land-use scene classification. *Int. J. Remote Sens.* **2016**, *37*, 2149–2167. [CrossRef]

32. Marmanis, D.; Datcu, M.; Esch, T.; Stilla, U. Deep learning earth observation classification using ImageNet pretrained networks. *IEEE Geosci. Remote Sens. Lett.* **2016**, *13*, 105–109. [CrossRef]

33. Cheng, G.; Ma, C.; Zhou, P.; Yao, X.; Han, J. Scene classification of high resolution remote sensing images using convolutional neural networks. In Proceedings of the IEEE International Geoscience Remote Sensing Symposium (IGARSS), Beijing, China, 10–15 July 2016; pp. 767–770. [CrossRef]

34. Zhao, L.; Zhang, W.; Tang, P. Analysis of the inter-dataset representation ability of deep features for high spatial resolution remote sensing image scene classification. *Multimed. Tools Appl.* **2018**. [CrossRef]

35. Xia, G.S.; Hu, J.; Hu, F.; Shi, B.; Bai, X.; Zhong, Y.; Zhang, L.; Lu, X. AID: A benchmark data set for performance evaluation of aerial scene classification. *IEEE Trans. Geosci. Remote Sens.* **2017**, *55*, 3965–3981. [CrossRef]

36. Sabour, S.; Frosst, N.; Hinton, G.E. Dynamic routing between capsules. In Proceedings of the 31st Conference on Neural Information Processing Systems (NIPS), Long Beach, CA, USA, 4–9 December 2017; pp. 3859–3869.

37. Andersen, P.A. Deep Reinforcement learning using capsules in advanced game environments. *arXiv* **2018**, arXiv:1801.09597.

38. Afshar, P.; Mohammadi, A.; Plataniotis, K.N. Brain Tumor Type Classification via Capsule Networks. *arXiv* **2018**, arXiv:1802.10200.

39. Iqbal, T.; Xu, Y.; Kong, Q.; Wang, W. Capsule routing for sound event detection. *arXiv* **2018**, arXiv:1806.04699.

40. LaLonde, R.; Bagci, U. Capsules for object segmentation. *arXiv* **2018**, arXiv:1804.04241.

41. Deng, F.; Pu, S.; Chen, X.; Shi, Y.; Yuan, T.; Pu, S. Hyperspectral Image Classification with Capsule Network Using Limited Training Samples. *Sensors* **2018**, *18*, 3153. [CrossRef] [PubMed]

42. Xi, E.; Bing, S.; Jin, Y. Capsule Network Performance on Complex Data. *arXiv* **2017**, arXiv:1712.03480.

43. Jaiswal, A.; AbdAlmageed, W.; Natarajan, P. CapsuleGAN: Generative adversarial capsule network. *arXiv* **2018**, arXiv:1802.06167.

44. Neill, J.O. Siamese capsule networks. *arXiv* **2018**, arXiv:1805.07242.

45. Mobiny, A.; Nguyen, H.V. Fast CapsNet for lung cancer screening. *arXiv* **2018**, arXiv:1806.07416.

46. Kumar, A.D. Novel Deep learning model for traffic sign detection using capsule networks. *arXiv* **2018**, arXiv:1805.04424.

47. Li, Y.; Qian, M.; Liu, P.; Cai, Q.; Li, X.; Guo, J.; Yan, H.; Yu, F.; Yuan, K.; Yu, J.; et al. The recognition of rice images by UAV based on capsule network. *Clust. Comput.* **2018**. [CrossRef]

48. Qiao, K.; Zhang, C.; Wang, L.; Yan, B.; Chen, J.; Zeng, L.; Tong, L. Accurate reconstruction of image stimuli from human fMRI based on the decoding model with capsule network architecture. *arXiv* **2018**, arXiv:1801.00602.

49. Zhao, W.; Ye, J.; Yang, M.; Lei, Z.; Zhang, S.; Zhao, Z. Investigating capsule networks with dynamic routing for text classification. *arXiv* **2018**, arXiv:1804.00538.

50. Xiang, C.; Zhang, L.; Tang, Y.; Zou, W.; Xu, C. MS-CapsNet: A novel multi-scale capsule network. *IEEE Signal Process. Lett.* **2018**, *25*, 1850–1854. [CrossRef]

51. Simonyan, K.; Zisserman, A. Very deep convolutional networks for large-scale image recognition. In Proceedings of the 2015 International Conference on Learning Representations (ICLR), San Diego, CA, USA, 7–9 May 2015; pp. 1–14.

52. Russakovsky, A.; Deng, J.; Su, H. ImageNet large scale visual recognition challenge. *Int. J. Comput. Vis.* **2015**, *115*, 211–252. [CrossRef]

53. Cheng, G.; Han, J.; Lu, X. Remote sensing image scene classification: Benchmark and state of the art. *Proc. IEEE* **2017**, *105*, 1865–1883. [CrossRef]

54. Gong, X.; Xie, Z.; Liu, Y.; Shi, X.; Zheng, Z. Deep salient feature based anti-noise transfer network for scene classification of remote sensing imagery. *Remote Sens.* **2018**, *10*, 410. [CrossRef]

55. Chen, G.; Zhang, X.; Tan, X.; Cheng, Y.F.; Dai, F.; Zhu, K.; Gong, Y.; Wang, Q. Training small networks for scene classification of remote sensing images via knowledge distillation. *Remote Sens.* **2018**, *10*, 719. [CrossRef]

56. Zeng, D.; Chen, S.; Chen, B.; Li, S. Improving remote sensing scene classification by integrating global-context and local-object features. *Remote Sens.* **2018**, *10*, 734. [CrossRef]

57. Chen, J.; Wang, C.; Ma, Z.; Chen, J.; He, D.; Ackland, S. Remote sensing scene classification based on convolutional neural networks pre-trained using attention-guided sparse filters. *Remote Sens.* **2018**, *10*, 290. [CrossRef]

58. Zou, J.; Li, W.; Chen, C.; Du, Q. Scene classification using local and global features with collaborative representation fusion. *Inf. Sci.* **2018**, *348*, 209–226. [CrossRef]

59. Mahdianpari, M.; Salehi, B.; Rezaee, M.; Mohammadimanesh, F.; Zhang, Y. Very Deep Convolutional Neural Networks for Complex Land Cover Mapping Using Multispectral Remote Sensing Imagery. *Remote Sens.* **2018**, *10*, 1119. [CrossRef]

60. Krizhevsky, A.; Sutskever, I.; Hinton, G.E. ImageNet classification with deep convolutional neural networks. In Proceedings of the 26th Annual Conference on Neural Information Processing Systems, Harrahs and Harveys, Lake Tahoe, CA, USA, 3–8 December 2012; NIPS Foundation: La Jolla, CA, USA, 2012.

61. Szegedy, C.; Liu, W.; Jia, Y. Going deeper with convolutions. In Proceedings of the 2015 IEEE Conference on Computer Vision and Pattern Recognition (CVPR), Boston, MA, USA, 7–12 June 2015; pp. 1–9. [CrossRef]

62. Szegedy, C.; Vanhoucke, V.; Ioffe, S.; Shlens, J. Rethinking the inception architecture for computer vision. *arXiv* **2015**, arXiv:1512.00567.

63. Bian, X.; Chen, C.; Tian, L.; Du, Q. Fusing local and global features for high-resolution scene classification. *IEEE J. Sel. Top. Appl. Earth Obs. Remote Sens.* **2017**, *10*, 2889–2901. [CrossRef]

64. Anwer, R.M.; Khan, F.S.; van deWeijer, J.; Monlinier, M.; Laaksonen, J. Binary patterns encoded convolutional neural networks for texture recognition and remote sensing scene classification. *arXiv* **2017**, arXiv:1706.01171. [CrossRef]

65. Weng, Q.; Mao, Z.; Lin, J.; Guo, W. Land-Use Classification via Extreme Learning Classifier Based on Deep Convolutional Features. *IEEE Geosci. Remote Sens. Lett.* **2017**, *14*, 704–708. [CrossRef]

66. Qi, K.; Guan, Q.; Yang, C.; Peng, F.; Shen, S.; Wu, H. Concentric Circle Pooling in Deep Convolutional Networks for Remote Sensing Scene Classification. *Remote Sens.* **2018**, *10*, 934. [CrossRef]

67. Cheng, G.; Li, Z.; Yao, X.; Li, K.; Wei, Z. Remote Sensing Image Scene Classification Using Bag of Convolutional Features. *IEEE Geosci. Remote Sens. Lett.* **2017**, *14*, 1735–1739. [CrossRef]

68. Liu, Y.; Huang, C. Scene classification via triplet networks. *IEEE J. Sel. Top. Appl. Earth Obs. Remote Sens.* **2018**, *11*, 220–237. [CrossRef]

remote sensing

MDPI

Article

Applying High-Resolution Imagery to Evaluate Restoration-Induced Changes in Stream Condition, Missouri River Headwaters Basin, Montana

Melanie K. Vanderhoof *[ID] and Clifton Burt

U.S. Geological Survey, Geosciences and Environmental Change Science Center, P.O. Box 25046, DFC, MS980, Denver, CO 80225, USA; steeprockwall@gmail.com
* Correspondence: mvanderhoof@usgs.gov; Tel.: +1-303-236-1411

Received: 13 April 2018; Accepted: 7 June 2018; Published: 9 June 2018

Abstract: Degradation of streams and associated riparian habitat across the Missouri River Headwaters Basin has motivated several stream restoration projects across the watershed. Many of these projects install a series of beaver dam analogues (BDAs) to aggrade incised streams, elevate local water tables, and create natural surface water storage by reconnecting streams with their floodplains. Satellite imagery can provide a spatially continuous mechanism to monitor the effects of these in-stream structures on stream surface area. However, remote sensing-based approaches to map narrow (e.g., <5 m wide) linear features such as streams have been under-developed relative to efforts to map other types of aquatic systems, such as wetlands or lakes. We mapped pre- and post-restoration (one to three years post-restoration) stream surface area and riparian greenness at four stream restoration sites using Worldview-2 and 3 images as well as a QuickBird-2 image. We found that panchromatic brightness and eCognition-based outputs (0.5 m resolution) provided high-accuracy maps of stream surface area (overall accuracy ranged from 91% to 99%) for streams as narrow as 1.5 m wide. Using image pairs, we were able to document increases in stream surface area immediately upstream of BDAs as well as increases in stream surface area along the restoration reach at Robb Creek, Alkali Creek and Long Creek (South). Although Long Creek (North) did not show a net increase in stream surface area along the restoration reach, we did observe an increase in riparian greenness, suggesting increased water retention adjacent to the stream. As high-resolution imagery becomes more widely collected and available, improvements in our ability to provide spatially continuous monitoring of stream systems can effectively complement more traditional field-based and gage-based datasets to inform watershed management.

Keywords: beaver mimicry; beaver dam analogue; QuickBird; riparian; stream restoration; Worldview

1. Introduction

Remotely sensed imagery has been widely applied to characterize variability in surface-water extent across space and time [1,2]. The spatial resolution ($\geq$30 m) of commonly used sources of imagery (e.g., Landsat, MODIS, AVHRR), however, has limited our ability to remotely monitor river systems, except for large rivers (e.g., >40 m wide) [3] or rivers under flood conditions [4–7]. Yet remote monitoring of the spatial distribution of river stage and condition has several applications including enhancing our ability to predict and monitor flood events, informing the source and distribution of flow to downstream gaged points, helping monitor ungaged watersheds, predicting carbon dioxide emissions, and informing river management [8–11]. The rapidly increasing availability of multispectral, high-resolution imagery ($\leq$5 m resolution, Dove, RapidEye (Planet, San Francisco, CA, USA), Worldview-2, 3 (DigitalGlobe, Westminster, CO, USA)) provides increased opportunity to potentially monitor river systems across diverse watershed sizes and flow conditions.

Multiple sources of fine resolution imagery have been applied to aquatic systems. LiDAR [12,13] and synthetic aperture radar (SAR) imagery have been successfully used to map surface water and can be preferable in forested environments or during storm events under cloud cover [14–16]. Multispectral, high-resolution imagery has also been used effectively to map surface water [17,18]. Riverscape units including the active channel have primarily been mapped by digitizing very high-resolution multispectral satellite imagery or aerial imagery [19–21] or by applying geographical object-based image analysis (GEOBIA) methods [11,22,23]. A GEOBIA approach segments an image into homogenous objects prior to object classification. Such an approach can help account for the greater within-class spectral variability that can occur with high-resolution imagery, relative to moderate-resolution imagery [18]. In general, however, efforts to remotely monitor narrow, linear water features, such as rivers and streams, have lagged behind efforts to remotely monitor lakes and wetlands [24,25].

The Upper Missouri River Headwaters Basin in southwestern Montana faces increasingly uncertain water supplies attributable to high water demand for agricultural irrigation [26,27] and public water supply [28,29]. In addition, shifts in the timing of runoff and peak streamflow are predicted with increasing amounts of winter precipitation and a declining snowpack related to climate change [30–32]. Societal water demands as well as climate-induced shifts in streamflow can threaten habitat critical for fish and aquatic species [33,34]. These risks have raised interest in increasing the capacity of streams to respond to extreme events [35–37]. One approach that is growing in popularity is to slow runoff, absorb excess floodwater, and encourage groundwater recharge by enhancing natural water storage in stream channels, riparian areas, and floodplains [38,39]. One way to create natural water storage is using in-stream, channel spanning structures called beaver dam analogues (BDAs) [40–42]. Over time BDAs have been shown to slow water flow, encourage channel stability and riparian vegetation, activate side channels, and improve water quality and fish habitat [40,41,43]. Installing BDAs along a reach of stream can potentially increase spring overbank flow and elevate riparian water tables [43]. If water is a limiting factor in the riverscape, elevating near-surface soil moisture along stream channels can encourage riparian vegetation [44,45]. Depending on channel shape and river stage, slowing the flow of water can increase the stream surface area through ponding and temporary flooding [38,45]. Downstream from beaver dams, channels are more likely to be stable, with lower sediment loads and a lower range of stream discharge [45]. Although the installation of BDAs is becoming relatively common, analyses that evaluate their impact are limited. Most studies to date have focused on the impact of actual beaver dams that differ from BDAs in the magnitude of change and maintenance activity. In this study, we tested how multispectral high-resolution imagery can be used to monitor stream condition along four stream reaches in the Upper Missouri Headwaters Basin, and how pairs of images can potentially be used to monitor BDA stream restoration projects. Our research questions included:

1. What methodological approaches are most effective to map stream surface area using multispectral high-resolution imagery? And,
2. How can image pairs (e.g., pre- and post-restoration) be used to monitor changes in stream surface area and riparian greenness?

2. Methods

2.1. Study Area and Restoration Activities

The four stream sites examined in this study occur within the Upper Missouri Headwaters Basin in southwest Montana (Figure 1). Annual precipitation across the Headwaters Basin averages 565 mm yr^{-1}, while the annual temperature maximum and minimum average 10 °C and −3 °C, respectively (1981–2010) [46]. Across the basin herbaceous vegetation (35%) and shrub/scrub (20%) dominate the large river valleys while evergreen forest dominates the higher elevations (35%) [47]. Two restoration sites occurred along reaches separated by 4.8 stream km in Long Creek (Figure 1), which flows south into the Red Rock River in the Red Rock River Hydrological Unit (Red Rock HUC8). Land cover adjacent to Long Creek is dominated by herbaceous vegetation, shrub/scrub, and emergent herbaceous wetlands [47]. The third restoration site occurred along Alkali Creek, which flows northwest into Blacktail Deer Creek in the Beaverhead HUC8. This site showed evidence of beaver activity just upstream from the restoration site. The fourth restoration site occurred along Robb Creek that flows north into the Ruby River (Ruby River HUC8). Land cover adjacent to both Alkali Creek and Robb Creek is dominated by herbaceous vegetation and shrub/scrub habitat [47].

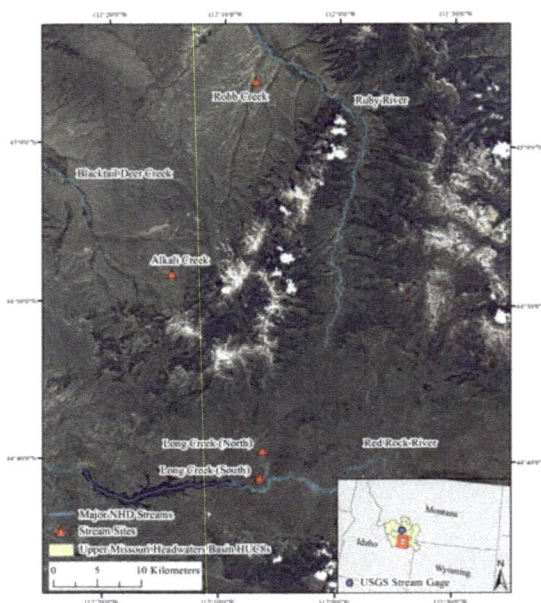

Figure 1. Distribution of the stream sites within the Upper Missouri Headwaters Basin. Background image is a Landsat 8 image (path 39, row 39, 9 June 2016). Location of the U.S. Geological Survey stream gage (Jefferson River, #06026500) is also shown relative to the restoration sites. NHD: National Hydrography Dataset, HUC8s: 8-digit Hydrological Units.

All restoration activities were developed and completed by the Nature Conservancy. A series of BDAs were installed in stream reaches at each restoration site. The structures were created from wooden posts installed vertically into the streambed across the channel with willow branches woven between posts. The structures collect organic material and sediment behind them, building up the stream bed height, ponding water upstream from the structures, stabilizing the channel and increasing connectivity with its floodplain [48]. The BDAs were accompanied by willow plantings along the stream to stabilize banks and cattle exclusions at most of the sites [48]. The design is cost effective as no heavy equipment is used and the in-stream structures are designed to be temporary [49]. BDAs were installed in two reaches of Long Creek (9 BDAs on the north reach and 7 BDAs on the south reach), a reach of Alkali Creek (6 BDAs) and a reach of Robb Creek (12 BDAs) (Table 1). The goal of the restorations at the Long Creek and Alkali Creek sites was to aggrade the streambed, improving hydrologic connectivity between the stream channel and associated floodplains. Along Robb Creek, the BDAs were designed to encourage reactivation of abandoned side channels. The time since restoration ranged from one to three years across the sites (Table 1).

Table 1. Characteristics of the restoration sites and a summary of the restoration activities performed at each site. A list of the high-resolution images representing pre- and post-restoration conditions across the four sites is also shown. Length refers to the stream distance from the most upstream to most downstream beaver dam analog (BDA).

Site	Elevation (m)	Slope (%)	Sinuosity	Width (Pre-Restoration, m)	BDAs (Length, m)	Willow Stakes	Restoration Date	Years Since Restoration
Alkali Creek	2249	2.1	2.1	1.6	6 (830)	~	16-Oct	1
Long Creek (North)	2033	0.9	2.7	3.5	9 (3857)	800	16-Aug	1
Long Creek (South)	2014	1	3.7	3.7	7 (2496)	2500	14-Sep	3
Robb Creek	1793	2.6	1.1	1.8	12 (1232)	2915	15-Nov	2

Site	Pre-Image Date	Jefferson River Discharge ($m^3\ s^{-1}$) (Daily Mean)	Pre-Image Source	Mean Off-Nadir View Angle	Post-Image Date	Jefferson River Discharge ($m^3\ s^{-1}$) (Daily Mean)	Post-Image Source	Mean Off-Nadir View Angle
Alkali Creek	30-Jun-14	126.9	Worldview-2	21.5	2-Aug-17	15.4	Worldview-3	17
Long Creek (North)	30-Jun-14	126.9	Worldview-2	21.9	20-Jun-17	140.5	Worldview-3	19
Long Creek (South)	30-Jun-14	126.9	Worldview-2	21.9	20-Jun-17	140.5	Worldview-3	19
Robb Creek	23-Jun-14	117.8	QuickBird-2	11.7	23-Jun-17	108.2	Worldview-2	28.3

2.2. Image Acquisition and Preprocessing

A total of six high-resolution images (2 m resolution) were acquired from DigitalGlobe (Westminster, CO, USA) via the NextView license for this analysis. These images included one QuickBird-2 image, three Worldview-2 images, and two Worldview-3 images (Table 1). "Pre-restoration" conditions were represented by images acquired during summer 2014, while "post-restoration" conditions were represented by images acquired during summer 2017. Using historical (1895–2017) Palmer Hydrological Drought Index (PHDI) values, we found that the pre- and post-restoration image dates represented similar historical wetness conditions (37.6% relative to 41.8% PHDI); however, 2013, the year prior to the pre-restoration images experienced a drought which may have influenced stream conditions in 2014 (Figure 2). We also compared the stream discharge values on the date the image was collected using a USGS stream gage downstream from the four restoration sites (Jefferson River, USGS Gage #06026500) (Figure 1). Discharge was reasonably similar between the image dates (10% higher post-restoration for the Long Creek image dates and 8% lower post-restoration for the Robb Creek image dates), for all sites except Alkali Stream. The post-restoration image at this site was collected in August when discharge was much lower relative to the early summer period (Figure 2). We converted the image (processing Level 1) pixel values from Digital Numbers to top-of-atmosphere reflectance in PCI Geomatica. For each image the panchromatic and multispectral bands were orthorectified together using PCI Geomatica's 2014 OrthoEngine. National Agricultural Imagery Program (NAIP) images (1 m resolution) were used as reference images (Long Creek and Robb Creek—22 October 2015, Alkali Creek—3 August 2013) together with the U.S. Geological Survey's 10 m National Elevation Dataset (NED) [50]. Images were pan-sharpened to 50 cm resolution using PCI Geomatica's PANSHARP2 tool [51]. The panchromatic band width for QuickBird-2 extends across all four of the multispectral bands (blue, green, red, NIR); however, the panchromatic band for Worldview-2 and Worldview-3 extends across only five of the eight spectral bands (excludes the coastal band, NIR1 and NIR2 bands). Prior work has shown that applying pan-sharpening methods to bands outside of the panchromatic range can distort the values in these bands [52]. To account for this, we pan-sharpened only the bands overlapping the panchromatic band using the PANSHARP2 tool. Bands outside of the panchromatic band range were resampled to 50 cm resolution using cubic convolution.

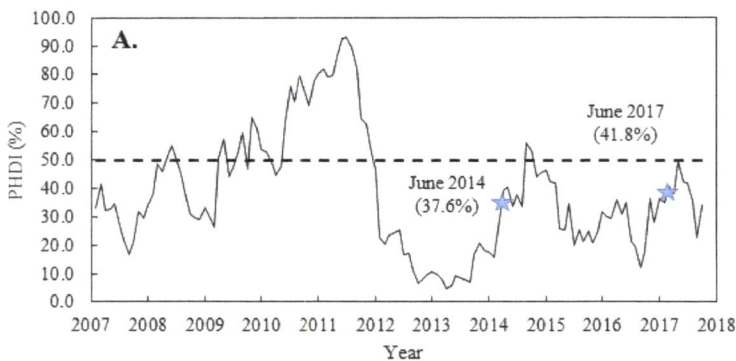

Figure 2. *Cont.*

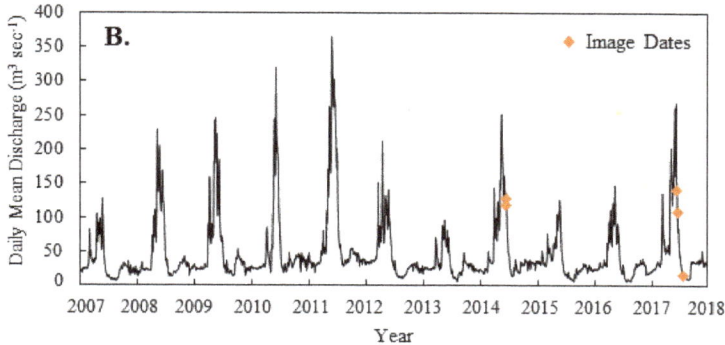

Figure 2. (A) The monthly Palmer Hydrological Drought Index (PHDI) values for southwestern Montana over the past 10 years, converted to percentages based on the historical record (1395–2017). Stars indicate the PHDI value at the time of the pre- and post-restoration images; (**B**) The daily mean discharge for the Jefferson River near Twin Bridges, Montana (USGS Gage #06026500) downstream of the four restoration sites. Diamonds indicate the discharge value at the time of the pre- and post-restoration images.

2.3. Object-Based Water Classification

We used the software eCognition (version 9.2.1, Trimble, Westminster, CO, USA) to process the high-resolution images into maps of surface-water extent. This software uses an object-oriented approach where an image is first segmented into objects representing meaningful features of the physical landscape, and the objects are then classified using user-defined rules and algorithms. Rules can be set in a hierarchical order as child rules under a parent process so that the ruleset can be automatically run in sequence. In this case, our objective was to segment, then classify objects into water and non-water where the objects of interest were the stream channels. Each pan-sharpened image was first clipped to the spatial area of interest, which included a minimum of 300 m stream length upstream from the restoration reach, the restoration reach and approximately 1 stream km downstream of the restored site (Figure 3). The stream length of the restoration reach (from the upstream to the downstream BDA) ranged from 830 m at Alkali Creek to 3.8 km at Long Creek (North) (Table 1).

To segment each image into objects we focused the segmentation along edges or sharp contrasts in the image. To do this we first modified the panchromatic band. Within eCognition, the Edge Extraction Lee Sigma filter was applied to the panchromatic band to create a (1) bright edge layer and (2) dark edge layer from the original image. The dark edge layer was then added to and the bright edge layer subtracted from an inverted version of the panchromatic band to enhance the edge contrasts along streams. The edge-enhanced panchromatic band was then used with the pan-sharpened blue, green, red, and near-infrared bands to guide the initial image segmentation (scale = 50–100). This approach focused segmentation along stream boundaries while segmenting the image at larger scales, where scale refers to the maximum allowable heterogeneity within an object. The larger scale was desired so that individual trees and tree shadows, which are spectrally similar to water, were segmented into larger objects that contained multiple trees, tree shadows, and intervening vegetation (Figure 4A). After the initial image segmentation, the objects were classified using spectral indices applied in a hierarchical rule-based approach. Our goal was to classify all objects as (1) water; (2) vegetation; (3) soil; or (4) shadow. Objects that were not spectrally similar enough to fit in any of these categories were classified temporarily as (1) water candidates (i.e., potentially water) or (2) unclassified. Objects classified as water candidates or unclassified using the rule-based approach were then re-segmented at a finer scale (scale = 15) to create smaller and more spectrally pure objects (Figure 4B).

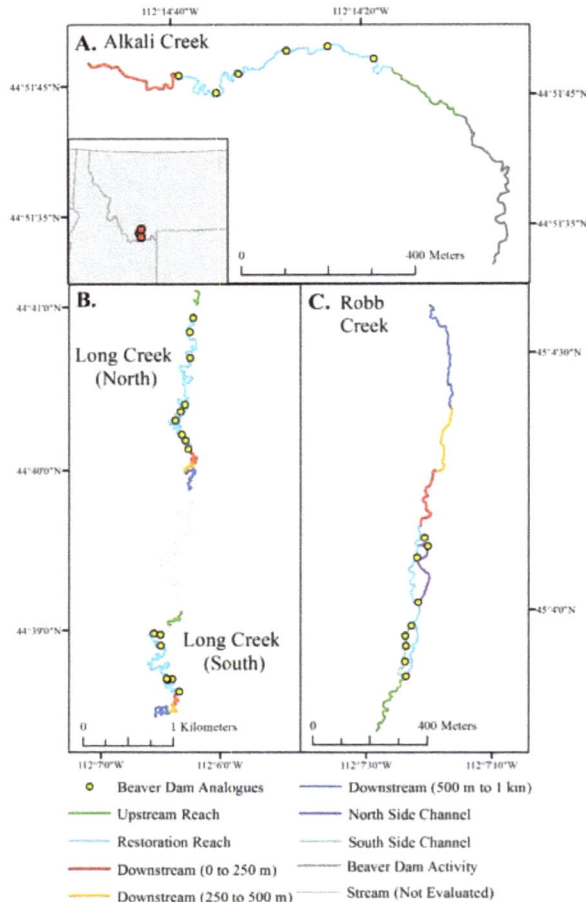

Figure 3. A schematic of (**A**) Alkali Creek, (**B**) Long Creek (North and South), and (**C**) Robb Creek, indicating the reaches along each of the stream sites that were analyzed, relative to the distribution of the beaver dam analogues. Flow direction can be determined from the relative location of the upstream and downstream reaches.

Several spectral indices were included in our hierarchical rule-based approach (Table 2). Worldview-2 and 3 images (8 spectral bands) provide data from several bands not available in Landsat TM, ETM+ or QuickBird-2 including a coastal band, red edge band and two separate near infrared bands (NIR1 and NIR2), offering opportunities for unique band combinations [53]. The Normalized Difference Water Index (NDWI) [54] and the Worldview Water Index (WWI) [55] (Table 2) were used as the primary means to identify water objects. Objects were classified as water when either the NDWI or WWI object values were greater than zero. Objects were identified as vegetation when the objects had high Enhanced Vegetation Index (EVI) [56] values, using the coefficients generally adopted [57], or when they showed both a high Normalized Difference Vegetation Index (NDVI) [58] as well as a minimal difference between NDVI and EVI. Objects were classified as soil when the Green-Red

Vegetation Index (GRVI) [59] value of objects was less than zero or where brightness values, derived from the panchromatic band, were high (Table 2). Shadow objects were identified using the normalized difference between the coastal band and blue band. For QuickBird-2 images in which the coastal band was not available this index was adapted to the normalized difference between the blue band and green band. A low brightness threshold, derived from the panchromatic band, was also used to identify shadowed areas. Finally, water candidates (i.e., potentially water) were identified using a series of indices. The rules to identify objects as water candidates were applied as child rules below the rules identifying objects as soil or vegetation. Indices used to identify water candidates included the NDWI using the coastal band instead of the green band [60] and panchromatic brightness, both of which were effective at identifying deeper water, as well as several novel band combinations including the Red Edge NDWI and Red Edge WWI both of which were helpful in mapping shallow water, particularly where sandy soils were visible below the water, and the Normalized Difference Coastal Red Edge Index (NDCREI) which was helpful in identifying turbid water. A list of the indices used, and the band combinations are shown in Table 2.

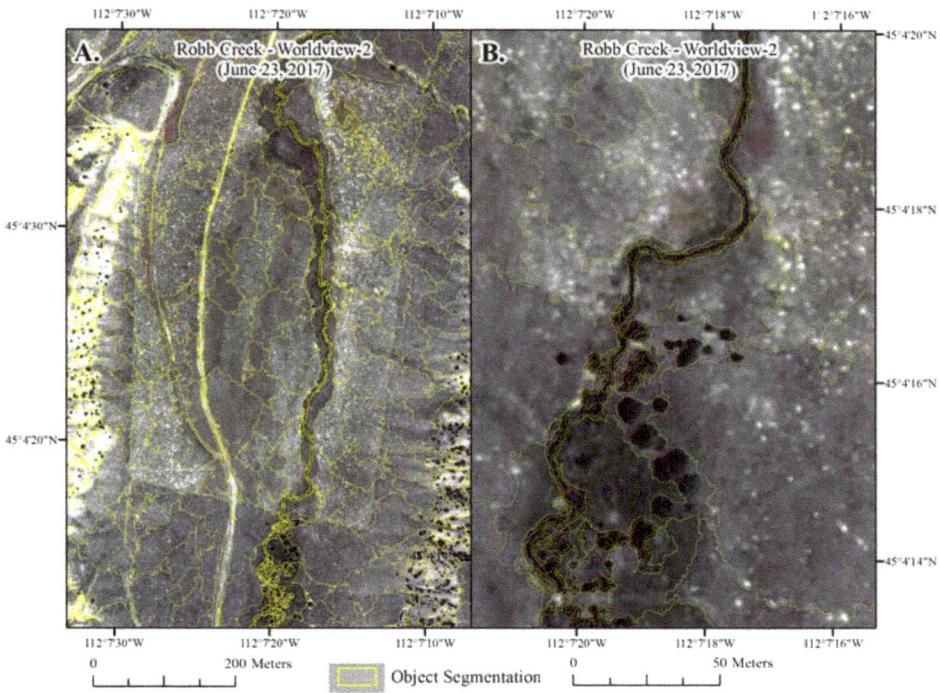

Figure 4. Examples from Robb Creek, Montana, of segmenting an image into objects of interest. Segmentation approaches that focus on detecting edges as well as iterative segmentation at different scales can allow objects to vary in size from large, upland objects (**A**) to narrow, small objects that follow the stream (**B**); DigitalGlobe Copyright 2017.

Table 2. Spectral indices used in object-based and pixel-based classification of surface water, as well as the characterization of riparian greenness. NIR: near infrared.

Index	Equation	Purpose
Normalized Difference Water Index (NDWI)	(Green − NIR1)/(Green + NIR1)	stream surface water area
Worldview Water Index (WWI)	(Coastal − NIR2)/(Coastal + NIR2)	stream surface water area
Panchromatic brightness		stream surface water area, bare ground, shadows
Enhanced Vegetation Index (EVI)	$2.5 \times$ (NIR1 − Red)/((NIR1 + 6) × (Red − 7.5) × (Blue + 1)	vegetation
Normalized Difference Vegetation Index (NDVI)	(NIR1 − Red)/(NIR1 + Red)	vegetation, riparian
NDVI and EVI difference	(NDVI − EVI)/(NDVI + EVI)	vegetation, riparian
Soil-Adjusted Vegetation Index (SAVI)	(NIR − red)/(NIR + red + L) × (1 + L), L = 0.5	vegetation, riparian
Green-Red Vegetation Index (GRVI)	(Green − Red)/(Green + Red)	bare ground
Worldview Shade Index	(Coastal − Blue)/(Coastal + Blue)	shadows (Worldview)
QuickBird Shade Index	(Blue − Green)/(Blue + Green)	shadows (QuickBird)
NDWI v2	(Coastal − NIR1)/(Coastal + NIR1)	deep water
Red Edge NDWI	(Red Edge − NIR1)/(Red Edge + NIR1)	shallow water
Red Edge WWI	(Red Edge − NIR2)/(Red Edge + NIR2)	shallow water
Normalized Difference Coastal Red Edge Index (NDCREI)	(Coastal − Red Edge)/(Coastal + Red Edge)	turbid water

The hierarchical rule-based approach of classification was initially applied to all objects (scale = 50–100). For objects that were classified as water candidates or unclassified by the initial rule-based approach, these objects were re-segmented to create smaller, more homogenous objects (scale = 15) and the rule-based approach was re-applied to this subset of smaller objects. For objects classified as water candidates or unclassified after both rounds of rule-based classifications, we applied a Random Forest classifier to determine if these remaining objects were water, vegetation, soil, or shadow. The Random Forest classifier was trained using the objects already classified (either using scale = 50–100 or scale = 15) by the rule-based approach. Bootstrap iterations (n = 500) were run using all indices shown in Table 2, the individual band values, and the band standard deviations of the objects as independent variables.

Following the classification of all objects, in images that contained dense riparian vegetation with shadows, we applied the "grow region" algorithm to the shadow class in eCognition. This step reclassified the objects neighboring shadow objects as shadow candidates. We repeated the process as needed. More inclusive shadow thresholds were then applied to the shadow candidates and the objects were converted to shadow if they were within the thresholds. The object-based elliptic fit shape attribute was also used to classify individual tree shadows for trees that occurred near the stream. Although a similar segmentation and image classification approach was applied to all high-resolution images, as is common in the eCognition environment, a trial-and-error approach to segmentation and image classification was used [61] so that segmentation scale, index thresholds and rulesets were not identical across images. A flowchart showing our eCognition methods from image segmentation to an output of water and non-water is shown in Figure 5.

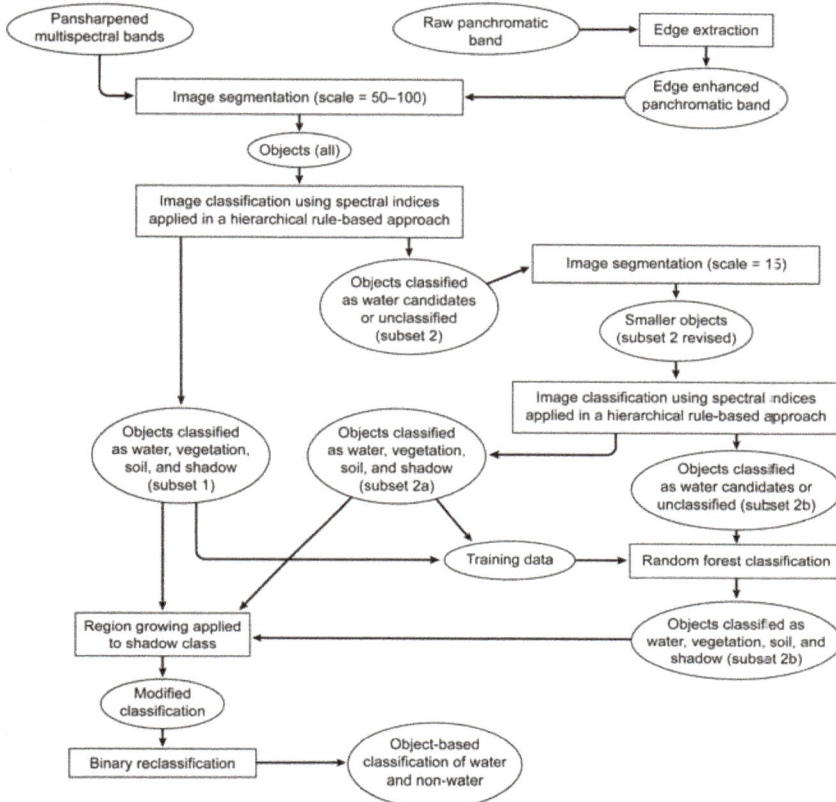

Figure 5. A flowchart showing the order of steps taken to process each image to water and non-water using an object-based approach. Inputs and outputs are shown in ovals while processing steps are shown in rectangles.

2.4. Pixel-Based Water Classification

Because a GEOBIA approach can be time-intensive and site specific, we were interested in comparing the performance of eCognition outputs with outputs producing using simplistic, single, spectral index thresholds. Using the pan-sharpened TOA reflectance values, we calculated the (1) NDWI [54]; (2) WWI [55]; and (3) the panchromatic brightness value. Brightness was calculated as the pixel value of the panchromatic band, a grayscale image of portions of the electromagnetic spectrum (Worldview-2, 3 (450–800 nm) and QuickBird-2 (450–900 nm)) (e.g., [62]). These three indices were selected as they were most prominently used to identify water objects in the eCognition image processing approach. The spectral index values of the validation points, described in Section 2.5, were used to guide the threshold selection. The rasters were thresholded to water and non-water using the maximum Youden's index, which maximizes the difference between the true positive rate and the false positive rate from the ROC curve and provides an optimal threshold independent from class prevalence [63,64]. The Youden's Index optimal threshold and corresponding AUC were calculated for each of the pixel-based outputs.

2.5. Stream Surface Area Validation

At each of the four sites, a stream line was manually delineated along each stream reach. The stream line was buffered (above and below the streamline, 1 m, 1.5 m, 1.8 m, 0.75 m for Alkali Creek, Long Creek (North), Long Creek (South), and Robb Creek, respectively) so that the total buffered area represented the average stream width. Points were randomly selected within the buffered stream area to represent water points (n = 200). The stream line was then buffered by 50 m and 200 points were randomly selected within the buffered area to represent non-water points. The 400 validation points per image were visually inspected using the raw pan-sharpened image to confirm status (water or non-water). Accuracy metrics calculated included overall accuracy, omission error, commission error, Dice coefficient, and relative bias. Omission and commission errors were calculated for the category of water. The Dice coefficient is the conditional probability that if one classifier (product or reference data) identifies a pixel as water, the other one will as well, integrating errors of omission and commission [65,66]. The relative bias provides the proportion that water is underestimated (negative bias) or overestimated (positive bias). Accuracy metrics were calculated for each of the pixel-based and object-based stream surface area outputs and presented by site, year, and methodology.

2.6. Changes in Stream Surface Area

To evaluate changes in stream surface area, our goal was to select the most accurate pair of stream surface area maps per site. This was determined using both the accuracy statistics as well as a visual assessment of quality. For Long Creek (North) and Long Creek (South), we used the panchromatic brightness output (97.4 and 98.4% overall accuracy, respectively when averaged across the two years, Table 3). For Robb Creek, we used the eCognition outputs (96.6% overall accuracy, averaged across the two years, Table 3), and for Alkali Creek, we merged the eCognition outputs with the panchromatic brightness outputs so that if water was identified by either output it was included as water. This was necessary only for Alkali Creek as the stream segment immediately downstream from the restoration was narrow (~1 to 1.5 m wide) and not adequately mapped using the eCognition output alone. Each output was edited manually to remove errors of commission. All outputs were converted to polygons and projected to WGS 1984 UTM zone 12 N prior to calculating area (m^2) in ArcGIS 10.3 (ESRI, Redlands, CA, USA).

All surface water continuous with the stream centerline was included in stream surface area calculations, while waterbodies that were disconnected from the stream centerline were excluded from the stream surface area calculations. The total stream surface area was calculated for (1) the restoration reach, which extended from the upstream BDA structure to the most downstream BDA structure; (2) a reach extending upstream from the restoration reach (stream length of 300 m), and three reaches extending downstream from the restoration reach including; (3) a stream length of 0 m to 250 m downstream; (4) a stream length of 250 m to 500 m downstream; and (5) a stream length of 500 m to 1 km downstream from the restored reach. A schematic showing the distribution of these reaches at each site is shown in Figure 3. The stream length evaluated upstream and downstream from the restoration reach was limited by the extent of the pre- and post-restoration images. The same five reaches were used in analysis of the 2014 and 2017 images. At Robb Creek, the BDAs were aimed at reactivating side channels so each of the side channels were considered separately from the main channel (Figure 3). At Alkali Creek, beaver activity upstream from the restoration site influenced conditions in the upstream reach, so that the reach containing extensive beaver activity was analyzed separately (Figure 3). In addition, dense riparian vegetation approximately 300 m downstream from the restoration site in Alkali Creek limited our analysis of downstream area, so that results were presented only for a single downstream reach (0 m to 300 m) (Figure 3).

Table 3. A comparison of the accuracy of methods to map stream extent across sites and image dates. Errors are presented for the accuracy of mapping stream surface area extent. All methods listed are pixel-based except for eCognition which is object-based. WWI: Worldview Water Index; NDWI: Normalized Difference Water Index; WV2: Worldview-2; WV3: Worldview-2; QB2: QuickBird-2; OE: omission error; CE: commission error; OA: overall accuracy; DC: Dice coefficient; RB: relative bias; AUC: area under curve; NIR: near-infrared.

Site and Method	Image Year/Sensor	Youden's Index Threshold	AUC	OE (%)	CE (%)	OA (%)	DC (%)	RB (%)
Alkali Creek								
WWI (coastal − NIR2)/(coastal + NIR2)	2014, WV2	−0.211	0.75	52.5	14.0	69.9	61.2	−44.8
NDWI (green − NIR)/(green + NIR)	2014, WV2	−0.316	0.60	68.5	13.1	63.4	46.2	−63.8
Panchromatic brightness	2014, WV2	0.080	0.91	3.5	12.7	91.3	91.7	10.5
eCognition	*2014, WV2*	~	~	10.5	0.6	94.5	94.2	−10.0
WWI (coastal − NIR2)/(coastal + NIR2)	2017, WV3	−0.176	0.91	24.5	6.2	85.3	83.7	−19.5
NDWI (green − NIR)/(green + NIR)	2017, WV3	−0.277	0.68	56.0	9.3	69.8	59.3	−51.5
Panchromatic brightness	*2017, WV3*	0.123	0.97	2.5	6.3	95.5	95.6	4.0
eCognition	2017, WV3	~	~	10.0	1.6	94.3	94.0	−8.5
Long Creek (North)								
WWI (coastal − NIR2)/(coastal + NIR2)	2014, WV2	−0.225	0.94	15.0	4.0	90.8	90.2	−11.5
NDWI (green − NIR)/(green + NIR)	2014, WV2	−0.389	0.92	36.5	0.8	81.5	77.4	−36.0
Panchromatic brightness	2014, WV2	0.079	0.98	6.0	1.6	96.3	96.2	−4.5
eCognition	*2014, WV2*	~	~	4.5	2.1	96.8	96.7	−2.5
WWI (coastal − NIR2)/(coastal + NIR2)	2017, WV3	−0.183	0.98	5.5	3.1	95.8	95.7	−2.5
NDWI (green − NIR)/(green + NIR)	2017, WV3	−0.326	0.97	9.0	2.2	94.5	94.3	−7.0
Panchromatic brightness	*2017, WV3*	0.102	1.00	2.5	0.5	98.5	98.5	−2.0
eCognition	2017, WV3	~	~	4.0	0.5	97.8	97.7	−3.5
Long Creek (South)								
WWI (coastal − NIR2)/(coastal + NIR2)	2014, WV2	−0.083	0.98	4.0	3.5	96.3	95.2	−0.5
NDWI (green − NIR)/(green + NIR)	2014, WV2	−0.274	0.98	6.5	4.6	94.5	94.4	−2.0
Panchromatic brightness	*2014, WV2*	0.076	0.99	2.0	1.5	98.3	98.2	−0.5
eCognition	2014, WV2	~	~	1.5	6.6	95.8	95.9	5.5
WWI (coastal − NIR2)/(coastal + NIR2)	2017, WV3	−0.055	0.99	2.0	3.4	97.3	97.3	1.5
NDWI (green − NIR)/(green + NIR)	2017, WV3	−0.234	0.96	10.5	3.8	93.0	92.7	−7.0
Panchromatic brightness	*2017, WV3*	0.081	0.99	2.0	1.0	98.5	98.5	−1.0
eCognition	*2017, WV3*	~	~	2.0	1.0	98.5	98.5	−1.0
Robb Creek								
WWI (coastal − NIR2)/(coastal + NIR2)	2014, QB2	~	~	~	~	~	~	~
NDWI (green − NIR)/(green + NIR)	2014, QB2	−0.278	0.74	62.5	2.6	68.3	54.2	−61.5
Panchromatic brightness	2014, QB2	0.137	0.94	4.0	7.2	94.3	94.3	3.5
eCognition	*2014, QB2*	~	~	6.5	2.1	95.8	95.7	−4.5
WWI (coastal − NIR2)/(coastal + NIR2)	2017, WV2	−0.243	0.96	16.0	4.5	90.0	89.4	−12.0
NDWI (green − NIR)/(green + NIR)	2017, WV2	−0.299	0.70	36.5	8.6	78.8	74.9	−30.5
Panchromatic brightness	*2017, WV2*	0.103	0.97	0.5	4.8	97.3	97.3	4.5
eCognition	2017, WV2	~	~	2.5	3.0	97.3	97.3	0.5

The challenge in evaluating the effects of restoration is to separate change attributable to natural variation (interannual, seasonal, event) from change attributable to the restoration activities. To accomplish this, we assumed that variability in hydro-climatic conditions would propagate similarly at the scale of adjacent or nearby stream reaches. Therefore, the ratio between the condition (e.g., amount of water, greenness) of an upstream reach and downstream reach at T_1 should be equivalent to the ratio between the condition of the same upstream reach and same downstream reach at T_2.

$$\frac{O\ upstream\ reach_{T1}}{O\ downstream\ reach_{T1}} = \frac{O\ upstream\ reach_{T2}}{E\ downstream\ reach_{T2}} \tag{1}$$

where O refers to the observed value and E refers to the expected value. If we let the downstream reach represent the restoration reach, we can provide the values for the pre-restoration conditions (upstream and downstream reach values at T_1) and the value for the upstream reach at T_2 and solve for the "expected" value of the downstream reach (i.e., restoration reach) at T_2. We can then use the observed downstream reach value at T_2, to calculate the percent change from the expected value that we can attribute to the restoration activities:

$$\textit{Change attributed to restoration } (\%) = \frac{(O \textit{ downstream reach}_{T2} - E \textit{ downstream reach}_{T2})}{E \textit{ downstream reach}_{T2}} * 100 \quad (2)$$

Calculating change as a function of the expected value can help account for variability between the images due to natural variability in climate and stream discharge. Although the outputs used to analyze changes in stream surface area tended to show high accuracy, if we assume that error induced by the methodology is random or consistent across the image extent, this approach also allowed us to take into account between-image variability in the accuracy of the mapped stream extent. We also recognize that changes to the restoration reach can potentially propagate upstream (e.g., [67]), possibly influencing the relationship between the upstream and downstream reaches; however, due to image extents we were limited in how far upstream we could document stream surface area necessitating the above assumptions.

In addition to calculating changes in stream surface area at a reach scale, we also calculated changes in stream width, inundated length, and changes to stream surface area just upstream from each BDA. To calculate changes in stream width, 20 points were randomly selected along the stream centerline within each site and reach and stream width was measured manually using the pan-sharpened raw imagery and averaged to obtain a single mean stream width per reach. The same points were used in both years. Inundated stream length was calculated as the percent of the stream centerline mapped as water. To quantify local changes in stream surface area induced by the structures, we calculated the stream surface area immediately upstream of each of the structures that showed a visually evident change (from 2014 to 2017) in stream width. Stream surface area was calculated from the classified (edited) stream surface area used to quantify reach-scale changes. The local change in stream surface area was presented as the change relative to pre-restoration stream surface area. The localized change was observed over a variable stream length distance, but the observed effect averaged 26 m of stream length upstream from the installed BDAs.

2.7. Changes in Riparian Condition

Riparian greenness was evaluated using three vegetation indices, NDVI, EVI and the Soil-Adjusted Vegetation Index (SAVI) [65]. Multiple indices were included because vegetation indices are sensitive to several conditions including canopy geometry (trees versus herbaceous vegetation), soil properties, sun position, and cloudiness [68–70]. While NDVI is the most commonly used vegetation index, EVI can help eliminate atmospheric noise, and SAVI reduces the influence of soil by including a soil adjustment factor (L) [56,68]. Changes in riparian greenness, averaged across the three indices, were then evaluated, (1) along the restoration reach (from the upstream BDA to the downstream BDA); (2) 0 m to 250 m downstream from the restoration reach; (3) 250 m to 500 m downstream from the restoration reach; and (4) 500 m to 1 km downstream from the restoration reach. At Robb Creek, the BDAs were designed to reactivate side channels so the side channels were considered separately from the main channel. We also tested how the effect changed as the buffer from the main channel increased from 10 m to 20 m from the channel centerline.

It was critical to control between-image differences in greenness not related to the restoration, therefore changes in greenness between the pre- and post-restoration riparian corridors were corrected using the difference in greenness across reference areas. Three reference polygons were selected at each site (ranging from 0.2 ha to 2.5 ha in size) representing: (1) herbaceous photosynthetic riparian vegetation upstream from the restoration site; (2) an upland patch dominated by photosynthetic grasses; and (3) an upland patch dominated by non-photosynthetic grasses. We avoided areas that appeared to show a difference in grazing intensity between the two image dates. The greenness values for each of the reference polygons were averaged to obtain a reference greenness value for each site and date. Because herbaceous vegetation is more sensitive to interannual change than riparian tree species, the reference polygons included a mix of riparian and upland vegetation samples. Additionally, because changes may have occurred for upstream vegetation in response to the restoration measures, we used a mix of riparian and non-riparian patches. Even at sites with willow plantings, short-term changes (one to three years) in riparian condition can be expected to primarily result from growth of herbaceous species, a change attributable to increased water availability in the shallow subsurface areas adjacent to the stream. The "reference greenness" values were seen as equivalent to the role of the upstream reach values when analyzing changes to stream surface area. The same analysis used to evaluate restoration effects on stream surface area above (Equations (1) and (2)) were applied here to evaluate restoration-induced changes to riparian condition.

3. Results

3.1. Accuracy of Stream Delineation Approaches

The accuracy of our estimates of stream surface area depended on both the stream width and the classification approach. Across all sites and classification approaches, the relative bias tended to be negative, indicating that the stream surface area, on average, was underestimated. Long Creek (South), which showed a stream width averaging 3 m to 3.5 m and Long Creek (North), which showed a stream width averaging 2.5 to 3.5 m wide, showed more consistent accuracy statistics across the approaches tested relative to the other two sites. The eCognition output and Panchromatic brightness consistently performed the best with errors of omission for water ranging from 2% to 6% and errors of commission ranging from 0.5% to 7%. In contrast, the NDWI tended to show higher errors of omission, with omission errors ranging from 7% to 37% and commission errors ranging from 1% to 5% (Table 3). Overall accuracy and dice coefficients were >95% except for NDWI outputs and one of the four WWI outputs (Table 3).

As stream width decreased at the Alkali Creek (averaged 1 m to 2.5 m) and Robb Creek (averaged 1.5 m to 2 m) sites, the accuracy of published indices (NDWI and WWI) was relatively poor. Using these indices errors of omission, for instance, ranged from 16% to 69% along the two streams. However, panchromatic brightness and eCognition outputs maintained relatively strong accuracy even as stream width decreased with errors of omission and commission across the two sites and years ranging from 3% to 11% and 2% to 13%, respectively, and the corresponding overall accuracy and Dice coefficient ranging from 91% to 97%.

It was also evident that classification accuracy for a given method can be inconsistent over time. Examples of this are shown in Figure 6, in which we can compare the outputs for each classification approach and year along the Alkali Creek site. Figure 6 also demonstrated that when surface water is a minority cover type across the image extent, an output can visually appear to be relatively noisy but statistically show a relatively low amount of calculated commission error. An example of this is the panchromatic brightness output for 2014 in which an error of commission of 13% created a visually "noisy" output. The visualized variability in the surface-water extent across methods justifies the need to analyze changes in stream surface area using the most accurate method possible and including a manual editing component as time allows so that uncertainty in surface-water extent does not obscure "true changes" in stream condition.

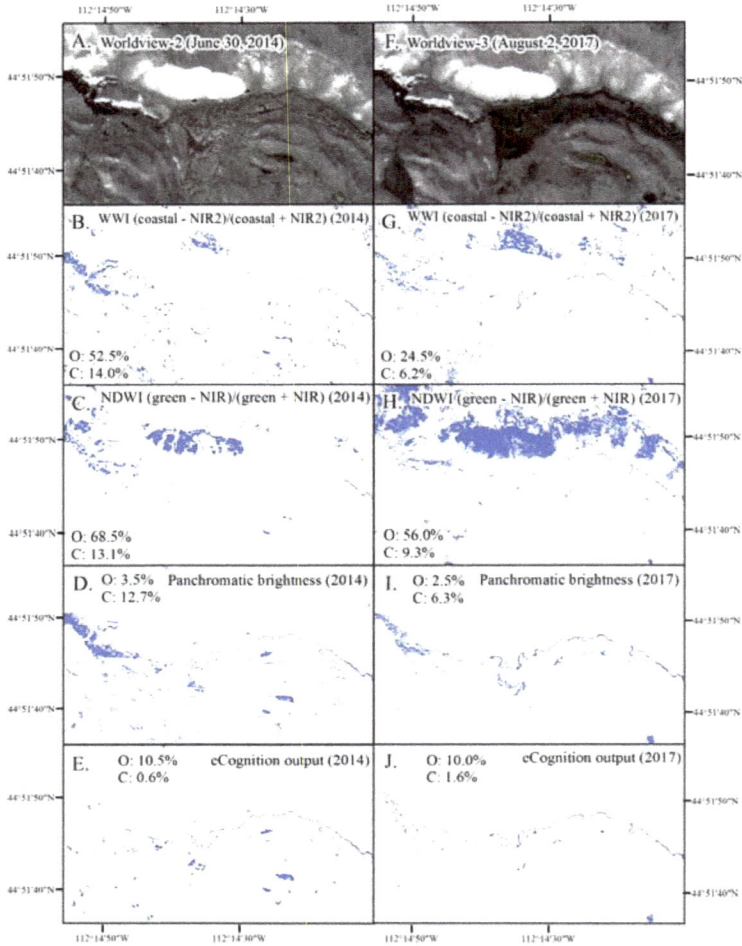

Figure 6. A comparison of mapped surface water extent for a reach of Alkali Creek in which challenges included shaded riparian vegetation (left) and narrow stretches of stream (center). Comparisons include, natural color images (**A,F**); the Worldview Water Index (WWI, **B,G**); the Normalized Difference Wetness Index (NDWI, **C,H**); the panchromatic brightness (**D,I**); and the eCognition output (**E,J**). O: omission error; C: commission error Copyright DigitalGlobe, 2014, 2017.

3.2. Changes to Stream Condition

The installation of BDAs resulted in proximal changes (e.g., increases in surface water immediately upstream from the in-stream structures) as well as changes at the scale of the restoration reach and downstream reaches. Increases in stream surface area immediately upstream from structures or in reactivated side channels were observed at all four sites. Increases in total stream surface area along the restoration reach were observed at three of the four sites, while a decrease in stream surface area downstream from the restoration reach was observed at all four sites (Table 4).

Table 4. Change to stream surface area. Local storage increase refers to a net change in the surface area of water upstream from individual beaver dam analogues (BDAs). Because the BDAs at Robb Creek served a primary purpose of redirecting water toward side channels, local storage is not shown.

Stream and Reach	Length (m)	Area (2014, m²)	Area (2017, m²)	Change (%)	Change Relative to Expected (%)	Inundated Stream Length (2014, %)	Inundated Stream Length (2017, %)	Surface Water Width (2014, m)	Surface Water Width (2017, m)	Local Storage Increase (Mean Per BDA) (m²)
Alkali Creek										
Upstream Beaver Activity	592	9665.3	2564.3	−73.5		97	96	2.4	1.9	
Upstream Reach	200	648.5	630.0	**−2.9**		94	99	2.3	3.2	
Restoration reach	830	1311.5	1595.5	21.7	25.2	87	86	1.6	1.8	302.9 (50.5)
Downstream (0 to 300 m)	300	594.5	403.8	−32.1	−30.1	37	63	1.4	1.1	
Long Creek (North)										
Upstream Reach	300	643.3	589.7	**−8.3**		75	79	2.4	2.8	
Restoration reach	3857	12,713.0	11,080.0	−12.8	−4.9	91	90	3.5	3.6	170.5 (18.9)
Downstream (0 to 250 m)	250	673.0	561.4	−16.6	−9.0	91	97	2.8	2.3	
Downstream (250 to 500 m)	250	1059.0	819.8	−22.6	−15.6	98	98	3.6	3.5	
Downstream (500 m to 1 km)	500	1492.5	1183.9	−20.7	−13.5	92	93	3.0	2.7	
Long Creek (South)										
Upstream Reach	300	981.0	1028.4	**4.8**		100	100	2.9	3.6	
Restoration reach	2496	10,424.5	12,434.3	19.3	13.8	98	100	3.7	5.1	746.7 (106.7)
Downstream (0 to 250 m)	250	780.3	653.8	−16.2	−20.1	100	100	2.7	3.0	
Downstream (250 to 500 m)	250	1030.4	900.6	−12.6	−16.6	100	100	3.5	3.5	
Downstream (500 m to 1 km)	500	2213.4	2209.8	−0.2	−4.8	100	100	3.0	3.3	
Robb Creek										
Upstream Reach	300	490.4	501.3	**2.2**		99	95	1.8	1.8	
Main Stem	691	2100.4	1566.2	−25.4	−27.1	92	86	1.8	2.3	
Restored Side Channel (South)	284	77.8	362.8	366.6	356.4	8	39	0.0	1.2	
Restored Side Channel (North)	257	292.5	531.1	81.6	77.6	40	73	1.5	2.5	
Downstream (0 to 250 m)	250	863.3	651.0	−24.6	−26.2	78	84	1.8	2.0	
Downstream (250 to 500 m)	250	704.6	495.2	−29.7	−31.2	21	49	1.6	1.8	
Downstream (500 m to 1 km)	500	1092.8	1117.4	2.3	0.0	79	70	1.7	1.7	

At Alkali Creek approximately one year post-restoration, we observed a net increase of 303 m^2 in stream surface area immediately upstream from installed BDAs (Figure 7). This amount of water represented 19% of the total stream surface area along the restoration reach in 2017. After controlling for differences in stream surface area attributable to interannual variability, we observed a 25% increase in stream surface area attributable to the restoration activities, and a 30% decrease in stream surface area downstream from the restoration reach (Table 4). These changes in stream surface area were matched by corresponding changes in stream width (Table 4). A complicating factor at this site was that beaver activity was present from 200 m to 800 m upstream from the restoration reach. Upstream ponding resulting from natural beaver dams was substantial in the 2014 image but total stream surface water along the beaver impacted reach (200 m to 800 m upstream) decreased 74% by 2017 (Table 4, Figure 8A).

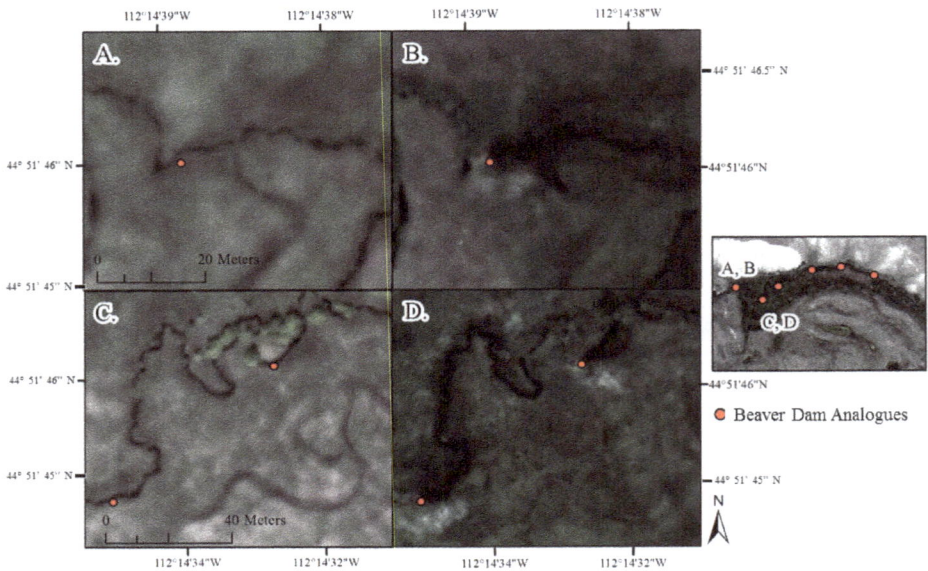

Figure 7. Retention of water upstream of beaver dam analogues along Alkali Creek are visible by comparing pre-restoration stream reaches (**A,C**) with post-restoration stream reaches (**B,D**). Copyright DigitalGlobe 2014, 2017.

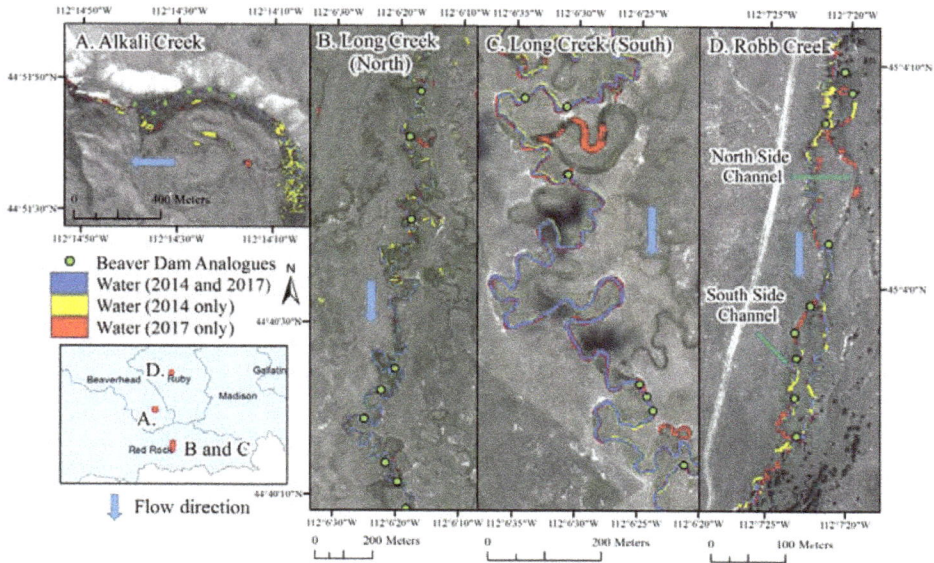

Figure 8. Change in stream surface area between the most upstream and most downstream beaver dam analogues (BDAs) along (**A**) Alkali Creek; (**B**) Long Creek (North); (**C**) Long Creek (South); and (**D**) Robb Creek. Copyright DigitalGlobe, 2014.

The goal of the restoration along Robb Creek was to reactivate two side channels. At approximately two years post-restoration, we observed a reactivation of the side channels that included a 78% increase in stream surface area along the northern side channel and a 356% increase in stream surface area along the southern side channel (Table 4, Figure 8). Correspondingly, we observed a 27% decrease in stream surface area along the main stream stem, and a decrease in stream surface area downstream from the restoration reach (0 m to 500 m downstream) (Table 4).

Long Creek (South) is 4.8 stream km downstream from Long Creek (North) and was the first restored of the four sites. At the time of the post-restoration image the site was approximately three years post-restoration. We observed a substantial amount of water stored upstream of many of the structures (net increase of water surface area of 747 m^2) (Figure 9), which represented 6% of the total stream surface area along the restoration reach in 2017. We observed a 14% increase in stream surface area along the restoration reach and a corresponding decrease in stream surface area (-17% to -20%) from 0 m to 500 m downstream from the restoration reach (Table 4). Long Creek (North) at one-year post-restoration was the only site where we did not observe an increase in stream surface area along the restoration reach, but instead observed a minor decrease of 5% in stream surface area. We also observed the smallest increase in stream surface area upstream from the BDA structures (net increase of proximal water surface area of 171 m^2), relative to the other sites evaluated (Figure 9). Similar to other sites, however, Long Creek (North) showed a decrease in stream surface area (-9% to -16%) through 1 km downstream from the restoration reach and an associated decrease in mean stream width (Table 4, Figure 8).

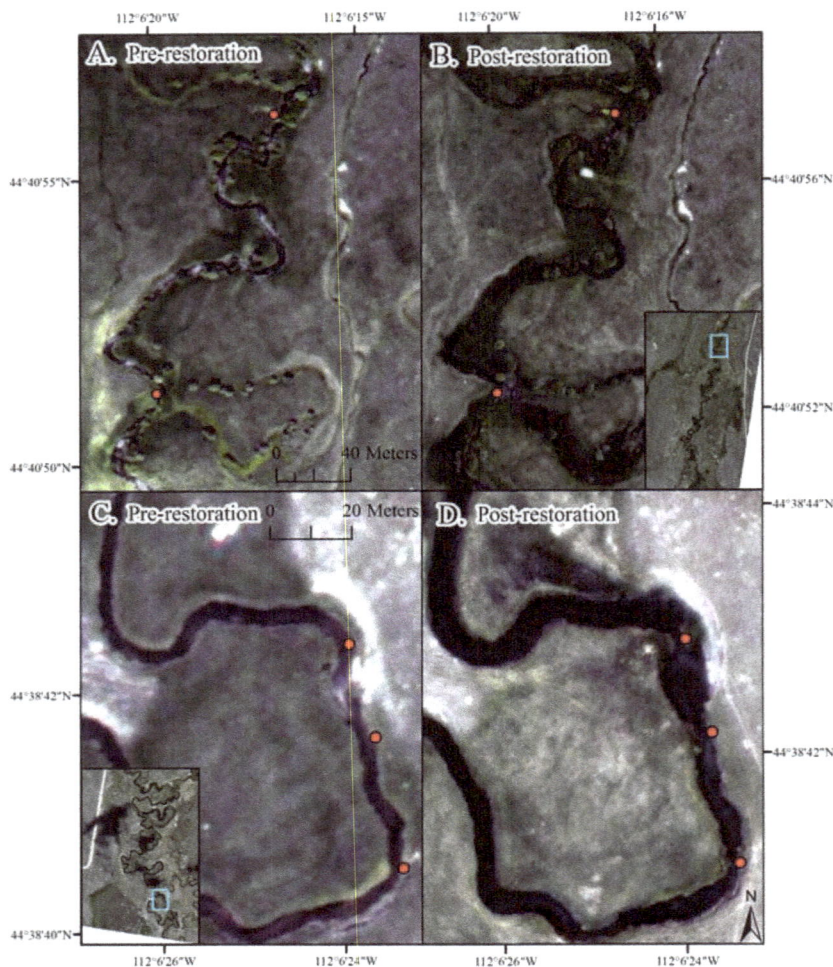

Figure 9. Visually apparent changes with the installation of the beaver dam analogues varied along Long Creek. Within Long Creek (North) changes from pre- to post-restoration included retention of water upstream from structures as well as the reactivation of abandoned side channels (**A,B**); Along Long Creek (South) changes included widening of the stream as well as retention of water upstream from structures (**C,D**). Copyright DigitalGlobe 2014, 2017.

3.3. Changes to Riparian Condition

At each site, the spectral greenness index used (NDVI, EVI or SAVI) showed a relatively minor but inconsistent influence on the reported change in greenness, after controlling for between-year differences in greenness. EVI, for example, showed a higher percent change in greenness relative to NDVI and SAVI along Alkali Creek and Robb Creek, but not along the Long Creek sites (Table 5). However, we also found that the directionality of change to riparian condition post-restoration at each of the sites did not depend on the greenness index used. Increases in riparian greenness along the restoration reach were observed at three of the four sites, while changes to riparian greenness downstream from the restoration reach were less consistent.

Table 5. Percent change in riparian greenness between pre- (2014) and post-restoration (2017) images after controlling for between-image differences in greenness not related to the restoration action. Greenness was evaluated using the Soil-Adjusted Vegetation Index (SAVI), Enhanced Vegetation Index (EVI) and Normalized Difference Vegetation Index (NDVI). Indices were evaluated using a 10 m, 15 m, and 20 m buffer from the stream. Changes along the restored reaches are in bold. DS: downstream; N: North; S: South.

Index (Buffer)	SAVI (%) (10 m)	SAVI (%) (15 m)	SAVI (%) (20 m)	EVI (%) (10 m)	EVI (%) (15 m)	EVI (%) (20 m)	NDVI (%) (10 m)	NDVI (%) (15 m)	NDVI (%) (20 m)	Average (%) (10 m)	Average (%) (15 m)	Average (%) (20 m)
Alkali Creek												
Restoration reach	18.2	15.5	13.3	24.6	21.1	18.3	17.4	15.4	13.6	**20.1**	**17.3**	**15.1**
0 to 250 m DS	12.2	9.5	7.0	16.6	13.3	10.3	7.9	5.6	3.1	12.2	9.5	6.8
250 to 500 m DS	29.6	30.5	29.1	37.8	38.6	37.1	7.3	7.5	7.1	24.9	25.5	24.4
500 to 1 km DS	25.2	27.3	26.3	33.2	35.2	34.1	6.6	7.5	6.6	21.7	23.3	22.3
Long Creek (North)												
Restoration reach	8.6	8.0	7.8	6.0	5.4	5.2	10.5	9.3	8.7	**8.4**	**7.6**	**7.2**
0 to 250 m DS	14.3	14.7	13.4	11.9	12.2	10.8	17.2	16.3	15.1	14.5	14.4	13.1
250 to 500 m DS	1.2	-1.2	1.2	-1.9	-4.2	-1.7	8.2	5.2	6.4	2.5	-0.1	2.0
500 to 1 km DS	-18.6	-16.9	-16.5	-22.7	-20.8	-20.4	-6.0	-5.5	-5.3	-15.8	-14.4	-14.1
Long Creek (South)												
Restoration reach	-0.4	-1.0	-0.2	-4.4	-4.7	-4.0	-2.3	-3.5	-3.1	**-2.4**	**-3.1**	**-2.4**
0 to 250 m DS	-0.6	0.6	-0.3	-4.7	0.3	-4.1	-5.5	-4.4	-5.4	-3.6	-1.2	-3.3
250 to 500 m DS	-1.5	-1.9	-1.9	-5.6	-5.6	-5.7	-7.0	-7.6	-8.3	-4.7	-5.0	-5.3
500 to 1 km DS	0.7	-0.5	-0.3	-3.3	-4.3	-4.3	-5.5	-7.0	-7.6	-2.7	-3.9	-4.1
Robb Creek												
Main Stem	3.5	4.5	4.3	4.8	5.8	5.6	2.3	3.2	3.3	**3.5**	**4.5**	**4.4**
Side Stem (N)	-0.4	1.4	2.1	1.0	2.7	3.6	3.2	4.1	4.2	**1.3**	**2.7**	**3.3**
Side Stem (S)	21.6	22.8	22.1	23.4	24.9	24.1	16.4	17.5	16.9	**20.5**	**21.7**	**21.0**
0 to 250 m DS	-5.6	-3.8	-2.5	-4.6	-2.8	-1.5	-6.5	-5.0	-3.9	-5.6	-3.9	-2.6
250 to 500 m DS	-13.8	-13.4	-13.3	-12.1	-12.0	-11.9	-10.3	-10.7	-10.8	-12.1	-12.0	-12.0
500 to 1 km DS	-11.2	-11.0	-10.4	-10.2	-10.0	-9.5	-10.8	-11.0	-10.6	-10.7	-10.7	-10.2

Along Alkali Creek, we observed an average increase in greenness of 20% using a 10 m stream buffer, declining to a 15% increase using a 20 m stream buffer (Table 5). Although we had observed a decrease in stream surface area downstream from this restoration site, we found an increase in greenness that persisted through 1 km downstream from the restoration reach. Defining reference conditions required particular attention along Alkali Creek because the post-restoration image was late summer (August 2, 2017) when water is more limited. This is evident in the contrast observed in the NDVI between the riparian area and uplands in the 2017 Alkali Creek image (Figure 10). Along Robb Creek, minor increases in greenness were observed along the main stem and northern side channel (<5%), while a substantial green-up was observed along the southern side channel that showed an increase in greenness of 21% (Figure 10). The decrease in stream surface area observed downstream of this restoration site was found to co-occur with a decrease in greenness, which ranged from a 4% to 12% decrease in greenness through 1 km downstream of the restoration (Table 4). Although Long Creek (North) showed a minor decrease in stream surface area within the restoration reach, we observed an increase in greenness of 7 to 8% along the restoration reach and an increase in greenness of 13 to 15% just downstream of the restoration reach (0 to 250 m) (Figure 11). In contrast, while we observed clear changes in stream surface area along Long Creek (South), changes in greenness were minimal (<5% change) (Table 5, Figure 11).

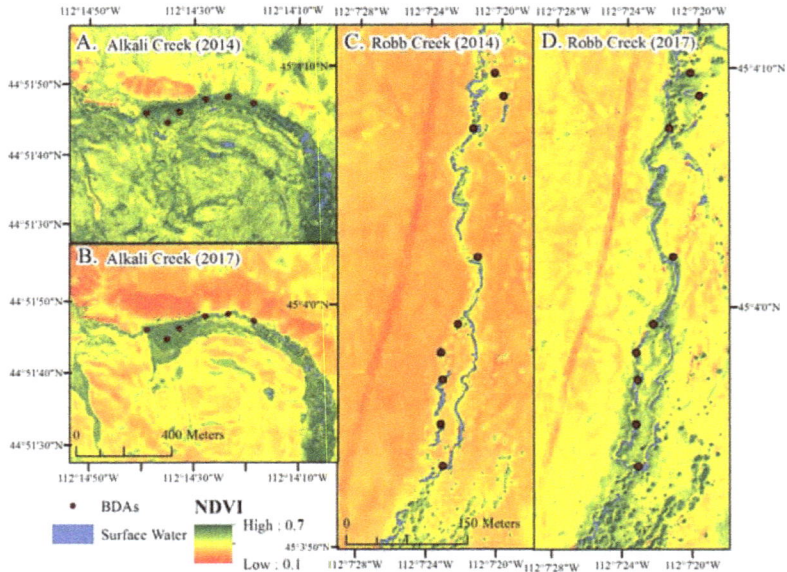

Figure 10. Change in riparian greenness along (**A**) Alkali Creek (2014), (**B**) Alkali Creek (2017), (**C**) Robb Creek (2014), and (**D**) Robb Creek (2017) from pre- to post-restoration conditions. NDVI: Normalized difference vegetation index; BDAs: beaver dam analogues.

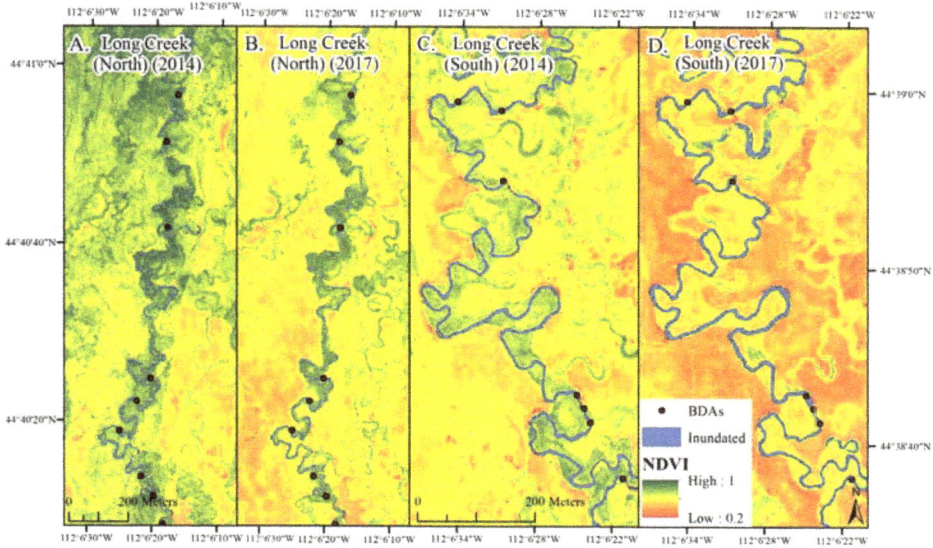

Figure 11. Change in riparian greenness along (**A**) Long Creek (North, 2014), (**B**) Long Creek (North, 2017), (**C**) Long Creek (South, 2014), and (**D**) Long Creek (South, 2017) from pre- to post-restoration conditions. NDVI: Normalized difference vegetation index; BDAs: beaver dam analogues.

4. Discussion

Long-term trends in the degradation of riparian and stream habitat are common across the western United States [71,72]. Satellite imagery has the potential to provide spatially continuous monitoring of stream extent and condition, which can complement point-based field efforts and stream gage data, and better inform stream management in response to degradation. However, only a limited number of studies have attempted to apply satellite imagery to streams, particularly smaller streams (<5 m wide). We found that pan-sharpened high-resolution imagery can be used to effectively monitor streams as narrow as 1.5 m wide. We tested sites where stream width was <1 m and found the results too poor to include. Panchromatic brightness consistently outperformed more established indices such as the NDWI. However, in experimenting with the classification of stream surface area we found that spectrally mixed portions of a stream can be more challenging to identify using this approach, for instance, portions of the stream showing high turbidity, high chlorophyll levels or bright sands can be missed using brightness alone. In addition, vegetation shadows, common in riparian areas, can be erroneously mapped as inundated areas, creating substantial errors of commission. Alternatively, while far more time consuming, a GEOBIA approach was able to greatly reduce errors of commission outside of the stream area by increasing object size with distance from stream and adding region growing to the shadow class within heavily shadowed riparian areas. While the index- and object-based approaches showed distinct advantages, we found that regardless of efforts accuracy results could be uneven across years. This is a major challenge in change detection analysis in which uneven accuracy over time, potentially attributable to differences in sensors, the off-nadir view angle, as well as the time of day that the image was collected or variability in local hydro-climatic conditions, can obscure change attributable to human-caused degradation or restoration [73]. We sought to minimize this source of error by (1) using only our most accurate outputs and further manually editing these outputs prior to analyzing changes in stream surface area; and (2) calculating the change as a function of the change from the expected value, therefore controlling, to the extent possible, for change due to image quality or variability in hydro-climatic conditions. This aspect of remote sensing change analysis, however, remains a challenge. Additionally, as the method that produced the highest overall accuracy varied also across sites, it was evident that the appropriate processing approach to minimize uneven error over time will vary across sites depending on the amount and type of riparian vegetation (herbaceous or tree), stream width, as well as water depth and clarity, a finding supported by others [74,75].

Improving techniques to map stream surface area with commercial high-resolution imagery offers opportunities to remotely monitor changes in key aspects of stream condition induced by flood or drought events, shifts in local land uses, or in-stream restoration activities. However, it is important to clarify that the stream data gleaned from a remote sensing analysis is intrinsically different from the data a hydrologist typically uses. For example, at the Alkali Stream site, differences in the seasonality of the image pair (June vs. August) meant that the downstream discharge in the Jefferson River was much lower at the August date relative to the June date (15 m^3 s^{-1} compared to 127 m^3 s^{-1}); however, the stream surface area upstream from the restoration reach along Alkali Creek was only 3% less at the August date relative to the June date. This contrast clarifies that a remote sensing analysis is not necessarily capturing changes in stream discharge, which are better measured with stream gages, but instead providing a spatially continuous dataset of changes in stream surface area, specifically stream width and the creation or change to riparian wetlands, which could in turn, impact downstream stream discharge.

Relying on image pairs, however, or only two points in time, can limit our understanding regarding seasonally specific effects. For example, local stakeholders are interested in the effects of BDAs on streams not just after snowmelt, but in particular during the late summer period in which water availability can be limited [27,31]. Because very few DigitalGlobe images have been collected and archived across southwestern Montana we were restricted to the early summer period and were therefore unable to evaluate the impact of the BDAs during this late summer period. This limitation in image timing means we were unable to observe how changes in stream surface area or riparian condition documented near the start of the growing season influenced conditions near the end of the growing season. However, this limitation will likely be reduced in the near future. Sentinel-2 (10–20 m resolution), launched in June 2015, was too coarse for the streams evaluated, but in the future, could be used to regularly monitor the condition of rivers >10 m wide. CubeSats, such as those launched by Planet (San Francisco, CA, USA), also show high potential for improved monitoring of stream condition at more frequent intervals. Obstacles to the widespread application of CubeSats for monitoring stream condition, however, include the limited number of spectral bands (blue, green, red, near-infrared), the cost of Planet imagery, as well as challenges in calibrating reflectance and georeferencing between satellites e.g., [76]. As these technical obstacles are overcome, satellite imagery can be more commonly used to monitor streams in a spatially continuous manner.

We found separating the influence of weather relative to the influence of human-induced change particularly challenging in the riparian areas. Trends in riparian condition could be very sensitive to how reference conditions were defined. In part, this is because we might expect that the magnitude of interannual variability may be inconsistent across areas dominated by trees versus herbaceous vegetation. Additionally, reference areas can be influenced by forces of change independent of weather patterns such as variability in grazing intensity. The use of image pairs, and the time-frame at which the projects were considered (one to three years post-restoration), also limited our ability to evaluate the success of the activities with a lagged response, such as the tree planting effort, which takes more time to establish and can act to further influence riparian-stream interactions over time [44,77]. Riparian and wetland herbaceous species, however, can respond very quickly to changes in riparian soil moisture [71,77]. Therefore, changes in riparian greenness can be used to indirectly assess short-term changes to the stream hydrologic processes [77–79]. Because changes in riparian greenness were detected even when using a 20 m stream buffer, it may be possible to monitor riparian condition for streams <5 m wide using moderate resolution satellites, such as Landsat or Sentinel-2. As riparian trees grow larger and become more established, they can also impact our ability to monitor stream surface area. For instance, stream length at most sites was found to be <100% inundated, a finding that can be attributed to overhanging vegetation or hyporheic flow masking stream water.

Because our analysis relied on just two high-resolution images per site, we view this analysis as an example of the capabilities and potential of using high-resolution imagery to monitor stream condition and not a conclusion on the impact of specific restoration methods (e.g., BDAs, riparian planting) on streams. However, despite these concerns, we saw proximal evidence of increases in stream surface area upstream from structures at all four sites and evidence of either increases in stream surface area or riparian greenness along the restoration reach at all four sites. We also saw a decrease in stream surface area downstream from the restoration reach at all four sites. These findings suggest that the restoration activities have induced increases in instream water storage, at least during the early summer period. Uneven findings regarding the change induced by BDAs could be due to several factors including stream size, pre-restoration conditions [41], restoration goals, restoration installation (including extent and age), and site specifics, all of which can influence how a stream responds to change through time. However, these findings suggest that high-resolution imagery can provide a spatially continuous understanding of how narrow (<5 m wide) streams respond to restoration projects.

5. Conclusions

As high-resolution, multispectral imagery becomes more frequently collected and available, regular monitoring of stream surface area and condition in response to local or watershed-based changes will become increasingly feasible. However, mapping narrow, linear water features that are subject to rapid changes in water depth and turbidity may require approaches independent from those widely applied to map wetlands and lakes [1,2,80]. We found that utilizing bands unique to Worldview-2 and 3 in an eCognition framework can produce accurate results and minimal errors of commission down to a stream width of approximately 1.5 m. By comparing images pre- and post-restoration across multiple sites in the Upper Missouri River Headwaters Basin, we were able to quantify proximal and reach-scale changes in stream surface area and riparian greenness in response to the installation of multiple BDAs at each site. In general, the installation of BDAs appeared to create an increase in stream surface area immediately upstream from many of the structures. These proximal changes tended to result in reach-scale increases in stream surface area and riparian greenness along the restoration reach as well as decreases in stream surface area for reaches just downstream (through 500 m) from the restoration reach at most of the sites. The consistency of the directional changes to stream surface area across the sites, despite differing patterns in discharge lower in the watershed between-image pairs, suggests that we were able to account for hydro-climatic variability. Restoring degraded streams can positively influence in-stream habitat, water quality as well as water quantity across a watershed [34,37,38]. Monitoring the impacts of stream restoration projects, including resource-efficient structures such as BDAs, can improve and inform site selection and expectations for future stream restoration efforts.

Author Contributions: M.K.V. designed the study, performed the data analysis, and wrote the manuscript. C.B. led the image processing effort.

Acknowledgments: This project was funded by a U.S. EPA Region 8 grant, entitled "Building drought resiliency and watershed prioritization using natural water storage techniques" and through the associated interagency agreement (DW-014-92475401-0). This project was also supported by the USGS Land Resources, Land Change Science Program. All restoration actions were developed and implemented by the Nature Conservancy. We thank Tina Laidlaw and Ayn Schmit with U.S. EPA Region 8 as well as Laurie Alexander, Heather Golden, Jay Christensen, and Charles Lane with U.S. EPA, Office of Research and Development for their support and leadership in acquiring the funds and initial project development. We thank Nathan Korb of the Nature Conservancy and Jeff Burrell of the Wildlife Conservation Society for their invaluable help in site selection and knowledge of the field conditions, and Todd Hawbaker, Laurie Alexander, and the anonymous reviewers for their vital comments on earlier versions of the manuscript. Following publication, the data related to this publication will be published in the ScienceBase catalog (doi:10.5066/P9F9618G). Any use of trade, product, or firm names is for descriptive purposes only and does not imply endorsement by the U.S. Government.

Conflicts of Interest: The authors declare no conflicts of interest.

References

1. Smith, L.C. Satellite remote sensing of river inundation area, stage, and discharge: A review. *Hydrol. Process.* **1997**, *11*, 1427–1439. [CrossRef]
2. Alsdorf, D.E.; Rodriguez, E.; Lettenmaier, D.P. Measuring surface water from space. *Rev. Geophys.* **2007**, *45*. [CrossRef]
3. Wang, J.J.; Lu, X.X.; Liew, S.C.; Zhou, Y. Retrieval of suspended sediment concentrations in large turbid rivers using Landsat ETM+: An example from the Yangtze River, China. *Earth Surf. Process. Landf.* **2009**, *34*, 1082–1092. [CrossRef]
4. Wang, Y.; Colby, J.D.; Mulcahy, K.A. An efficient method for mapping flood extent in a coastal floodplain using Landsat TM and DEM data. *Int. J. Remote Sens.* **2002**, *23*, 3681–3696. [CrossRef]
5. Qi, S.; Brown, D.G.; Tian, Q.; Jiang, L.; Zhao, T.; Bergen, K.M. Inundation extent and flood frequency mapping using Landsat imagery and Digital Elevation Models. *GISci. Remote Sens.* **2009**, *46*, 101–127. [CrossRef]
6. Chen, Y.; Huang, C.; Ticchurst, C.; Merrin, L.; Thew, P. An evaluation of MODIS daily and 8-day composite products for floodplain and wetland inundation mapping. *Wetlands* **2013**, *33*, 823–835. [CrossRef]
7. Ogilvie, A.; Belaud, G.; Belenne, C.; Bailly, J.S.; Bader, J.C.; Oleksiak, A.; Ferry, L.; Martin, D. Decadal monitoring of the Niger Inner Delta flood dynamics using MODIS optical data. *J. Hydrol.* **2015**, *523*, 368–383. [CrossRef]
8. Schumann, G.; Di Baldassarre, G.; Alsdorf, D.E.; Bates, P.D. Near real-time flood wave approximation on large rivers from space: Application to the River Po, Northern Italy. *Water Resour. Res.* **2010**, *46*, W05601. [CrossRef]
9. Allen, G.H.; Pavelsky, T.M. Patterns of river width and surface area revealed by the satellite-derived North American River Width data set. *Geophys. Res. Lett.* **2015**, *42*, 395–402. [CrossRef]
10. Hotchkiss, E.R.; Hall, R.O., Jr.; Sponseller, R.A.; Butman, D.; Klaminder, J.; Laudon, H.; Rosvall, M.; Karisson, J. Sources of and processes controlling CO_2 emissions change with the size of streams and rivers. *Nat. Geosci.* **2015**, *8*, 696–699. [CrossRef]
11. Demarchi, L.; Bizzi, S.; Piegay, H. Hierarchical object-based mapping of riverscape units and in-stream mesohabitats using LiDAR and VHR imagery. *Remote Sens.* **2016**, *8*, 97. [CrossRef]
12. Lang, M.W.; McCarty, G.W. Lidar intensity for improved detection of inundation below the forest canopy. *Wetlands* **2009**, *29*, 1166–1178. [CrossRef]
13. Wu, Q.; Lane, C.R. Delineating wetland catchments and modeling hydrologic connectivity using lidar data and aerial imagery. *Hydrol. Earth Syst. Sci.* **2017**, *21*, 3579–3595. [CrossRef]
14. Clewley, D.; Whitcomb, J.; Moghaddam, M.; McDonald, K.; Chapman, B.; Bunting, P. Evaluation of ALOS PALSAR data for high-resolution mapping of vegetated wetlands in Alaska. *Remote Sens.* **2015**, *7*, 7272–7297. [CrossRef]
15. Hess, L.L.; Melack, J.M.; Affonso, A.G.; Barbosa, C.; Gastil-Buhl, M.; Novo, E.M.L.M. Wetlands of the lowland Amazon basin: Extent, vegetative cover, and dual-season inundated area as mapped with JERS-1 synthetic aperture radar. *Wetlands* **2015**, *35*, 745–756. [CrossRef]
16. Schlaffer, S.; Chini, M.; Dettmering, D.; Wagner, W. Mapping wetlands in Zambia using seasonal backscatter signatures derived from ENVISaT ASaR time series. *Remote Sens.* **2016**, *8*, 402. [CrossRef]
17. White, D.C.; Lewis, M.M. A new approach to monitoring spatial distribution and dynamics of wetlands and associated flows of Australian Great Artesian Basin springs using QuickBird satellite imagery. *J. Hydrol.* **2011**, *408*, 140–152. [CrossRef]
18. Whiteside, T.G.; Bartolo, R.E. Mapping aquatic vegetation in a tropical wetland using high spatial resolution multispectral satellite imagery. *Remote Sens.* **2015**, *7*, 11664–11694. [CrossRef]
19. Liébault, F.; Piegay, H. Assessment of channel changes due to long term bedload supply decrease, Roubion River, France. *Geomorphology* **2001**, *36*, 167–186. [CrossRef]
20. Bollati, I.M.; Pellegrini, L.; Rinaldi, M.; Duci, G.; Pelfini, M. Reach-scale morphological adjustments and stages of channel evolution: The case of the Trebbia River (northern Italy). *Geomorphology* **2014**, *221*, 176–186. [CrossRef]
21. Toone, J.; Rice, S.P.; Piégay, H. Spatial discontinuity and temporal evolution of channel morphology along a mixed bedrock-alluvial river, upper Drôme River, southeast France: Contingent responses to external and internal controls. *Geomorphology* **2014**, *205*, 5–16. [CrossRef]

22. Belletti, B.; Dufour, S.; Piégay, H. What is the relative effect of space and time to explain the braided river width and island patterns at a regional scale? *River Res. Appl.* **2013**, *31*, 1–15. [CrossRef]
23. Bertrand, M.; Piégay, H.; Pont, D.; Liébault, F.; Sauquet, E. Sensitivity analysis of environmental changes associated with riverscape evolutions following sediment reintroduction: Geomatic approach on the Drôme River network, France. *Int. J. River Basin Manag.* **2013**, *11*, 19–32. [CrossRef]
24. Marcus, W.A.; Fonstad, M.A. Optical remote mapping of rivers at sub-meter resolutions and watershed extents. *Earth Surf. Process. Landf.* **2008**, *33*, 4–24. [CrossRef]
25. Jiang, H.; Feng, M.; Zhu, Y.; Lu, N.; Huang, J.; Xiao, T. An automated method for extracting rivers and lakes form Landsat imagery. *Remote Sens.* **2014**, *6*, 5067–5089. [CrossRef]
26. Goklany, I.M. Comparing 20th century trends in U.S. and global agricultural water and land use. *Water Int.* **2002**, *27*, 321–329. [CrossRef]
27. Schaible, G.D.; Aillery, M.P. *Water Conservation in Irrigated Agriculture: Trends and Challenges in the Face of Emerging Demands*; EIB-99; U.S. Department of Agriculture, Economic Research Service: Washington, DC, USA, 2012.
28. Hansen, A.J.; Rasker, R.; Maxwell, B.; Rotella, J.J.; Johnson, J.D.; Parmenter, A.W.; Langner, U.; Cohen, W.B.; Lawrence, R.L.; Kraska, P.V. Ecological causes and consequences of demographic change in the new west. *Bioscience* **2002**, *52*, 151–162. [CrossRef]
29. Gude, P.H.; Hansen, A.J.; Rasker, R.; Maxwell, B. Rates and drivers of rural residential development in the Greater Yellowstone. *Landsc. Urban Plan.* **2006**, *77*, 131–151. [CrossRef]
30. Pederson, G.T.; Gray, S.T.; Woodhouse, C.A.; Betancourt, J.L.; Fagre, D.B.; Littell, J.S.; Watson, E.; Luckman, B.H.; Graumlich, L.J. The Unusual Nature of Recent Snowpack Declines in the North American Cordillera. *Science* **2011**, *333*, 332–335. [CrossRef] [PubMed]
31. Pederson, G.T.; Betancourt, J.L.; McCabe, G.J. Regional patterns and proximal causes of the 60 recent snowpack decline in the Rocky Mountains, U.S. *Geophys. Res. Lett.* **2013**, *40*, 1811–1816. [CrossRef]
32. U.S. Bureau of Reclamation. *Climate Change Analysis for the Missouri River Basin*; Technical Memorandum No. 86-68210-2012-03; U.S. Bureau of Reclamation: Washington, DC, USA, 2012.
33. Lemly, A.D.; Kingsford, R.T.; Thomson, J.R. Irrigated agriculture and wildlife conservation: Conflict on a global scale. *Environ. Manag.* **2000**, *25*, 485–512. [CrossRef] [PubMed]
34. Isaak, D.J.; Wollrab, S.; Horan, D.; Chandler, G. Climate change effects on stream and river temperatures across the northwest U.S. from 1980–2009 and implications for salmonid fishes. *Clim. Chang.* **2012**, *113*, 499–524. [CrossRef]
35. Ziemer, L.S.; Kendy, E.; Wilson, J. Ground water management in Montana: On the road from beleaguered to science-based policy. *Public Land Resour. Law Rev.* **2006**, *76*, 75–97.
36. Jones, H.P.; Hole, D.G.; Zavaleta, E.S. Harnessing nature to help people adapt to climate change. *Nat. Clim. Chang.* **2012**, *2*, 504–509. [CrossRef]
37. Gartner, T.; Mulligan, J.; Schmidt, R.; Gunn, J. *Natural Infrastructure*; World Resources Institute: Washington, DC, USA, 2013; Volume 56, p. 18.
38. Acreman, M.; Holden, J. How Wetlands Affect Floods. *Wetlands* **2013**, *33*, 773–786. [CrossRef]
39. Montana Department of Natural Resources and Conservation. *The 2015 Montana State Water Plan*; Montana Department of Natural Resources and Conservation: Helena, MT, USA, 2015; 20p.
40. Kemp, P.S.; Worthington, T.A.; Langford, T.E.L.; Tree, A.R.J.; Gaywood, M.J. Qualitative and quantitative effects of reintroduced beavers on stream fish. *Fish Fish.* **2012**, *13*, 158–181. [CrossRef]
41. Pollock, M.M.; Beechie, T.J.; Wheaton, J.M.; Jordan, C.E.; Bouwes, N.; Weber, N.; Volk, C. Using Beaver Dams to Restore Incised Stream Ecosystems. *Bioscience* **2014**, *64*, 279–290. [CrossRef]
42. Bouwes, N.; Weber, N.; Jordan, C.E.; Saunders, C.; Tattam, I.A.; Volk, C.; Wheaton, J.M.; Pollock, M.M. Ecosystem experiment reveals benefits of natural and simulated beaver dams to a threatened population of steelhead (*Oncorhynchus mykiss*). *Sci. Rep.* **2016**, *6*, 1–12. [CrossRef] [PubMed]
43. Hill, A.R.; Duval, T.P. Beaver dams along an agricultural stream in southern Ontario, Canada: Their impact on riparian zone hydrology and nitrogen chemistry. *Hydrol. Process.* **2009**, *23*, 1324–1336. [CrossRef]
44. Knopf, F.L.; Johnson, R.R.; Rich, T.; Samson, F.B.; Szaro, R.C. Conservation of riparian systems in the United States. *Wilson Bull.* **1988**, *100*, 272–284.
45. Gurnell, A.M. The hydrogeomorphological effects of beaver dam-building activity. *Prog. Phys. Geogr.* **1998**, *22*, 167–189. [CrossRef]

46. PRISM Climate Group, Oregon State University. Available online: http://prism.oregonstate.edu (accessed on 10 July 2012).

47. Homer, C.; Dewitx, J.; Yang, L.; Jin, S.; Danielson, P.; Xian, G.; Coulston, J.; Herold, N.; Wickham, J.; Megown, K. Completion of the 2011 National Land Cover Database for the conterminous United States—Representing a decade of land cover change information. *Photogramm. Eng. Remote Sens.* **2015**, *81*, 345–354.

48. Podolak, K.; Kelsey, R.; Harris, S.; Korb, N. Why the Nature Conservancy is Restoring Streams by Acting Like a Beaver. Cool Green Science, Nature Conservancy. Available online: https://blog.nature.org/science (accessed on 22 January 2018).

49. Pollock, M.M.; Lewallen, G.; Woodruff, K.; Jordan, C.E.; Castro, J.M. (Eds.) *The Beaver Restoration Guidebook: Working with Beaver to Restore Streams, Wetlands, and Floodplains*; version 1.0; United States Fish and Wildlife Service: Portland, OR, USA, 2015; 189p. Available online: http://www.fws.gov/oregonfwo/ToolsForLandowners/RiverScience/Beaver.asp (accessed on 22 January 2018).

50. Gesch, D.; Oimoen, M.; Greenlee, S.; Nelson, C.; Steuck, M.; Tyler, D. The National Elevation Dataset. *Photogramm. Eng. Remote Sens.* **2002**, *68*, 5–11.

51. Zhang, Y. Problems in the fusion of commercial high-resolution satellite as well as LANDSAT 7 images and initial solutions. In *GeoSpatial Theory, Processing and Applications*; ISPRS: Ottawa, ON, Canada, 2002; Volume 34, Part 4.

52. Li, H.; Jing, L.; Tang, Y. Assessment of pansharpening methods applied to WorldView-2 imagery fusion. *Sensors* **2017**, *17*, 89. [CrossRef] [PubMed]

53. Marchisio, G.; Pacifici, F.; Padwick, C. On the relative predictive value of the new spectral bands in the WorldView-2 sensor. In Proceedings of the 2010 IEEE International Geoscience and Remote Sensing Symposium (IGARSS), Honolulu, HI, USA, 25–30 July 2010; pp. 2723–2726.

54. McFeeters, S.K. The use of the Normalized Difference Water Index (NDWI) in the delineation of open water features. *Int. J. Remote Sens.* **1996**, *17*, 1425–1432. [CrossRef]

55. Wolf, A.F. Using Worldview-2 Vis-NIR multispectral imagery to support land mapping and features extraction using normalized difference index ratios. In *Proceedings SPIE 8390, Algorithms and Technologies for Multispectral, Hyperspectral, and Ultraspectral Imagery XVIII*; SPIE: Baltimore, MD, USA, 2012; p. 83900N.

56. Liu, H.Q.; Huete, A.R. A feedback based mpoiodification of the NDVI to minimize canopy background and atmospheric noise. *IEEE Trans. Geosci. Remote Sens.* **1995**, *33*, 457–465.

57. Huete, A.R.; Liu, H.Q.; Batchily, K.; YanLeeuwen, W. A comparison of vegetation indices global set of TM images for EOS-MODIS. *Remote Sens. Environ.* **1997**, *59*, 440–451. [CrossRef]

58. Tucker, C.J. Red and photographic infrared linear combinations for monitoring vegetation. *Remote Sens. Environ.* **1979**, *8*, 127–150. [CrossRef]

59. Falkowski, M.J.; Gessler, P.E.; Morgan, P.; Hudak, A.T.; Smith, A.M.S. Characterizing and mapping forest fire fuels using ASTER imagery and gradient modeling. *For. Ecol. Manag.* **2005**, *217*, 129–146. [CrossRef]

60. Jawak, S.D.; Luis, A.J. A spectral index ratio-based Antarctic land-cover mapping using hyperspatial 8-band WorldView-2 Imagery. *Polar Sci.* **2013**, *7*, 18–38. [CrossRef]

61. Myint, S.W.; Gober, P.; Brazel, A.; Grossman-Clarke, S.; Weng, Q. Per-pixel vs. Object-based classification of urban land cover extraction using high spatial resolution imagery. *Remote Sens. Environ.* **2011**, *115*, 1145–1161. [CrossRef]

62. Stumpf, R.P.; Holderied, K.; Sinclair, M. Determination of water depth with high-resolution satellite imagery over variable bottom types. *Limnol. Oceanogr.* **2003**, *48*, 547–556. [CrossRef]

63. Youden, W.J. Index for rating diagnostic tests. *Cancer* **1950**, *3*, 32–35. [CrossRef]

64. Lopez-Raton, M.; Rodriguez-Alvarez, M.X. Package "OptimalCutpoints". Available online: http://cran.r-project.org/web/packages/OptimalCutpoints/OptimalCutpoints.pdf (accessed on 11 May 2018).

65. Fleiss, J.L. *Statistical Methods for Rates and Proportions*, 2nd ed.; John Wiley & Sons: New York, NY, USA, 1981.

66. Forbes, A.D. Classification-algorithm evaluation: Five performance measures based on confusion matrices. *J. Clin. Monit.* **1995**, *11*, 189–206. [CrossRef] [PubMed]

67. Liro, M. Conceptual model for assessing the channel changes upstream from dam reservoir. *Quaest. Geogr.* **2014**, *33*, 61–74. [CrossRef]

68. Huete, A.R. A soil-adjusted vegetation index (SAVI). *Remote Sens. Environ.* **1988**, *25*, 295–309. [CrossRef]

69. Baret, F.; Guyot, G. Potentials and limits of vegetation indices for LAI and APAR assessment. *Remote Sens. Environ.* **1991**, *35*, 161–173. [CrossRef]

70. Qi, J.; Huete, A.R.; Moran, M.S.; Chehbouni, A.; Jackson, R.D. Interpretation of vegetation indices derived from multi-temporal SPOT images. *Remote Sens. Environ.* **1993**, *44*, 89–101. [CrossRef]

71. Stromberg, J.C. Restoration of riparian vegetation in the south-western United States: Importance of flow regimes and fluvial dynamism. *J. Arid Environ.* **2001**, *49*, 17–34. [CrossRef]

72. Richardson, D.M.; Holmes, P.M.; Esler, K.J.; Galatowitsch, S.M.; Stromberg, J.C.; Kirkman, S.P.; Pysek, P.; Hobbs, R.J. Riparian vegetation: Degradation, alien plant invasions, and restoration prospects. *Divers. Distrib.* **2007**, *13*, 126–139. [CrossRef]

73. Mouillot, F.; Schultz, M.G.; Yue, C.; Cadule, P.; Tansey, K.; Ciais, P.; Chuvieco, E. Ten years of global burned area products from spaceborne remote sensing—A review: Analysis of user needs and recommendations for future developments. *Int. J. Appl. Earth Obs. Geoinf.* **2014**, *26*, 64–79. [CrossRef]

74. Galvão, L.S.; Filho, W.P.; Abdon, M.M.; Novo, E.M.M.L.; Silva, J.S.V.; Ponzoni, F.J. Spectral reflectance characterization of shallow lakes from the Brazilian Pantanal wetlands with field and airborne hyperspectral data. *Int. J. Remote Sens.* **2003**, *24*, 4093–4112. [CrossRef]

75. Tyler, A.N.; Svab, E.; Preston, T.; Presing, M.; Kovacs, W.A. Remote sensing of the water quality of shallow lakes: A mixture modelling approach to quantifying phytoplankton in water characterized by high-suspended sediment. *Int. J. Remote Sens.* **2006**, *27*, 1521–1537. [CrossRef]

76. McCabe, M.F.; Aragon, B.; Houborg, R.; Mascaro, J. CubeSats in hydrology: Ultrahigh-resolution insights into vegetation dynamics and terrestrial evaporation. *Water Resour. Res.* **2017**. [CrossRef]

77. Rood, S.B.; Gourley, C.R.; Ammon, E.M.; Heki, L.G.; Klotz, J.R.; Morrison, M.L.; Mosley, D.; Scoppettone, G.G.; Swanson, S.; Wagner, P.L. Flows for floodplain forests: A successful riparian restoration. *BioScience* **2003**, *53*, 647–656. [CrossRef]

78. Stromberg, J.C.; Lite, S.J.; Rychener, T.J.; Levick, L.R.; Dixon, M.D.; Watts, J.M. Status of the riparian ecosystem in the Upper San Pedro River, Arizona: Application of an assessment model. *Environ. Monit. Assess.* **2006**, *115*, 145–173. [CrossRef] [PubMed]

79. Jones, K.B.; Edmonds, C.E.; Slonecker, E.T.; Wickham, J.D.; Neale, A.C.; Wade, T.G.; Riiters, K.H.; Kepner, W.G. Detecting changes in riparian habitat conditions based on patterns of greenness change: A case study from the Upper San Pedro River Basin, USA. *Ecol. Indic.* **2008**, *8*, 89–99. [CrossRef]

80. Bizzi, S.; Demarchi, L.; Grabowski, C.; Weissteiner, C.J.; Van de Bund, W. The use of remote sensing to characterize hydromorphological properties of European rivers. *Aquat. Sci.* **2016**, *78*, 57–70. [CrossRef]

remote sensing

MDPI

Article

Seabed Mapping in Coastal Shallow Waters Using High Resolution Multispectral and Hyperspectral Imagery

Javier Marcello [1,*] , Francisco Eugenio [1], Javier Martín [2] and Ferran Marqués [3]

1   Instituto de Oceanografía y Cambio Global, IOCAG, Universidad de las Palmas de Gran Canaria, ULPGC, Parque Científico Tecnológico Marino de Taliarte, s/n, Telde, 35214 Las Palmas, Spain; francisco.eugenio@ulpgc.es
2   Departamento de Física, Universidad de las Palmas de Gran Canaria, ULPGC, 35017 Las Palmas, Spain; javier.martin@ulpgc.es
3   Signal Theory and Communications Department, Universitat Politecnica de Catalunya BarcelonaTECH, 08034 Barcelona, Spain; ferran.marques@upc.edu
*   Correspondence: javier.marcello@ulpgc.es; Tel.: +34-928-457365

Received: 18 June 2018; Accepted: 28 July 2018; Published: 2 August 2018

Abstract: Coastal ecosystems experience multiple anthropogenic and climate change pressures. To monitor the variability of the benthic habitats in shallow waters, the implementation of effective strategies is required to support coastal planning. In this context, high-resolution remote sensing data can be of fundamental importance to generate precise seabed maps in coastal shallow water areas. In this work, satellite and airborne multispectral and hyperspectral imagery were used to map benthic habitats in a complex ecosystem. In it, submerged green aquatic vegetation meadows have low density, are located at depths up to 20 m, and the sea surface is regularly affected by persistent local winds. A robust mapping methodology has been identified after a comprehensive analysis of different corrections, feature extraction, and classification approaches. In particular, atmospheric, sunglint, and water column corrections were tested. In addition, to increase the mapping accuracy, we assessed the use of derived information from rotation transforms, texture parameters, and abundance maps produced by linear unmixing algorithms. Finally, maximum likelihood (ML), spectral angle mapper (SAM), and support vector machine (SVM) classification algorithms were considered at the pixel and object levels. In summary, a complete processing methodology was implemented, and results demonstrate the better performance of SVM but the higher robustness of ML to the nature of information and the number of bands considered. Hyperspectral data increases the overall accuracy with respect to the multispectral bands (4.7% for ML and 9.5% for SVM) but the inclusion of additional features, in general, did not significantly improve the seabed map quality.

Keywords: benthic mapping; seagrass; airborne hypespectral imagery; Worldview-2; atmospheric correction; sunglint correction; water column correction; dimensionality reduction techniques; SVM classification; linear unmixing

---

1. Introduction

Coastal ecosystems are essential because they support high levels of biodiversity and primary production, but their complexity and high spatial and temporal variability make their study particularly challenging. Seagrasses are extremely important marine angiosperms (flowering plants) with a worldwide distribution. Seagrass meadows are among the most productive ecosystems in the world, which help protect the shoreline from soil erosion, serve as a refuge area for other species, and absorb carbon from the atmosphere [1,2]. Thus, seagrasses are essential, and their preservation in a sustainable manner needs the appropriate management tools. In this sense, satellite remote sensing is a cost-effective solution that has many advantages, compared to traditional techniques, like airborne photography with photo-interpretation or in-situ measurements (binomic maps from oceanographic ships). This way, satellite remote sensing is becoming a fundamental technology for the monitoring of benthic habitats (e.g., seagrass meadows) in shallow waters, as it provides periodic and synoptic data at different spatial scales and spectral resolutions [3].

Seafloor mapping using satellite remote sensing is a complex and challenging task, as optical bands have limited water penetration capability and the best channels to reach the seafloor (shorter wavelengths) suffer from higher atmospheric distortion. Hence, the signal recorded at the sensor level coming from the seabed is very low, even in clear waters [4,5]. Towards the goal of mapping benthic habitats at high spatial resolution and achieving a reasonable accuracy, the use of hyperspectral (HS) imagery can be considered as an alternative to multispectral (MS) data. Unfortunately, high spatial hyperspectral sensors onboard satellites are not yet available and, in consequence, high-resolution data from airborne or drone HS sensors are the only options to collect HS data to map complex benthic habitats environments.

To map the seafloor, the use of high-resolution remote sensing is promising but requires the application of different geometric and radiometric corrections. Specifically, the removal of the atmospheric absorption and scattering and the sunglint effect over the sea surface are essential preprocessing steps. In addition, the water column disturbance can be corrected; however, it is a very complex issue in coastal areas due to the variability of the scattering and absorption in the water column, the bottom type, and the water depth [6].

Regarding the removal of the atmospheric effects, correction approaches can be basically grouped into physical radiative transfer models and empirical methods exclusively considering information obtained from the image scene itself [7]. Many scene-based empirical approaches have been developed to remove atmospheric effects from multispectral and hyperspectral imaging data [8–11]. Concerning the physical models, they are more advanced, complex and based on simulations of the conditions of the atmosphere from its physical-chemical characteristics and the day and time of acquisition of the image. At the present time, there are a number of model-based correction algorithms, for example MODerate resolution atmospheric TRANsmission (MODTRAN), Atmosphere CORrection Now (ACRON), Fast Line-of-sight Atmospheric Analysis of Spectral Hypercubes (FLAASH), High-accuracy Atmospheric Correction for Hyperspectral Data (HATCH), Atmospheric and Topographic CORrection (ATCOR), or Second Simulation of a Satellite Signal in the Solar Spectrum (6S) [12–14]. Some of these algorithms include more advanced features, such as spectral smoothing, topographic correction, and adjacency effect correction.

On the other hand, the removal of sunglint is necessary for the reliable retrieval of bathymetry and seafloor mapping in shallow-water environments. Deglinting techniques have been developed for low-resolution open waters and also for high-resolution coastal applications [15]. In general, algorithms use the near-infrared (NIR) channel to eliminate sunglint assuming that water reflectivity in the NIR band is negligible [16]. This assumption is usually correct, except when turbidity is high, or the seabed reflectance is important, which can occur in very shallow areas [6].

Concerning the water column correction, Lyzenga [17] proposed the depth invariant index (DII), an image-based method to decrease the water column attenuation effect. This correction technique has been applied in previous works, due to its simplicity, with different degrees of success [18–21]. On the other hand, in the last decades, some radiative transfer models have been proposed, but they are more complex, and the difficulty of accurately measuring some in-situ water parameters can limit their applicability [22–25].

Once preprocessing algorithms have been applied, classification techniques can be used to generate the seabed maps. Classification is one of the most active areas of research in the field of remotely sensed image processing. For example, the classification of hyperspectral imagery is a challenging task because of the imbalance among the high dimensionality of the data and the limited amount of available training samples, as well as the implicit spectral redundancy. For this reason, specific approaches have been developed, like random forests, support vector machines (SVMs), deep learning or logistic regressions [26]. Unmixing techniques have also attracted the attention of the hyperspectral community. Unmixing algorithms separate the pixel spectra into a collection of constituent pure spectral signatures, named endmembers, and the corresponding set of fractional abundances, representing the percentage of each endmember that is present in the pixel [27].

Recent research to create seabed maps using remote sensing imagery has been mainly devoted to map coral reefs [28–34] or seagrass meadows [3,18,35–40]. Commonly, these studies address very shallow, clear and calm waters, and very dense vegetal species (i.e., *Posidonia oceanica*). As a continuation of our preliminary study [14], in this work, hyperspectral and multispectral imagery have been used to compare the benefits of each type of data to map the seafloor in a complex coastal area where submerged green aquatic vegetation meadows have low density, are relatively located at considerable depths (5 to 20 meters), where the sea surface is usually not completely calm due to persistent local surface winds, and, consequently, where very few bands reach an acceptable signal-to-noise ratio. Hence, a thorough analysis has been performed to obtain a robust methodology to produce accurate benthic habitat maps. To achieve this goal, different corrections, object-oriented, and pixel-based classification approaches have been considered, and diverse feature extraction strategies have also been tested. In summary, contributions are presented regarding the best correction techniques, feature extraction methods and classification approaches in such a challenging scenario. Moreover, a comparative assessment of the benefits of satellite multispectral and airborne hyperspectral imagery is included to map the seafloor in complex coastal zones.

2. Materials and Methods

2.1. Study Area

The Maspalomas Natural Reserve (Gran Canaria, Spain) is an important coastal-dune ecosystem covering approximately 4 km^2 and with a high touristic pressure, visited by more than 2 million people each year. The marine vegetation in the coastal fringe is basically composed of seagrass beds. The most abundant seagrass species is *Cymodocea nodosa* but, more recently, the *Caulerpa prolifera* green algae has also become dominant in this area. Figure 1 shows the geographic location of Maspalomas.

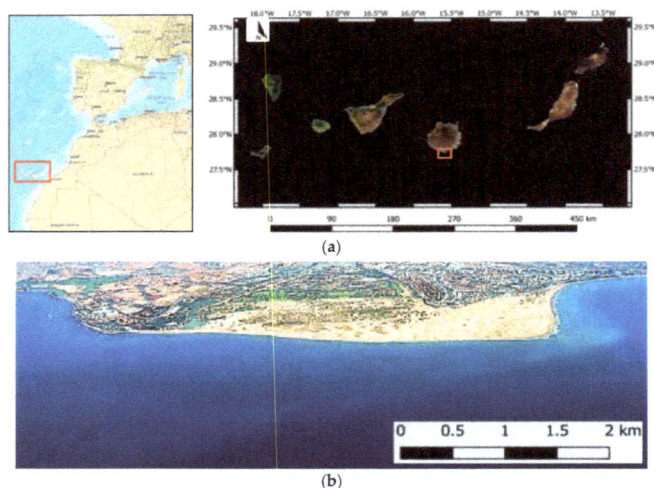

(a)

(b)

Figure 1. Maspalomas area: (**a**) Geographic location; (**b**) Panoramic view of the study area (scale is approximate).

2.2. Multisensor Remotely Sensed Data

A flight campaign was performed on June 2, 2017 and data were collected at 2.5 m resolution with the Airborne Hyperspectral Scanner (AHS). The AHS sensor (developed by ArgonST, USA) is operated by the Spanish Aerospace Institute (INTA) onboard the CASA 212-200 Paternina. AHS incorporates an 80-band imaging radiometer covering the range from 0.43 to 12.8 μm. In our study, only the first 20 channels were selected, covering the visible and near-infrared (NIR) spectrum from 0.434 to 1.015 μm with 12-bits of radiometric resolution [41]. Additionally, a WorldView-2 (WV-2) image collected on January 17, 2013 was considered for this study. Its sensor has a radiometric resolution of 11 bits and a spatial resolution, at the nadir, of 1.8 m for the 8 multispectral bands (0.40–1.04 μm). The panchromatic wideband was not used because this channel only provides information about the seabed in the very first meters of depth and, in consequence, pansharpening algorithms are not effective.

Table 1 includes the spectral characteristics of both sensors and Figure 2a,b show the Worldview and AHS imagery for the area of interest processed in this work, respectively.

Table 1. Airborne Hyperspectral Scanner (AHS) and Worldview-2 spectral channels.

Sensor	Spectral Band	Wavelength (nm)	Bandwidth (nm)
AHS	Visible and Near-IR (20 channels)	434–1015	28–30
WV-2	Coastal Blue	400–450	47.3
	Blue	450–510	54.3
	Green	510–580	63.0
	Yellow	585–625	37.4
	Red	630–690	57.4
	Red-edge	705–745	39.3
	Near-IR 1	770–895	98.9
	Near-IR 2	860–1040	99.6
	Panchromatic	450–800	284.6

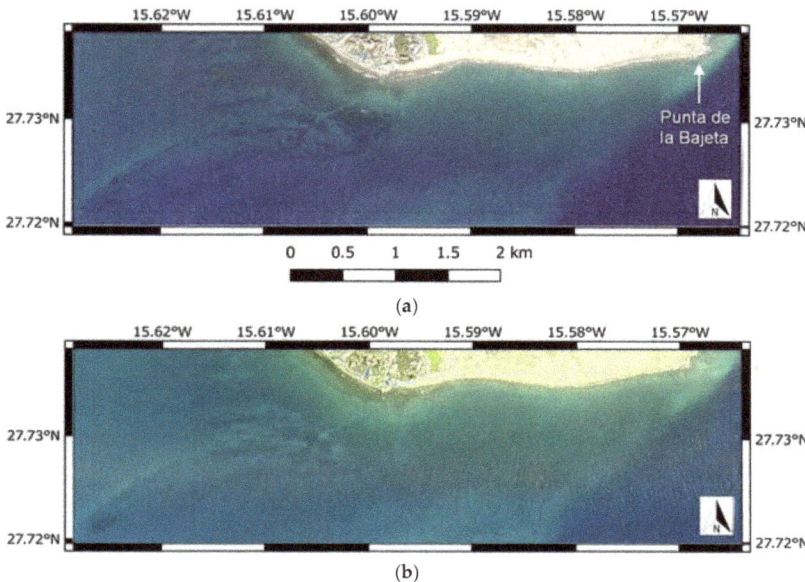

(a)

(b)

Figure 2. *Cont.*

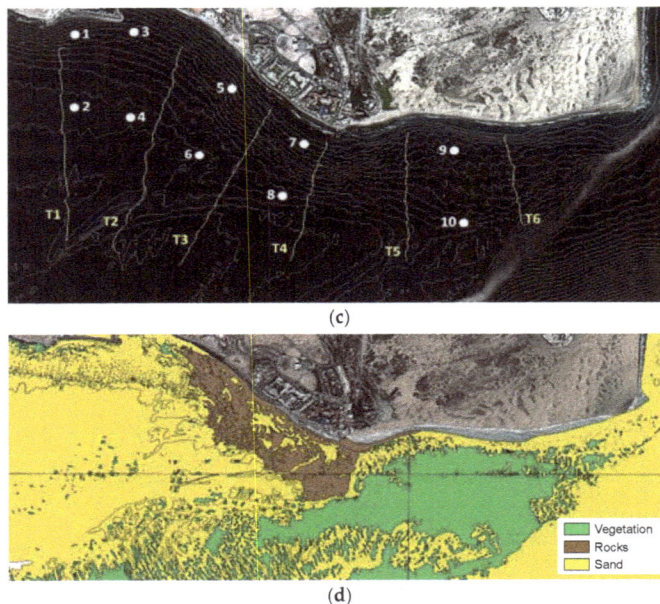

Figure 2. Color composite images after logarithmic stretching: (**a**) Worldview-2 of January 17, 2013 (channels 5-3-2) and (**b**) AHS of June 2, 2017 (channels 8-5-2); (**c**) Ship transects and sampling sites during the field campaigns of June 2, 2017 and June 4, 2015 (isobaths included at 1 m steps); (**d**) Reference benthic map 2013 [42].

2.3. In-Situ Measurements

In-situ data were acquired simultaneously to the AHS campaign. A total of 6 transects were performed, measuring the bathymetry using an ecosounder Reson Navisound 110 and recording images of the seafloor using two different video cameras (Neptune and Go Pro Hero 3+). Precise geographic and temporal information was provided by a differential GPS receiver model Trimble DSM132. Ten additional sites visited during 2015 were also used in the analysis, providing bathymetry and video records from the Go Pro camera. The variation of the sea level due to tides and waves was obtained from a nearby calibrated tide gauge.

Figure 2c presents the real ship transects during the 2017 campaign, as well as the sites monitored in 2015 (marked by yellow dots). Isobaths are also included, and the sites and equidistant transects are perpendicular to the shore to get the maximum information of the area at different depths up to 20 m.

Finally, to assess the accuracy of the benthic maps derived from HS and MS imagery, apart from the accurate information from the in-situ transect and sampling sites, a reference map providing global information for the whole area was desirable. Unfortunately, very limited cartography is available and the map considered (Figure 2d) is the reference map available of the Maspalomas coast, which is part of a Spanish coastline eco-cartographical study from a series of maritime engineering and marine ecology studies structured in a GIS [42].

This information was used just as a coarse reference, but the quantitative validation only considered the precise information measured during the campaigns of 2015 and 2017.

The data used in this study correspond to different years (2013 to 2017). However, the seabed of Maspalomas is in a fairly stable area with the exception of Punta de la Bajeta (marked in Figure 2a), which is the zone with the greatest topographic and sedimentary variability due to storms from the southwest.

The seafloor images recorded during the field campaign (see Figure 3 for examples) show the complexity to discriminate habitat classes because they are usually mixed. Especially, submerged green aquatic vegetation meadows grow on sandy bottoms, and they have low density and mainly live between 5 and 25 m depth. In addition, rocks are partially covered by algae making automated classifications more challenging.

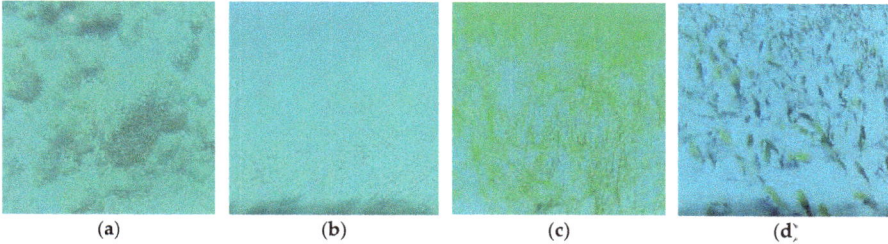

Figure 3. Seafloor classes: (**a**) Rocks; (**b**) Sand; (**c**) *Cymodocea nodosa*; (**d**) *Caulerpa prolifera*.

2.4. Mapping Methodology

The overall processing protocol to generate the seabed maps is presented in Figure 4. Different inputs to the classification algorithm were analyzed to select the best methodology. Following, there is a detailed description of the different steps involved.

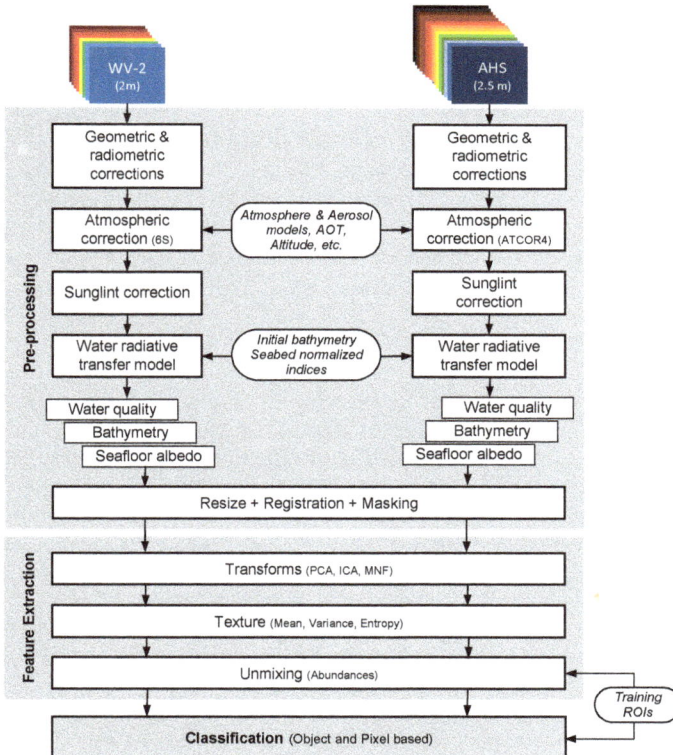

Figure 4. Flowchart of the processing methodology to generate benthic maps.

2.4.1. Multisensor Imagery Corrections

- Multispectral Satellite Imagery

DigitalGlobe owns and operates a world-class constellation of high resolution, high accuracy Earth imaging satellites. The acquired WV-2 image is the level 2 ortho-ready product that has the geometric correction implemented and a horizontal accuracy specification of 5 meters or better [43].

In coastal areas, radiometric and atmospheric corrections have proven to be a crucial step in the processing of high-resolution satellite images due to the low signal-to-noise ratio at sensor level. A comparative evaluation of advanced atmospheric models (FLAASH, ATCOR, and 6S) in the coastal area of Maspalomas using high-resolution WorldView-2 data showed that the 6S algorithm achieved the highest accuracy when the corrected reflectance was compared to field spectroradiometer data (RMSE of 0.0271 and bias of −0.0217) [14]. Hence, we have applied the 6S correction model but adapted to the WorldView-2 spectral response and to the particular scene geometry and date of the image. 6S is a radiative transfer model that generates the constants $(x_a, x_b$ and $x_c)$ to estimate the surface (BOA: Bottom Of Atmosphere) reflectance by [14,44,45]:

$$\rho'_{BOA} = \frac{(x_a L_{TOA}) - x_b}{1 + (x_c((x_a L_{TOA}) - x_b))} \tag{1}$$

corrected to consider the adjacency effect by:

$$\rho_{BOA} = \rho'_{BOA} + \frac{\tau_{odif}}{\tau_{odir}} [\rho'_{BOA} - \bar{\rho}] \tag{2}$$

where L_{TOA} is the radiance measured by the sensor (TOA: Top Of Atmosphere), ρ'_{BOA} is the initially corrected surface reflectivity by the 6S model, ρ_{BOA} is the surface reflectivity taking into account the adjacency effect, τ_{odif} and τ_{odir} are the diffuse and direct transmittances and $\bar{\rho}$ is the average reflectivity contribution from the pixel background [44].

Moreover, specular reflection of solar radiation is a serious disturbance for water quality, bathymetry, and benthic mapping in shallow-water environments. We applied the method suggested by References [15,16] to remove sunglint. Therefore, regions of the image having sunglint were selected, preferably areas of deep water. For each visible channel, all the pixels from these regions were included in a linear regression between the NIR against each visible band. Therefore, the reflectance of each pixel in the visible band i $(\rho_{BOA,i})$ can be deglinted $\left(\rho^{DG}_{BOA,i}\right)$ using the following equation:

$$\rho^{DG}_{BOA,i} = \rho_{BOA,i} - b_i * (\rho_{BOA,NIR} - MIN_{NIR}) \tag{3}$$

where b_i is the slope of the regression line for band i, $\rho_{BOA,NIR}$ is the reflectance of the NIR channel and MIN_{NIR} corresponds to the minimum reflectance value of the NIR image.

Improvements in the glint removal algorithm were performed [46] because, for WorldView-2, all the sensor bands are not recording the energy at the same precise time. In addition, the deglinting process, in the presence of considerable waves, alters the spectral content of the image. To overcome this inconvenience, a histogram matching was applied to statistically equalize each channel after and before deglinting. Finally, to remove the foam caused by the waves (whitecaps), pixels achieving reflectance values above a threshold were replaced by interpolation [46].

Water column correction is a very complex matter due to the variability of the bottom type, water depth, and water attenuation (scattering and absorption in the water column). Lyzenga proposed a simpler depth invariant index [17] but, in the last decades, different water models have been developed. In this work, a radiative transfer model for coastal waters was implemented [14]. The assumption is that the water leaving radiance is not only caused by the water column (IOPs: water Inherent Optical Properties), but it is also affected by the seafloor albedo and its corresponding bathymetry. In particular, the Lee et al. [47] exponential expression was used, that is an improved version of the formulation proposed by Reference [22]. The model can be expressed by:

$$r_{rs}^m(\lambda) \approx r_{rs,\infty}(\lambda)\left(1 - e^{-[\frac{1}{\mu_s^{sw}} + \frac{D_u^c}{\mu_v^{sw}}]k_d z}\right) + \frac{\rho_{alb}(\lambda)}{\pi}e^{-[\frac{1}{\mu_s^{sw}} + \frac{D_u^b}{\mu_v^{sw}}]k_d z} \qquad (4)$$

where $r_{rs}^m(\lambda)$ is the modeled reflectivity, $r_{rs,\infty}(\lambda)$ corresponds to the reflectivity below the sea surface for deep waters generated by the IOPs, $\rho_{alb}(\lambda)$ represents the seafloor reflectivity (albedo), z stands for the bathymetry, k_d is the water diffuse attenuation coefficient, μ_s^{sw} and μ_v^{sw} allow the corrections in the sun and sensor trajectories for the downwards and upwards directions, and, finally, D_u^c and D_u^b are light diffusion ascending factors due to the water column and the bottom reflectivity, respectively.

As proposed in Reference [48], the Fully Constrained Linear Unmixing method (FCLU) was used to model the seabed albedo. The albedo is obtained as the sum of the products of the abundances of the p pure benthic elements (ab_i) by their albedo for each wavelength $(em_i(\lambda))$:

$$\rho_{alb}(\lambda) = \sum_i^P ab_i * em_i(\lambda) \qquad (5)$$

Due to the limitations of the sensitivity of the multispectral sensors, we decided to use the three most significant and separable spectra of the seabed coverage in the area (sand, rocks and green vegetation). The radiative transfer modeling was adapted to multispectral sensors by integrating the result of the monochromatic model for the bandwidths of the 6 first channels of WV-2.

- Airborne Hyperspectral Imagery

Regarding the AHS geometric accuracy, the airborne inertial system achieves a final angular precision of 0.008° for the roll and the pitch, and 0.015° for the true heading; and a 12 channels GPS receiver provides trajectory location with accuracies of 5 to 10 cm [41].

For the hyperspectral data, atmospheric and illumination corrections were performed by INTA using the ATCOR4 model [41,49]. ATCOR4 is the ATCOR specific version for hyperspectral airborne data, and it is based on the MODTRAN-5 radiative transfer code.

The multispectral channels of the AHS sensor are somewhat narrower than the WV2 channels, providing, in the visible range, 12 multispectral channels instead of the 6 of the WV2. The greater number of channels implies greater spectral information and, therefore, an assumed greater sensitivity in the classification of the benthic classes. In any case, the methodology carried out for the elimination of the specular solar brightness, and the radiative transfer modeling is identical to that applied to WV-2.

Finally, land and deep water masks were applied to the corrected bands after both sensors have been homogenized in terms of resolution and have been spatially aligned using a large database of singular and well-distributed ground control points. Seafloor is difficult to properly monitor over 20 m depth in this area using satellite or airborne remote sensing imagery; therefore, a mask was applied at the 20 m isobath.

2.4.2. Feature Extraction

In addition to the spectral channels provided by each sensor, we also studied the inclusion of additional information to check the possible increase in classification performance. Therefore, we considered adding, to the multispectral or hyperspectral bands, components after Principal Component Analysis (PCA), Independent Component Analysis (ICA) and Minimum Noise Fraction (MNF) transforms; textural features to enrich the spatial information, and abundance maps extracted from linear unmixing techniques. Next, the feature extraction techniques are explained in more detail.

- Image Transforms

In hyperspectral imagery, the high number of narrow bands requires an increase in the number of training pixels to maintain a minimum statistical confidence and functionality for classification. This problem, known as the Hughes' phenomenon [50] or the curse of dimensionality, can be addressed by overcoming data redundancy. Several transforms have been proposed in the last decades to extract reliable information, reducing redundancy and noise. Traditional feature-extraction techniques, mainly applied to reduce the dimensionality of hyperspectral data, are PCA, ICA, and MNF [51].

PCA uses an orthogonal transformation to convert a set of correlated bands into a new set of uncorrelated components [52]. PCA is frequently applied for dimensionality reduction because it retains most of the information of the original data in the first principal components. In consequence, computation times decrease, and the Hughes' phenomenon is avoided, and, accordingly, the reduction of the considered components allows obtaining more precise thematic maps [53].

ICA decomposes the set of image bands into a linear combination of independent source signals [54]. ICA not only decorrelates second-order statistics, like PCA does but also decreases higher-order dependencies, generating a new set of components as independent as possible. It is an alternative approach to PCA for dimensionality reduction.

MNF Rotation is a linear transformation to segregate noise in the data and to reduce the computational requirements for subsequent processing [55]. This MNF transformation orders the new components so that they maximize the signal-to-noise ratio, rather than the information content [56]. MNF, apart from reducing the dimensionality, can be used to remove noise from data by performing a forward transform, by determining which bands contain the coherent information, by examining the images and eigenvalues, and then applying the inverse transform using only the appropriate bands, or adding as well the filtered or smoothed noisy bands.

- Texture Maps

We analyzed the inclusion of spatial information into the classification to improve the separability between the classes. There is a wide variety of texture measurements and, in this experiment, the parameters used were derived from the Gray-Level-Co-occurrence Matrix (GLCM) [57]. The main idea of the GLCM measurements is that the texture information contained in an image is based on the adjacency relationship between gray levels in the image. The relationship of the occurrence frequencies between pixel pairs can be calculated reliably for specific directions and distances between them.

After a preliminary analysis, we observed that most of the texture maps, derived from the co-occurrence matrix, were very similar (variance, entropy, dissimilarity, etc.). Therefore, to avoid the inclusion of redundant information, only the mean and variance parameters were finally used. Instead of applying the GLCM to each original band and, consequently, to have a large dataset of redundant texture information, the GLCM was only applied to the first component of the PCA and MNF transforms.

- Abundance Maps

In general, many pixels in the image represent a mixture of spectral signatures from different classes (e.g., seagrass and sand). In this context, mixing models estimate the contribution (abundance) of each endmember (pure class) to the total reflectance of each pixel. Linear mixing models provide adequate results in a large number of applications [27], and while non-linear models can be more precise, they require detailed information about the geometry and physical properties of the objects, which is not usually available and, thus, hinders its usefulness.

There are different strategies for the selection of pure pixels. In this work, as supervised classification techniques were applied, the training regions selected for each class were used to get the spectral signatures of each endmember.

After the application of unmixing techniques, an abundance map for each class to be discriminated was generated. Each map represents the percentage of contribution of a class to the total reflectance value of the pixel. In consequence, maps have values between 0 and 1 and, if a pixel is a mixture of different classes, the abundance of each class is obtained.

2.4.3. Classification

We used pixel and object-based supervised classifiers [51,58,59]. Specifically, Maximum Likelihood (ML), Spectral Angle Mapper (SAM), and Support Vector Machine (SVM) algorithms were assessed.

ML classification is one of the most common supervised techniques used with remotely sensed data [51]. ML considers that each class can be modeled as a normal distribution, allowing to describe that class by a Gaussian probability function, from its vector of means and covariance matrix. ML assigns the pixel to the class that maximizes the probability function.

The SAM classifier [52] compares the similarity between two spectra from their angular deviation, assuming that they form two vectors in an n-dimensional space (n being the number of bands). This algorithm measures the similarity as a function of the angle both vectors form in such a space.

SVM [60] is a machine learning algorithm to discriminate two classes by fitting an optimal hyperplane to separate the training samples of each class. The samples closest to the decision boundary are the so-called support vectors. SVM has been efficiently applied to classify both linear and nonlinearly separable classes applying a kernel function into a higher dimensional space, whose new data distribution allows better fitting of a linear hyperplane. Although deep learning approaches are becoming popular [61], they require large training datasets, and that is a great inconvenience in many operational applications. In addition, a recent assessment comparing advanced classification methods (SVMs, random forests, neural networks, deep convolutional neural networks, logistic regression-based techniques, and sparse representation-based classifiers) demonstrated that SVM is widely used because of its accuracy, stability, automation, and simplicity [26]. After a detailed review of SVM literature [18,62–64] and many tests conducted using the SVM algorithm with high-resolution imagery [65,66], the Gaussian radial basis function kernel $K(x_i, x_j)$ was selected for the SVM classification and their parameters were properly adjusted:

$$K(x_i, x_j) = e^{\left(-\frac{\|x_i - x_j^2\|}{2\sigma^2}\right)} \tag{6}$$

For the segmentation and merging steps in the object based classification (OBIA), after testing different combinations, using diverse features and number of bands, AHS channels 1 to 3 and WV-2 channel 1 were used. During the segmentation process, the image is divided into homogeneous regions according to several parameters (band weights, scale, color, shape, texture, etc.) defined by the operator, with the objective of creating the suitable object borders. We tested a multiresolution segmentation approach [67] and an algorithm based on watershed segmentation and merging stages [68]. An over-segmented image was preferred, and as soon as a suitable segmentation was attained, the classifier was applied.

Finally, the classification accuracy for each possible combination of input data was estimated using independent test regions of interest (ROIs) located in the image and computing the kappa coefficient, the confusion matrix, and its derived measures.

3. Results and Discussion

After the correction of each dataset (see Figure 4), three supervised classifiers were applied to different combinations of input data. All the analysis was performed for HS and MS imagery (AHS and WV-2, respectively) at the complex area of Maspalomas.

As seafloor reflectivity is very weak, and following the steps of Figure 4, precise preprocessing algorithms were applied to correct limitations in the sensor calibration, solar illumination geometry, viewing effects, as well as the atmospheric, sunglint, and water column disturbances. In this sense, geometric, radiometric and atmospheric corrections were performed. As specified, 6S was selected to model the atmosphere and to remove the absorption and scattering effects in the multispectral image [14], and ATCOR4 for the hyperspectral data [49]. This selection took into account the results of a previous validation campaign comparing real sea surface reflectance recorded by a field spectroradiometer (ADS Fieldspec 3) and the reflectance estimated for WV-2 data using different models. Next, deglinting algorithms were applied to eliminate the solar glint and whitecaps over both datasets. Finally, the seafloor albedo was generated applying the radiative transfer model described in Section 2.4.1. As shown in Figure 5, for a small area, the improvement is considerable, especially for the AHS data, as some areas were severely affected by de-sunglint.

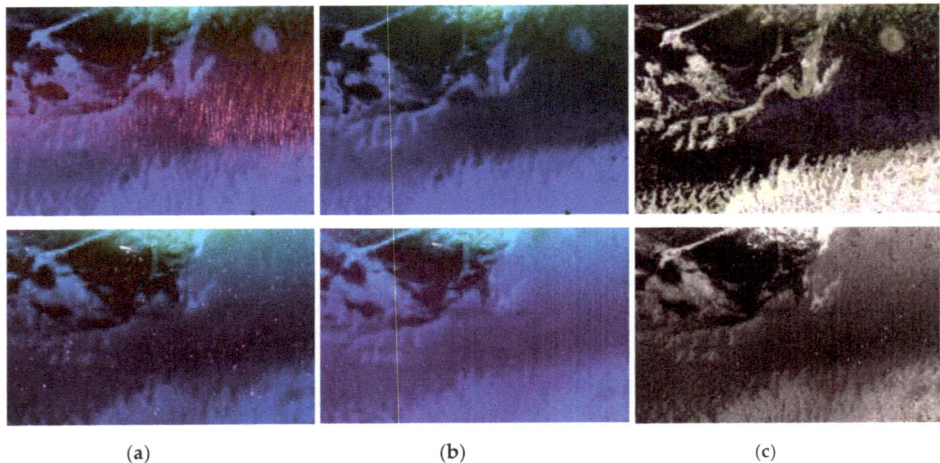

(a) (b) (c)

Figure 5. Color composite images, for Airborne Hyperspectral Scanner (AHS) (up) and Worldview-2 (WV-2) (down), after: (**a**) atmospheric correction; (**b**) sunglint removal; (**c**) water column correction.

A preliminary analysis was performed to find the most suitable corrected imagery to address the classification problem. Specifically, images obtained after the different pre-processing steps were assessed to identify the more reliable data source for the mapping production. The following thematic classes were considered: sand (yellow), rocks (brown) and *Cymodocea* or *Caulerpa* (green). Using the information from the ship transects and sampling sites (Figure 2c), sets of training and validation regions were generated including regions of each class at five-meter step depths from 5 to 20 m. Approximately, 3000 and 6000 pixels per class were selected for the training and test ROIs, respectively. The class pair separability (Jeffries-Matusita distance [52]) in the bands ranges from 1.218 to 1.693 for WV-2 and between 1.802 and 1.985 for AHS. These values corroborate a better discrimination capability of HS data as more spectral richness is available.

Table 2 presents the results of applying the three supervised classifiers to the data after the atmospheric, sunglint, and water column corrections. The same independent training and validation regions of interest were used in all the experiments. As expected, the airborne hyperspectral imagery allows a better classification than the satellite multispectral data (92.01% with respect to 88.66%). It can be appreciated that the best overall accuracy was achieved after the deglinting step. The water column removal did not improve the seafloor mapping, even after applying a complex radiative model. Even providing adjusted water IOPs and bathymetry values, the modeling of the background albedo by linear mixing of benthic classes in this complex area does not seem adequate for the subsequent classification. The very low reflectivity of the coastal bottom, which usually contributes less than 1% of the radiation observed by the sensor, produces errors in the adjustment of the abundances of the modeled pure benthic elements. Clearly, the model considered has to be further improved. As indicated, the water column modelling in coastal areas is complex and depends on the water quality parameters, as well as the bathymetry and the type of seabed. For this reason, this preprocessing is not always considered and some studies demonstrate that better results are not always achieved [18,20,40]. Finally, regarding the classification algorithm, SVM is the most appropriate approach for AHS but Maximum Likelihood works better with WV-2.

Figure 6 shows examples of seafloor maps generated for the AHS sensor, with SVM, and for the WV-2 data using the ML classifier. Comparing the results with the reference benthic map (Figure 2d) and the available video records from the ship transects, higher accuracy can be noted for AHS and using the imagery after the atmospheric and sunglint correction (middle row). Excessive amount of submerged vegetation is identified for WV-2 and some rocks incorrectly appear on the right side when these pixels should be labeled as vegetation.

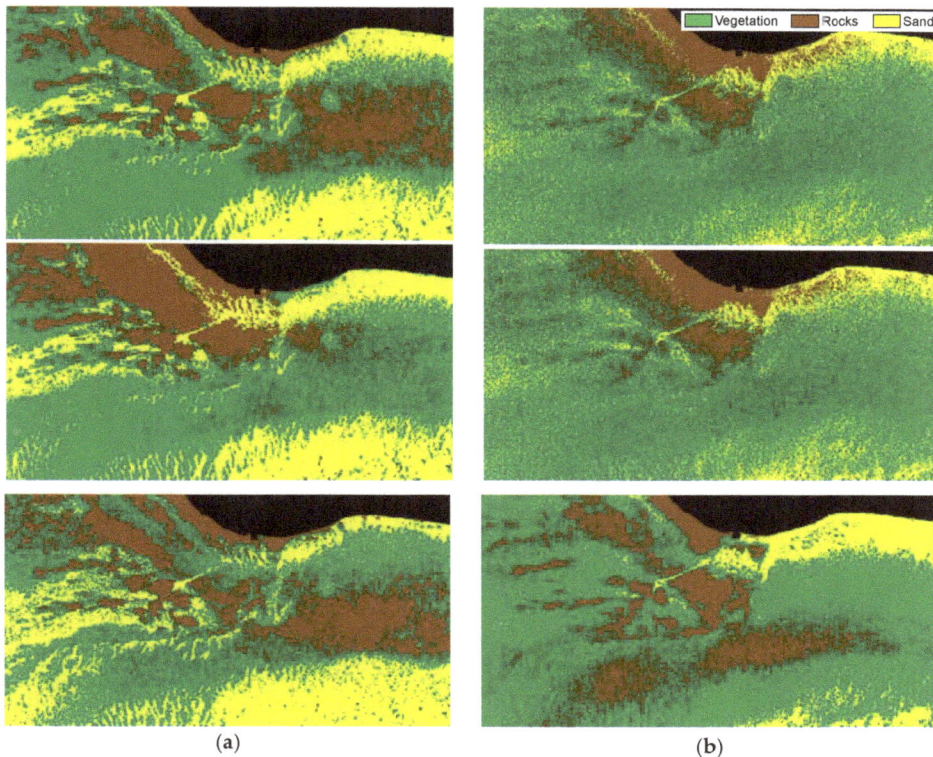

(a)   (b)

Figure 6. Seafloor maps after the atmospheric correction (up), sunglint removal (middle) and water column correction (down), for: (**a**) AHS using Support Vector Machine (SVM); (**b**) WV-2 using maximum likelihood (ML).

To improve the previous seabed cartography, a detailed feature extraction and classification assessment was only performed using the preprocessed data after the atmospheric and sunglint correction stages.

As stated in Section 2.4.2, to improve the benthic maps, additional information was obtained using feature extraction techniques. In particular, PCA, ICA and MNF were applied to the corrected spectral bands. In the analysis, the classifier performance was assessed including the complete new set of components after these transforms and, in addition, the best components were also tested discarding noisy bands. Figure 7 shows the first bands of each transform, as well as the original spectral bands as a reference (the remaining bands were not displayed as they are too noisy). Regarding the spectral channels, we can appreciate that only shorter wavelengths (first bands) can reach the seafloor and, in consequence, even dealing with hyperspectral data only a few channels are really valuable to map benthic habitats up to a depth of 20 m. On the other hand, PCA, ICA and MNF provide useful information in the first four components. The true color image and false color composites using the first three components are also included, and it is possible to check the worse behavior of ICA and the noise removal effect of MNF.

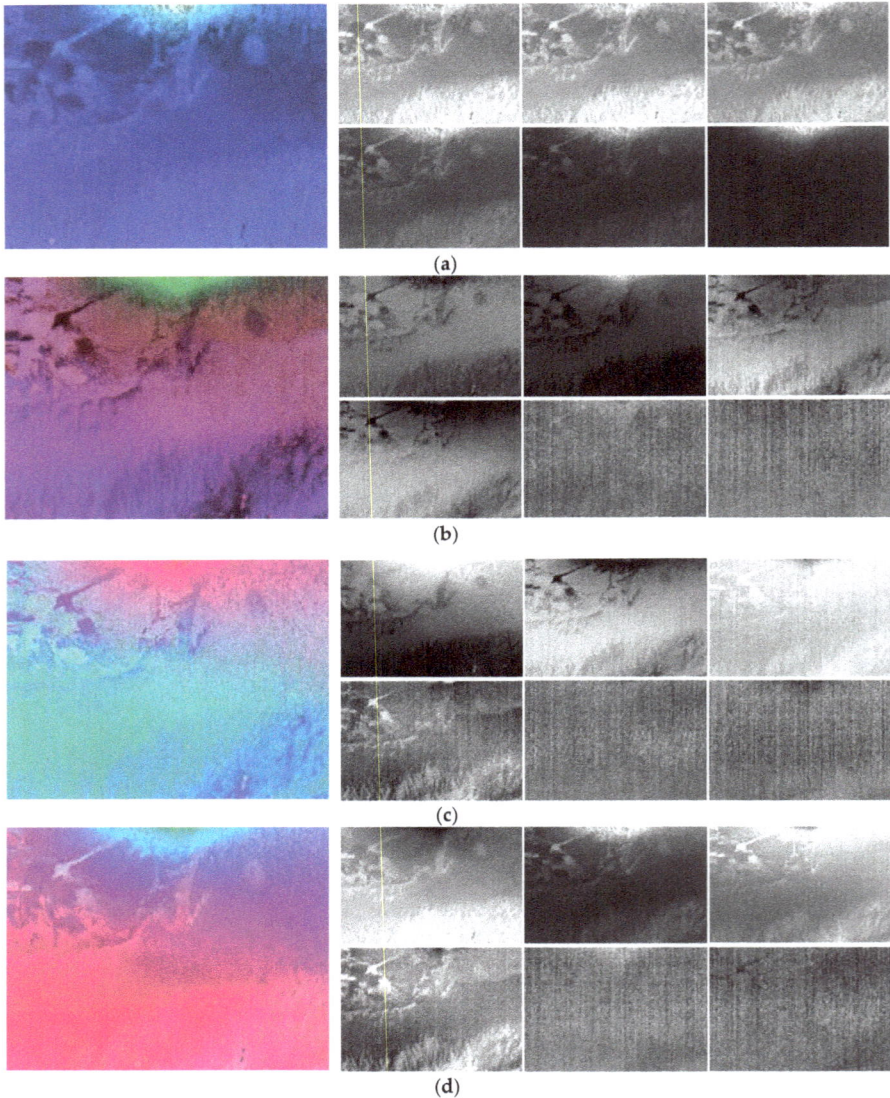

Figure 7. AHS color composite (RGB for the original bands and the 3 first components for the transforms) and the first bands for each transform: (**a**) original bands; (**b**) Principal Component Analysis (PCA); (**c**) Independent Component Analysis (ICA); (**d**) Minimum Noise Fraction (MNF).

Table 2. Overall accuracy (%) of Maximum Likelihood (ML), Support Vector Machine (SVM) and Spectral Angle Mapper (SAM) for Airborne Hyperspectral Scanner (AHS) and Worldview-2 (WV-2), and the different input combinations after each correction stage (AC: Atmospheric correction, SC: Sunglint Correction and WCC: Water Column Correction. Best accuracies marked in bold).

Sensor	Input	ML	SVM	SAM
AHS	AC	88.87	91.34	58.13
	AC+SC	**91.81**	**92.01**	58.35
	AC+SC+WCC	82.42	84.66	40.44
WV-2	AC	88.08	74.66	54.68
	AC+SC	88.66	80.63	**58.37**
	AC+SC+WCC	76.76	69.17	45.76

Additional textural parameters and abundance maps after unmixing were also inputted to the classifiers as auxiliary information.

Table 3 summarizes the AHS and WV-2 accuracy results of each classifier for the following input combinations:

- Spectral bands after atmospheric and sunglint corrections.
- Components after the application of three-dimensionality reduction techniques (PCA, ICA, and MNF). The complete dataset and a reduced number of bands or components were both tested.
- Abundance maps of each class after the application of linear unmixing techniques.
- Texture information (mean and variance) extracted from the first PCA/MNF component.

Table 3. Overall accuracy (%) of ML, SVM, and SAM for AHS and WV, and the different input combinations (Best accuracies marked in bold).

Sensor	Input	ML	SVM	SAM	Average
AHS	Bands (21)	91.81	92.01	58.35	80.72
	Bands 1-8 (8)	**93.77**	84.56	57.36	78.56
	PCA (21)	91.81	94.48	47.39	77.89
	PCA 1-4 (4)	93.25	92.39	50.54	78.73
	ICA (21)	91.81	85.57	29.58	68.99
	ICA 1-4 (4)	88.61	79.29	40.92	69.61
	MNF (21)	91.81	90.60	36.59	73.00
	MNF 1-4 (4)	93.57	90.11	39.08	74.25
	LU_ab (3)	90.63	73.33	48.57	70.84
	B+LU_ab (24)	92.20	90.04	58.35	80.20
	B+Text_PCA1 (23)	91.30	97.29	58.35	**82.31**
	B+Text_MNF1 (23)	92.30	85.90	58.35	78.85
	OBIA Bands (21)	85.70	**97.36**	**61.51**	81.52
	Average	**91.43**	88.69	49.61	76.58
WV-2	Bands (8)	88.66	80.63	58.37	75.89
	Bands 1-3 (3)	85.48	79.97	52.86	72.77
	PCA (8)	88.66	80.91	69.13	79.57
	PCA 1-4 (4)	87.60	82.27	68.79	79.55
	ICA (8)	88.66	70.90	58.26	72.61
	ICA 2-5 (4)	76.44	71.72	34.40	60.85
	MNF (8)	88.66	80.91	53.44	74.34
	MNF 1-4 (4)	88.34	80.52	53.31	74.06
	LU_ab(3)	87.70	70.16	74.39	77.42
	B+LU_ab (11)	**88.71**	81.16	**74.38**	81.42
	B+Text_PCA1 (10)	87.50	81.44	58.75	75.90
	B+Text_MNF1 (10)	88.41	78.45	57.57	74.81
	OBIA Bands (8)	82.27	**91.66**	64.55	79.49
	Average	**86.70**	79.28	59.86	75.28

Pixel-based classification was applied to the previous options and, finally, object-based classification was applied to the spectral bands.

With respect to the sensors, we can appreciate that AHS provides better accuracy than WV-2, as expected, mainly due to the availability of additional bands and a better radiometric resolution. Specifically, a major improvement is attained for SVM (mean accuracy increase of 9.5%) than for ML (4.7% average increase).

Concerning the classification algorithms, SAM did not work properly because, even being more insensitive to variations of the bathymetry, classes are spectrally overlapped, and only very few bands are useful due to the water column attenuation. SVM is the algorithm achieving the best accuracy, but the simpler and faster ML demonstrates good performance and, in many cases, better than SVM (average results in Table 3 confirm it). Actually, Figure 8 presents the comparative performance of both classifiers, and it can be appreciated that ML is more robust, providing more stable results regardless of the input information used or the number of bands considered. Specifically, the standard deviation (averaged for AHS and WV-2) of the overall accuracy for the different combinations is 2.9% for ML and 6.4% for SVM.

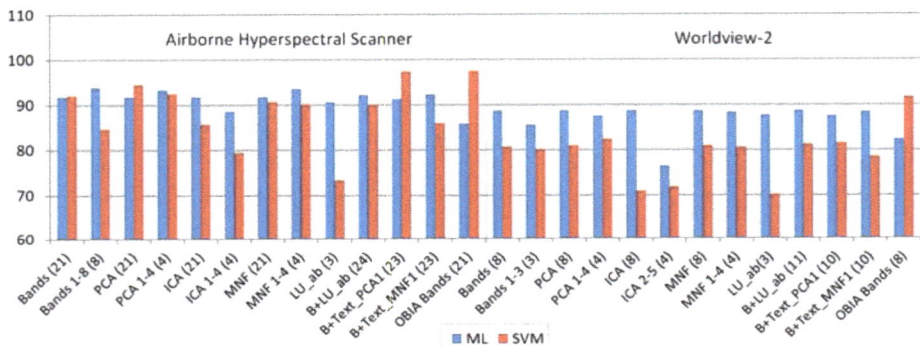

Figure 8. ML and SVM overall accuracies for both sensors and the different input combinations (LU: Linear Unmixing, B: Bands, Text: Texture. The number of input bands appears in parentheses).

PCA and MNF perform much better than ICA, but the improvement is, in general, negligible with respect to the original bands. Also, the reduction of the number of bands/components to avoid the Hughes' phenomenon is basically not increasing the classification accuracy except for ML and the hyperspectral data. The number of training pixels for each class is high enough (3000), and that could be a possible explanation.

The application of unmixing techniques before the classification did not improve the accuracy due to the small number of bands actually available. It can be appreciated the degraded performance of SVM when only the three abundances are considered in the classification scheme.

Finally, texture information is a feature that can be included in the final methodology as precision values in some circumstances increase the performance. Specifically, the improvement is more evident for SVM and using the texture information provided by the first component of PCA. For ML, texture generally does not provide a better accuracy of the benthic map.

It is important to highlight that results obtained by the object-based classification techniques (OBIA) are not always the best. Basically, OBIA only provides superior performance than pixel-based techniques for the SVM algorithm. However, results are quite dependent on the type of segmentation considered.

In general, the overall accuracies for ML and SVM are high as few classes are considered and the validation pixels chosen to numerically assess accuracy were selected in clear and central locations of each seabed type. In any case, the relative results between the different classifiers and input combinations displayed in Table 3 are reliable.

Figure 9 includes an example of the AHS and WV-2 segmentation for a specific area. AHS provides more detailed information and, in consequence, the number of objects increases.

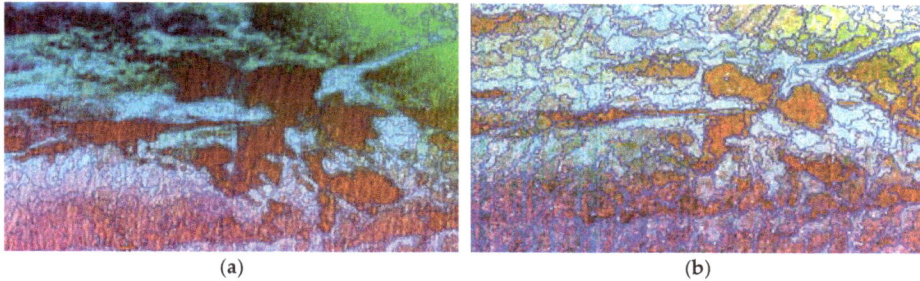

(a)   (b)

Figure 9. Objects after the segmentation for rocky and sandy areas of the seafloor: (a) AHS; (b) WV-2.

Figure 10 compares the best pixel-based seafloor maps generated by ML and SVM for the AHS image. A majority filter of 5 × 5 window size was applied to remove the salt and pepper effect. Both maps are very accurate, but ML overestimates vegetation (green) in some specific areas, while SVM the rocks (brown) in others. Finally, Figure 11 shows the best maps for each sensor obtained using the object-based classification with the SVM algorithm. Results are similar and, in general, match the available eco-cartographic map included in Figure 2d, except for the western side of the area. In any case, as indicated, this map was just considered a coarse reference and really ship transects T1 and T2 in Figure 2c demonstrate the existence of vegetation meadows in that area, in agreement with AHS and WV-2 maps. It is also important to highlight that a vulnerable and complex ecosystem was studied where the density of submerged green aquatic vegetation beds is quite low and, therefore, there is a considerable mixture of sand and plant contributions in each pixel of the image.

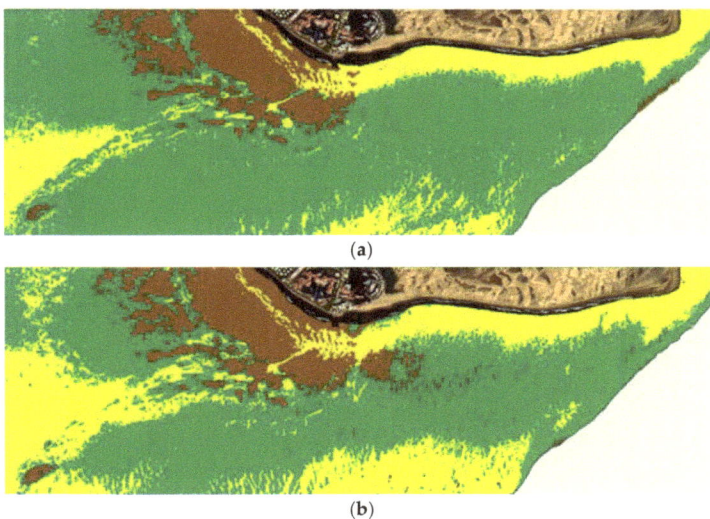

(a)

(b)

Figure 10. AHS pixel-based classification (majority 5 × 5): (a) ML using bands 1 to 8; (b) SVM using the bands plus the texture of the first principal component.

(a)

(b)

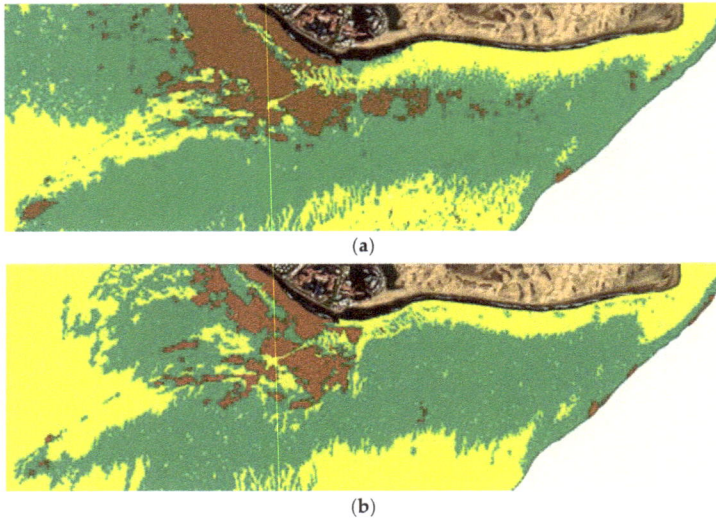

Figure 11. Object Based Classification (OBIA) (SVM): (**a**) AHS; (**b**) WV-2.

These methodologies will be shortly applied to generate precise benthic maps of natural protected ecosystems in other vulnerable coastal ecosystems. In addition, these will be applied to hyperspectral imagery recorded from drone platforms with the goal of discriminating between the different vegetation species.

4. Conclusions

A comprehensive analysis was performed to identify the best input dataset and to obtain a robust classification methodology to generate accurate benthic habitat maps. The assessment considered pixel-based and object-oriented classification methods in shallow waters using hyperspectral and multispectral data.

A vulnerable and complex coastal ecosystem was selected where the submerged green aquatic vegetation meadows to be classified are located at depths between 5 and 20 meters and have low density, implying the availability of very few spectral channels with information and a considerable mixing of spectral contributions in each image pixel.

Appropriate and improved atmospheric and sunglint correction techniques were applied to the HS and MS data. Next, a water radiative transfer model was also considered to remove the water column disturbances and to generate the seafloor albedo maps. A preliminary analysis was performed to identify the most suitable preprocessed imagery to be used for seabed classification. Three different supervised classifiers (maximum likelihood, support vector machines, and spectral angle mapper) were tested.

A detailed analysis of different feature extraction methods was performed with the goal to increase the discrimination capability of the classifiers. To our knowledge, the effect of three rotation transforms to generate benthic maps was assessed for the first time. Texture parameters were, as well, added to check whether spatial and context information improve classifications. Finally, the inclusion of abundance maps for each cover, obtained by the application of linear unmixing algorithms, was also considered but, given the small number of spectral bands actually reaching the seafloor, results were not fully satisfactory. The best results were produced by SVM and the OBIA approach. However, to generate benthic habitat maps, the simple ML has shown an excellent performance and superior stability and robustness than SVM (average overall accuracies over 3% and 7% for AHS and WV-2 data, respectively).

In summary, a robust methodology was identified, including the best correction techniques, feature extraction methods, and classification approaches, and it was successfully applied to multispectral and hyperspectral data in a complex coastal zone.

Author Contributions: Conceptualization, J.M. (Javier Marcello) and F.M.; Methodology, J.M. (Javier Marcello), F.E. and F.M.; Software, J.M. (Javier Martín); Validation, J.M. (Javier Marcello); Investigation, J.M. (Javier Marcello) and F.E.; Writing-Original Draft Preparation, J.M. (Javier Marcello); Writing-Review & Editing, F.E., J.M. (Javier Martín) and F.M.

Funding: This research was funded by Spanish *Agencia Estatal de Invetigación* (AEI) and by the *Fondo Europeo de Desarrollo Regional* (FEDER): project ARTEMISAT-2 (CTM2016-77733-R).

Acknowledgments: Authors want to acknowledge INTA (*Instituto Nacional de Técnica Aeroespacial*) for providing the AHS imagery. This work has been supported by the ARTEMISAT-2 (CTM2016-77733-R) project, funded by the Spanish AEI and FEDER funds.

Conflicts of Interest: The authors declare no conflict of interest.

References

1. Horning, E.; Robinson, J.; Sterling, E.; Turner, W.; Spector, S. *Remote Sensing for Ecology and Conservation*; Oxford University Press: New York, NY, USA, 2010; ISBN 978-0-19-921995-7.

2. Wang, Y. *Remote Sensing of Coastal Environments*; Taylor and Francis Series; CRC Press: Boca Raton, FL, USA, 2010; ISBN 978-1-42-009442-8.

3. Hossain, M.S.; Bujang, J.S.; Zakaria, M.H.; Hashim, M. The application of remote sensing to seagrass ecosystems: An overview and future research prospects. *Int. J. Remote Sens.* **2015**, *36*, 61–114. [CrossRef]

4. Lyons, M.; Phinn, S.; Roelfsema, C. Integrating Quickbird multi-spectral satellite and field data: Mapping bathymetry, seagrass cover, seagrass species and change in Moreton bay, Australia in 2004 and 2007. *Remote Sens.* **2011**, *3*, 42–64. [CrossRef]

5. Knudby, A.; Nordlund, L. Remote Sensing of Seagrasses in a Patchy Multi-Species Environment. *Int. J. Remote Sens.* **2011**, *32*, 2227–2244. [CrossRef]

6. Eugenio, F.; Marcello, J.; Martin, J. High-resolution maps of bathymetry and benthic habitats in shallow-water environments using multispectral remote sensing imagery. *IEEE Trans. Geosci. Remote Sens.* **2015**, *53*, 3539–3549. [CrossRef]

7. Rani, N.; Mandla, V.R.; Singh, T. Evaluation of atmospheric corrections on hyperspectral data with special reference to mineral mapping. *Geosci. Front.* **2017**, *8*, 797–808. [CrossRef]

8. Chavez, P.S. An improved dark-object subtraction technique for atmospheric scattering correction of multispectral data. *Remote Sens. Environ.* **1988**, *24*, 459–479. [CrossRef]

9. Chavez, P.S. Image-Based Atmospheric Corrections. Revisited and Improved. *Photogramm. Eng. Remote Sens.* **1996**, *62*, 1025–1036.

10. Bernstein, L.S.; Adler-Golden, S.M.; Jin, X.; Gregor, B.; Sundberg, R.L. Quick atmospheric correction (QUAC) code for VNIR-SWIR spectral imagery: Algorithm details. In Proceedings of the IEEE Workshop on Hyperspectral Image and Signal Processing: Evolution in Remote Sensing (WHISPERS), Shanghai, China, 4–7 June 2012.

11. Marcello, J.; Eugenio, F.; Perdomo, U.; Medina, A. Assessment of atmospheric algorithms to retrieve vegetation in natural protected areas using multispectral high resolution imagery. *Sensors* **2016**, *16*, 1624. [CrossRef] [PubMed]

12. Adler-Golden, S.M.; Matthew, M.W.; Bernstein, L.S.; Levine, R.Y.; Berk, A.; Richtsmeier, S.C.; Acharya, P.K.; Anderson, G.P.; Felde, G.; Gardner, J.; et al. Atmospheric Correction for Short-Wave Spectral Imagery based on MODTRAN4. In *Imaging Spectrometry V*; International Society for Optics and Photonics: Bellingham, WA, USA, 1999; Volume 3753.

13. Gao, B.-C.; Davis Curtiss, O.; Goetz, A.F.H. A review of atmospheric correction techniques for hyperspectral remote sensing of land surfaces and ocean colour. In Proceedings of the IEEE International Geoscience and Remote Sensing Symposium, Denver, CO, USA, 31 July–4 August 2006. [CrossRef]

14. Eugenio, F.; Marcello, J.; Martin, J.; Rodríguez-Esparragón, D. Benthic Habitat Mapping Using Multispectral High-Resolution Imagery: Evaluation of Shallow Water Atmospheric Correction Techniques. *Sensors* **2017**, *17*, 2639. [CrossRef] [PubMed]

15. Kay, S.; Hedley, J.; Lavender, S. Sun Glint Correction of High and Low Spatial Resolution Images of Aquatic Scenes: A Review of Methods for Visible and Near-Infrared Wavelengths. *Remote Sens.* **2009**, *1*, 697–730. [CrossRef]

16. Hedley, J.D.; Harborne, A.R.; Mumby, P.J. Simple and robust removal of sun glint for mapping shallow-water bentos. *Int. J. Remote Sens.* **2005**, *26*, 2107–2112. [CrossRef]

17. Lyzenga, D.R. Remote sensing of bottom reflectance and water attenuation parameters in shallow water using Aircraft and Landsat data. *Int. J. Remote Sens.* **1981**, *2*, 72–82. [CrossRef]

18. Traganos, D.; Reinartz, P. Mapping Mediterranean seagrasses with Sentinel-2 imagery. *Mar. Pollut. Bull.* **2017**. [CrossRef] [PubMed]

19. Manessa, M.D.M.; Haidar, M.; Budhiman, S.; Winarso, G.; Kanno, A.; Sagawa, T.; Sekine, M. Evaluating the performance of Lyzenga's water column correction in case-1 coral reef water using a simulated Wolrdview-2 imagery. In *Proceedings of IOP Conference Series: Earth and Environmental Science; IOP Publishing: Bristol;* IOP Publishing: Bristol, UK, 2016; Volume 47. [CrossRef]

20. Wicaksono, P. Improving the accuracy of Multispectral-based benthic habitats mapping using image rotations: The application of Principle Component Analysis and Independent Component Analysis. *Eur. J. Remote Sens.* **2016**, *49*, 433–463. [CrossRef]

21. Tamondong, A.M.; Blanco, A.C.; Fortes, M.D.; Nadaoka, K. Mapping of Seagrass and Other Bentic Habitat in Balinao, Pangasinan Using WorldView-2 Satellite Image. In Proceedings of the IEEE International Geoscience and Remote Sensing Symposium, Melbourne, Australia, 21–26 July 2013; pp. 1579–1582. [CrossRef]

22. Maritorena, S.; Morel, A.; Gentili, B. Diffuse reflectance of oceanic shallow waters: Influence of water depth and bottom albedo. *Limnol. Oceanogr.* **1994**, *39*, 1689–1703. [CrossRef]

23. Garcia, R.; Lee, A.; Hochberg, E.J. Hyperspectral Shallow-Water Remote Sensing with an Enhanced Benthic Classifier. *Remote Sens.* **2018**, *10*, 147. [CrossRef]

24. Loisel, H.; Stramski, D.; Dessailly, D.; Jamet, C.; Li, L.; Reynolds, R.A. An Inverse Model for Estimating the Optical Absorption and Backscattering Coefficients of Seawater From Remote-Sensing Reflectance Over a Broad Range of Oceanic and Coastal Marine Environments. *J. Geophys. Res. Oceans* **2018**, *123*, 2141–2171. [CrossRef]

25. Barnes, B.B.; Garcia, R.; Hu, C.; Lee, Z. Multi-band spectral matching inversion algorithm to derive water column properties in optically shallow waters: An optimization of parameterization. *Remote Sens. Environ.* **2018**, *204*, 424–438. [CrossRef]

26. Ghamisi, P.; Plaza, J.; Chen, Y.; Li, J.; Plaza, A. Advanced spectral classifiers for hyperspectral images. *IEEE Geosci. Remote Sens. Mag.* **2017**, *5*, 8–32. [CrossRef]

27. Bioucas, J.; Plaza, A.; Dobigeon, N.; Parente, M.; Du, Q.; Gader, P.; Chanussot, J. Hyperspectral unmixing overview: Geometrical, statistical, and sparse regression-based approaches. *IEEE J. Sel. Top. Appl. Earth Obs. Remote Sens.* **2012**, *5*, 354–379. [CrossRef]

28. Hedley, J.D.; Roelfsema, C.M.; Chollett, I.; Harborne, A.R.; Heron, S.F.; Weeks, S.; Skirving, W.J.; Strong, A.E.; Eakin, C.M.; Christensen, T.R.L.; et al. Remote Sensing of Coral Reefs for Monitoring and Management: A Review. *Remote Sens.* **2016**, *8*, 118. [CrossRef]

29. Roelfsema, C.; Kovacs, E.; Ortiz, J.C.; Wolff, N.H.; Callaghan, D.; Wettle, M.; Ronan, M.; Hamylton, S.M.; Mumby, P.J.; Phinn, S. Coral reef habitat mapping: A combination of object-based image analysis and ecological modelling. *Remote Sens. Environ.* **2018**, *208*, 27–41. [CrossRef]

30. Mohamed, H.; Nadaoka, K.; Nakamura, T. Assessment of Machine Learning Algorithms for Automatic Benthic Cover Monitoring and Mapping Using Towed Underwater Video Camera and High-Resolution Satellite Images. *Remote Sens.* **2018**, *10*, 773. [CrossRef]

31. Purkis, S.J. Remote Sensing Tropical Coral Reefs: The View from Above. *Annu. Rev. Mar. Sci.* **2018**, *10*, 149–168. [CrossRef] [PubMed]

32. Petit, T.; Bajjouk, T.; Mouquet, P.; Rochette, S.; Vozel, B.; Delacourt, C. Hyperspectral remote sensing of coral reefs by semi-analytical model inversion—Comparison of different inversion setups. *Remote Sens. Environ.* **2017**, *190*, 348–365. [CrossRef]

33. Zhang, C. Applying data fusion techniques for benthic habitat mapping and monitoring in a coral reef ecosystem. *ISPRS J. Photogramm. Remote Sens.* **2015**, *104*, 213–223. [CrossRef]
34. Leiper, I.A.; Phinn, S.R.; Roelfsema, C.M.; Joyce, K.E.; Dekker, A.G. Mapping Coral Reef Benthos, Substrates, and Bathymetry, Using Compact Airborne Spectrographic Imager (CASI) Data. *Remote Sens.* **2014**, *6*, 6423–6445. [CrossRef]
35. Roelfsema, C.; Kovacs, E.M.; Saunders, M.I.; Phinn, S.; Lyons, M.; Maxwell, P. Challenges of remote sensing for quantifying changes in large complex seagrass environments. *Estuar. Coast. Shelf Sci.* **2013**, *133*, 161–171. [CrossRef]
36. Baumstark, R.; Duffey, R.; Pu, R. Mapping seagrass and colonized hard bottom in Springs Coast, Florida using WorldView-2 satellite imagery. *Estuar. Coast. Shelf Sci.* **2016**, *181*, 83–92. [CrossRef]
37. Koedsin, W.; Intararuang, W.; Ritchie, R.J.; Huete, A. An Integrated Field and Remote Sensing Method for Mapping Seagrass Species, Cover, and Biomass in Southern Thailand. *Remote Sens.* **2016**, *8*, 292. [CrossRef]
38. Uhrin, A.V.; Townsend, P.A. Improved seagrass mapping using linear spectral unmixing of aerial photographs. *Estuar. Coast. Shelf Sci.* **2016**, *171*, 11–22. [CrossRef]
39. Valle, M.; Palà, V.; Lafon, V.; Dehouck, A.; Garmendia, J.M.; Borja, A.; Chust, G. Mapping estuarine habitats using airborne hyperspectral imagery, with special focus on seagrass meadows. *Estuar. Coast. Shelf Sci.* **2015**, *164*, 433–442. [CrossRef]
40. Zhang, C.; Selch, D.; Xie, Z.; Roberts, C.; Cooper, H.; Chen, G. Object-based benthic habitat mapping in the Florida Keys from hyperspectral imagery. *Estuar. Coast. Shelf Sci.* **2013**, *134*, 88–97. [CrossRef]
41. De Miguel, E.; Fernández-Renau, A.; Prado, E.; Jiménez, M.; Gutiérrez, O.; Linés, C.; Gómez, J.; Martín, A.I.; Muñoz, F. A review of INTA AHS PAF. *EARSeL eProc.* **2014**, *13*, 20–29.
42. Gesplan. *Plan Regional de Ordenación de la Acuicultura de Canarias. Tomo I: Memoria de Información del Medio Natural Terrestre y Marino. Plano de Sustratos de Gran Canaria*; Gobierno de Canarias: Las Palmas de Gran Canaria, Spain, 2013; pp. 1–344.
43. Digitalglobe. Accuracy of Worldview Products. White Paper. 2016. Available online: https://dg-cms-uploads-production.s3.amazonaws.com/uploads/document/file/38/DG_ACCURACY_WP_V3.pdf (accessed on 1 June 2018).
44. Vermote, E.; Tanré, D.; Deuzé, J.L.; Herman, M.; Morcrette, J.J.; Kotchenova, S.Y. *Second Simulation of a Satellite Signal in the Solar Spectrum—Vector (6SV)*; 6S User Guide Version 3; NASA Goddard Space Flight Center: Greenbelt, MD, USA, 2006.
45. Kotchenova, S.Y.; Vermote, E.F.; Matarrese, R.; Klemm, F.J. Validation of vector version of 6s radiative transfer code for atmospheric correction of satellite data. Parth radiance. *Appl. Opt.* **2006**, *45*, 6762–6774. [CrossRef] [PubMed]
46. Martin, J.; Eugenio, F.; Marcello, J.; Medina, A. Automatic sunglint removal of multispectral WV-2 imagery for retrieving coastal shallow water parameters. *Remote Sens.* **2016**, *8*, 37. [CrossRef]
47. Lee, Z.; Carder, K.L.; Mobley, C.D.; Steward, R.G.; Patch, J.S. Hyperspectral remote sensing for shallow waters: 2. Deriving bottom depths and water properties by optimization. *Appl. Opt.* **1999**, *38*, 3831–3843. [CrossRef] [PubMed]
48. Heylen, R.; Burazerović, D.; Scheunders, P. Fully constrained least squares spectral unmixing by simplex projection. *IEEE Trans. Geosci. Remote Sens.* **2011**, *49*, 4112–4122. [CrossRef]
49. Richter, R.; Schläpfer, D. Geo-atmospheric processing of airborne imaging spectrometry data. Part 2: Atmospheric/Topographic correction. *Int. J. Remote Sens.* **2002**, *23*, 2631–2649. [CrossRef]
50. Hughes, G. On the mean accuracy of statistical pattern recognizers. *IEEE Trans. Inf. Theory* **1968**, *14*, 55–63. [CrossRef]
51. Ibarrola-Ulzurrun, E.; Marcello, J.; Gonzalo-Martin, C. Assessment of Component Selection Strategies in Hyperspectral Imagery. *Entropy* **2017**, *19*, 666. [CrossRef]
52. Richards, J.A. *Remote Sensing Digital Image Analysis*, 5th ed.; Springer: Berlin, Germany, 2013; ISBN 978-3-54-029711-6.
53. Benediktsson, J.A.; Ghamisi, P. *Spectral-Spatial Classification of Hyperspectral Remote Sensing Images*; Artech House: Boston, MA, USA, 2015; ISBN 978-1-60-807812-7.
54. Li, C.; Yin, J.; Zhao, J. Using improved ICA method for hyperspectral data classification. *Arab. J. Sci. Eng.* **2014**, *39*, 181–189. [CrossRef]

55. Green, A.A.; Berman, M.; Switzer, P.; Craig, M.D. A transformation for ordering multispectral data in terms of image quality with implications for noise removal. *IEEE Trans. Geosci. Remote Sens.* **1988**, *26*, 65–74. [CrossRef]

56. Luo, G.; Chen, G.; Tian, L.; Qin, K.; Qian, S.E. Minimum noise fraction versus principal component analysis as a preprocessing step for hyperspectral imagery denoising. *Can. J. Remote Sens.* **2016**, *42*, 106–116. [CrossRef]

57. Haralick, R.; Shanmugam, K.; Dinstein, I. Textural features for image classification. *IEEE Trans. Syst. Man Cybern.* **1973**, *3*, 610–621. [CrossRef]

58. Tso, B.; Mather, P.M. *Classification Methods for Remotely Sensed Data*; Taylor and Francis Inc.: New York, NY, USA, 2009; ISBN 978-1-42-009072-7.

59. Li, M.; Zhang, S.; Zhang, B.; Li, S.; Wu, C. A review of remote sensing image classification technique: The role of spatio-contextual information. *Eur. J. Remote Sens.* **2014**, *47*, 389–411. [CrossRef]

60. Vapnik, V. *The Nature of Statistical Learning Theory*, 2nd ed.; Springer: Berlin, Germany, 1999; ISBN 978-1-47-573264-1.

61. Yu, X.; Wu, X.; Luo, C.; Ren, P. Deep learning in remote sensing scene classification: A data augmentation enhanced convolutional neural network framework. *GISci. Remote Sens.* **2017**, *54*, 741–758. [CrossRef]

62. Maulik, U.; Chakraborty, D. Remote Sensing Image Classification: A survey of support-vector-machine-based advanced techniques. *IEEE Geosci. Remote Sens. Mag.* **2017**, *5*, 33–52. [CrossRef]

63. Mountrakis, G.; Im, J.; Ogole, C. Support vector machines in remote sensing: A review. *ISPRS J. Photogramm. Remote Sens.* **2011**, *66*, 247–259. [CrossRef]

64. Pal, M.; Mather, P.M. Support vector machines for classification in remote sensing. *Int. J. Remote Sens.* **2005**, *26*, 1007–1011. [CrossRef]

65. Marcello, J.; Eugenio, F.; Marqués, F.; Martín, J. Precise classification of coastal benthic habitats using high resolution Worldview-2 imagery. In Proceedings of the IEEE Geoscience and Remote Sensing Symposium (IGARSS), Milan, Italy, 26–31 July 2015.

66. Ibarrola-Ulzurrun, E.; Gonzalo-Martín, C.; Marcello, J. Vulnerable land ecosystems classification using spatial context and spectral indices. In *Earth Resources and Environmental Remote Sensing/GIS Applications VIII, Proceedings of the SPIE Remote Sensing, Warsaw, Poland, 11–14 September 2017*; SPIE: Bellingham, WA, USA.

67. Baatz, M.; Schape, A. Multiresolution segmentation an optimization approach for high quality multi scale image segmentation. In Proceedings of the Angewandte Geographische Informations Verarbeitung XII, Wichmann Verlag, Karlsruhe, Germany, 30 June 2000.

68. Jin, X. Segmentation-Based Image Processing System. U.S. Patent 8,260,048, 4 September 2012.

remote sensing

MDPI

Article

Two-Step Urban Water Index (TSUWI): A New Technique for High-Resolution Mapping of Urban Surface Water

Wei Wu [1,2], Qiangzi Li [1,*], Yuan Zhang [1,*], Xin Du [1] and Hongyan Wang [1]

[1] Institute of Remote Sensing and Digital Earth, Chinese Academy of Sciences, No. 20 Datun Road, Chaoyang District, Beijing 100101, China; wuwei@radi.ac.cn (W.W.); duxin@radi.ac.cn (X.D.); wanghy@radi.ac.cn (H.W.)
[2] University of Chinese Academy of Sciences, Beijing 100049, China
*   Correspondence: liqz@radi.ac.cn (Q.L.); zhangyuan@radi.ac.cn (Y.Z.); Tel.: +86-10-6485-5094 (Q.L.)

Received: 19 September 2018; Accepted: 26 October 2018; Published: 29 October 2018

Abstract: Urban surface water mapping is essential for studying its role in urban ecosystems and local microclimates. However, fast and accurate extraction of urban water remains a great challenge due to the limitations of conventional water indexes and the presence of shadows. Therefore, we proposed a new urban water mapping technique named the Two-Step Urban Water Index (TSUWI), which combines an Urban Water Index (UWI) and an Urban Shadow Index (USI). These two subindexes were established based on spectral analysis and linear Support Vector Machine (SVM) training of pure pixels from eight training sites across China. The performance of the TSUWI was compared with that of the Normalized Difference Water Index (NDWI), High Resolution Water Index (HRWI) and SVM classifier at twelve test sites. The results showed that this method consistently achieved good performance with a mean Kappa Coefficient (KC) of 0.97 and a mean total error (TE) of 5.82%. Overall, classification accuracy of TSUWI was significantly higher than that of the NDWI, HRWI, and SVM (*p*-value < 0.01). At most test sites, TSUWI improved accuracy by decreasing the TEs by more than 45% compared to NDWI and HRWI, and by more than 15% compared to SVM. In addition, both UWI and USI were shown to have more stable optimal thresholds that are close to 0 and maintain better performance near their optimum thresholds. Therefore, TSUWI can be used as a simple yet robust method for urban water mapping with high accuracy.

Keywords: urban water mapping; water index; shadow detection; threshold stability

---

1. Introduction

Urban surface water such as rivers, lakes, reservoirs, and ponds, exerts a significant influence on urban ecosystem services [1] and local microclimates [2]. As a consequence of Land Use/Land Cover (LULC) and environmental changes and natural hazards, variations in urban surface water, may result in a series of ecological, climate, health, and socioeconomic problems, such as water supply shortages [3,4], biodiversity losses [5], aggravation of the urban heat island effect [6,7], and even outbreaks of waterborne infectious diseases [8]. These problems tend to be more prominent in cities with rapid urbanization [9–12]. With the development of urbanization, the urban space has been expanded, leading to the shrinking of water bodies. Meanwhile, frequent human activities may lead to the deterioration of water quality towards being turbid, stink, or black. Therefore, timely and accurate mapping of urban surface water is crucial for urban planning and disaster assessments [13,14].

Remote sensing techniques, with their advantages of large area coverage, integration, speed, and periodicity, have been widely used to delineate surface water and monitor surface water dynamics. Various methods have been proposed to identify surface water bodies, which can be divided into four types: thematic classification [15,16], spectral unmixing [17–19], single band thresholding [20,21], and spectral water index [22–24]. The last is the most widely used, due to its ease of use, relatively high mapping accuracy, and low computational expense [25]. Over the past few decades, many water indexes have been presented in the literature. McFeeters [23] proposed the first water index called the Normalized Difference Water Index (NDWI) with a default threshold of 0, which utilized the reflectance difference between water and vegetation and soil in the red and near-infrared (NIR) bands. To suppress the signals from buildings, Xu [24] replaced the NIR band with the shortwave infrared (SWIR) band in the NDWI formulation, creating the Modified Normalized Difference Water Index (MNDWI). Although the MNDWI shows high accuracy, it was still unable to suppress shadows. Therefore, Feyisa et al. [22] proposed an automated water extraction index (AWEI) and that has been demonstrated to be effective in different environments, particularly in mountainous areas with deep shadows.

However, mapping small and narrow urban surface water bodies requires the adoption of high-resolution images [26]. Most high-resolution images, such as Gaofen-2 (GF-2), IKONOS, RapidEye and Ziyuan-3 (ZY-3), have only visible and NIR bands and lack the bands necessary to compute most of the conventional water indexes designed for low- or medium-resolution images. This condition necessitates the development of a water index for fast and accurate mapping of urban surface water. Compared to water detection in rural areas, the complex urban setting poses great challenges for mapping water bodies with high-resolution imagery. Because urban water is greatly affected by human activities, it may contain high amounts of pollutants, such as suspended solids, high levels of nutrients, heavy metals, and sewage runoff. These pollutants make the spectral properties of urban surface water quite different from those of unpolluted water [27]. In addition, abundant shadows cast by tall buildings and trees are present in remotely sensed images in urban areas. Due to the similarity in spectral patterns between shadows and water, it is difficult to remove shadow noise from urban water maps.

NDWI and High Resolution Water Index (HRWI) are two water indexes commonly used for urban water extraction. However, NDWI tends to misclassify buildings and shadows as water when applied to high-resolution images [28]. HRWI is a water index that was proposed by Yao et al. [29]. HRWI also exhibits limited ability to distinguish shadows and should be applied together with a shadow detection model. Summaries of the aforementioned water indexes are presented in Table S1. To the best of our knowledge, a water index for high-resolution images with four standard bands that can effectively suppress all non-water pixels and extract urban water with high accuracy has not been proposed.

In this paper, a new urban water index, called the Two-Step Urban Water Index (TSUWI), was proposed to map urban water bodies from high-resolution imagery based on the full utilization of the spectral information from different objects in the visible and NIR bands. As one simple water index may not address all issues at the same time, the TSUWI combines two subindexes of an urban water index (UWI) and an urban shadow index (USI). The TSUWI proposed in this paper is expected to improve the accuracy of urban water mapping by suppressing the signal from artificial construction and shadows, and to be robust under various water conditions with stable thresholds and high accuracy.

2. Study Areas and Materials

2.1. Study Sites

Given the complex terrain and distinct climates over China, a total of twelve study sites with variable environmental conditions and diverse types of water bodies were selected to establish and validate the new urban water index. According to different application goals, these study sites were divided into two types: training sites and test sites.

Training sites were used for pure pixel selection to formulate the new urban water extraction index. Given that the features of urban surfaces are spatially variant, eight training sites characterized by different surface water types, climates, topographies, and urban development levels were therefore deliberately chosen. The sites were selected from eight cities around China: Chengdu, Guangzhou, Nanchang, Qingdao, Shanghai, Aksu, Lhasa and Shigatse (Figure 1). While covering all major challenging issues affecting the accuracy of urban water extraction, such as shadows, low-albedo buildings and black soil, these sites span across water bodies of different depths, turbidity levels, chemical compositions, and surface appearances, including rivers, lakes, reservoirs, pools and seas. The sites in Chengdu, Guangzhou, Nanchang, Qingdao and Shanghai are located in eastern China within a dense water network. The water bodies in Chengdu, Guangzhou, and Nanchang are typical inland waters and consist of several main rivers surrounded by many small ponds, regular or irregular lakes and artificial reservoirs. The main rivers in Chengdu are relatively narrow, while those in Nanchang are turbid with large amounts of fluid mud. Qingdao and Shanghai are coastal port cities containing both marine and inland water bodies. The site in Qingdao has harbors, tidal creeks, and a portion of the sea, and the main water body in Shanghai is a complicated mixture of suspended sediment and intrusive seawater because of its special location in the turbidity maximum zone of the Yangtze River Estuary. Aksu, Lhasa and Shigatse were selected to represent water bodies in China's western cities, which tend to be rare and shallow due to the arid or semiarid climates. The site in Aksu primarily has a narrow river and a large shallow lake. Both sites in Lhasa and Shigatse have a narrow river, but part of the river in Shigatse is semi-dry.

Test sites were selected to assess the accuracy and robustness of the TSUWI. The whole image at each test site were used to delineate the true water body boundaries for the assessments. Considering that the pure pixels sampled for the development of the TSUWI covered only an extremely small portion of the image, the aforementioned eight training sites were also used as test sites. To enhance the reliability of the assessments, another four test sites located in Fuzhou, Haerbin, Yinchuan, and Dongguan were added to constitute the set of test sites. Fuzhou, Haerbin, and Dongguan are located in Eastern China where there are plenty of water bodies, while Yinchuan is located in Western China where water bodies are scarce. These twelve test sites are distributed across different regions of China (Figure 1). The wide range of variability in water types and environmental conditions of the twelve test sites imposes great difficulty for accurately mapping urban water bodies, which makes these sites ideal test sites. Table 1 shows the detailed descriptions of these twelve study sites.

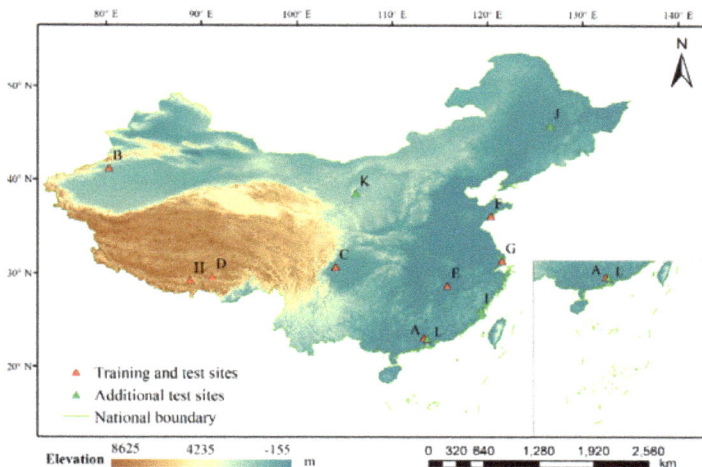

Figure 1. *Cont.*

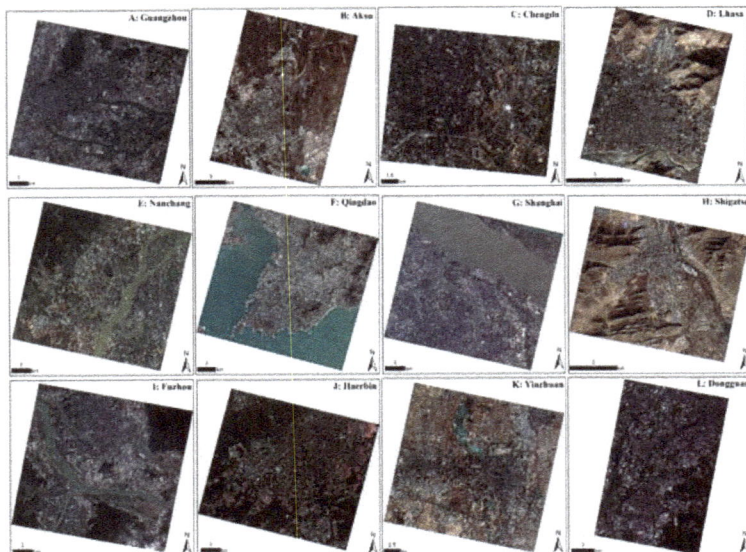

Figure 1. Locations of the twelve study sites and Gaofen-2 (GF-2) scenes in true color composites of red, green, and blue bands.

Table 1. Characteristics of the twelve study sites.

Study Sites	Main Water Types/Features	Location	Area (km²)	Topography	Climate
Guangzhou	River [c,t]/lake [c]/pond [c,t,e]	23.11°N, 113.33°E	574.9	Plain/hills	Subtropical oceanic monsoon
Aksu	River [c,n]/lake [s,t]	41.19°N, 80.31°E	171.7	Basin	Temperate continental arid
Chengdu	River [c,t,n]/pond [c,t]/reservoir [c]	30.64°N, 104.11°E	521.9	Plain	Subtropical humid monsoon
Lhasa	River [c,t]/pond [c]	29.67°N, 91.16°E	54.0	Mountain/plateau	Plateau mountain
Nanchang	River [t,w]/lake [c,t]/pond [c,t,e]/reservoir [c,t]	28.69°N, 115.82°E	516.2	Plain/hills	Subtropical humid monsoon
Qingdao	Sea [c]/tidal creek [t]/river [n,c,t]/pond [c,t]	36.13°N, 120.38°E	561.8	Plain/hills	Warm temperate monsoon
Shanghai	Harbor [w,t]/river [c,t]/lake [c,t]/pond [c,t]	31.36°N, 121.55°E	504.9	Plain	Subtropical oceanic monsoon
Shigatse	River [c,t]/lake [c,t]/pond [c,t]	29.25°N, 88.89°E	84.0	Mountain/plateau	Plateau mountain
Fuzhou	River [c,t,w]/lake [c,t]/pond [c,t]/reservoir [c]	26.01°N, 119.30°E	522.2	Basin/hills	Subtropical oceanic monsoon
Haerbin	River [t,w]/lake [c,t,e]/ponds [c,t,e]	45.71°N, 126.60°E	468.3	Plain	Temperate monsoon
Yinchuan	Lake [c,t]/river [c,n]/pond [c,t]	38.48°N, 106.24°E	525.1	Plain	Temperate continental
Dongguan	River [c,t]/lake [c,t]/pond [c,t,e]/reservoir [c]	22.98°N, 113.68°E	288.9	Plain/hills	Subtropical oceanic monsoon

Note: [c] means clear water, [t] means turbid water, [e] means eutrophic water, [n] means narrow water, and [w] means wide water.

2.2. GF-2 Imagery

As a civil land observation satellite with currently the highest resolution in China, GF-2 is equipped with two multispectral scanners and characterized by submeter spatial resolution, high positioning accuracy and rapid posture maneuverability. With a revisit cycle of 5 days and a swath width of 45 km, GF-2 is a critical data source for urban remote sensing applications. The basic characteristics of GF-2 satellite is shown in Table 2.

Twelve GF-2 images were ordered from the website of the China Center for Resources Satellite Data and Application (available at http://cresda.com/CN/index.shtml). One image was used for each site. When choosing images, all available data were inspected to avoid the influence of clouds on the water bodies. The GF-2 images contain one panchromatic band and four multispectral bands (comprised of blue, green, red, and near-infrared bands). All images were Level 1A products, which contain enough information for further image preprocessing, such as radiometric correction and geometric correction. The detailed information on these GF-2 images is presented in Table 3.

Table 2. Basic characteristics of GF-2 satellite. NIR: near-infrared.

Spectral Bands	Wavelength (μm)	Resolution (Nadir Point)	Swath Width	Side-Swing Ability	Revisit Cycle
Panchromatic	0.45–0.90	0.8 m			
Band1—Blue	0.45–0.52				
Band2—Green	0.52–0.59	3.2 m	45 km	±35°	5 days
Band3—Red	0.63–0.69				
Band4—NIR	0.77–0.89				

Table 3. Characteristics of the twelve study sites.

Study Sites	GF-2 Scene			Supplementary Reference Data
	Acquisition Date	Path	Row	
Guangzhou	4 November 2016	1016	185	Google Earth™ image acquired on 5 October/5 November/9 December 2016, ©Digital Globe
Aksu	29 February 2016	102	135	Google Earth™ image acquired on 17 April 2016, ©Digital Globe
Chengdu	21 March 2015	27	164	Google Earth™ image acquired on 11 February/21 March 2015, ©Digital Globe, CNES/Airbus
Lhasa	3 December 2016	63	167	Google Earth™ image acquired on 3 December 2016, ©Digital Globe
Nanchang	28 November 2016	1013	170	Google Earth™ image acquired on 24 September/1 December/31 December 2016, ©Digital Globe
Qingdao	16 February 2016	1006	149	Google Earth™ image acquired on 16 January 2016, ©Digital Globe
Shanghai	2 January 2015	999	162	Google Earth™ image acquired on 18 December 2014, and 24 January/18 February 2015, ©Digital Globe
Shigatse	12 January 2017	69	168	Google Earth™ image acquired on 21 May 2018, ©CNES/Airbus
Fuzhou	7 December 2016	1001	177	Google Earth™ image acquired on 21 January/1 March 2017, © Digital Globe, CNES/Airbus
Haerbin	10 September 2015	997	122	Google Earth™ image acquired on 19 June/9 July/16 September/24 October 2015, © Digital Globe
Yinchuan	4 January 2017	27	142	Google Earth™ image acquired on 30 October/2 November/ 13 November 2016, and 21 January 2017, ©Digital Globe
Dongguan	15 February 2017	1015	186	Google Earth™ image acquired on 12 February 2017, ©Digital Globe

2.3. Reference Data

The true water body boundaries of all twelve study sites were manually digitized on-screen to evaluate the accuracies of the extracted water surface. In consideration of the inevitable bias caused by the time span between GF-2 images and other data sources, the digitization was implemented on the GF-2 images, which was sharpened by the panchromatic band with higher spatial resolution. Water conditions are sometimes extremely complicated, and small urban water bodies adjacent to tall buildings, especially dark and quadrangle-shaped buildings, could easily be confused with building shadows due to their similar spectra and morphology. Google Earth^{TM} images, which were acquired on dates as close as possible to the GF-2 images, were supplied to assist with the visual interpretation by providing another different overview of the urban surfaces. Table 3 lists the detailed information about these supplementary reference data.

2.4. Image Preprocessing

The GF-2 images in the form of raw digital number (DN) values were calibrated to the top of atmosphere (TOA) reflectance via radiometric calibration. Atmosphere scattering and absorption could bring unexpected spectral bias, leading to significantly reduced image quality. As a consequence, an atmospheric correction was applied to the obtained TOA reflectance using the Fast Line-of-Sight Atmospheric Analysis of Spectral Hypercubes (FLAASH) module in ENVI v.5.3 [30]. Relative atmospheric parameters were determined via a lookup table [31], which is based on a seasonal-latitude surface temperature model. The initial visibility applied in this procedure was estimated using the aerosol optical depth (AOD) obtained from MODIS Terra aerosol products of version 6 [22].

Due to the effects of sensor tilt and terrain relief, orthorectification was carefully undertaken with the GF-2 images after atmospheric correction. Given that each GF-2 image contains Rational Polynomial Coefficient (RPC) information in the header file, this procedure was performed using the RPC orthorectification workflow in ENVI v5.3. To improve the precision of the geometric correction, ground control points (GPCs) with average root mean square (RMS) values of no more than 0.5 pixels were selected for each image to refine the RPCs, and the high-resolution digital elevation data (30 m) of ASTER GDEM v.2 [32] were supplied. The output pixel sizes for panchromatic band and multispectral bands are 1 m and 4 m, respectively. Afterwards, all orthorectified images were clipped to achieve higher urban water percentages and lower visual interpretation costs.

3. Methodology

3.1. Pure Pixel Selection

A dataset of pure pixel reflectance values of nine major urban land cover types was sampled from the GF-2 multispectral images of eight training sites. The urban land cover types are bright soil, black soil, bright built, dark built, vegetation, asphalt, light shadow, dark shadow, and water. These pure pixels were utilized to examine the spectral differences between water and other land cover types and act as samples fed into a linear Support Vector Machine (SVM) model for index coefficient training, aiming to design an urban water index that accurately distinguishes water from other urban surfaces. This new index is expected to be robust against various water type changes within complex urban environments. Therefore, as discussed in Section 2.1, eight training sites located in different cities across China, including a wide range of water types and all the major interference factors, were used to extract pure pixels.

Pure pixels were generated by manual digitization of the GF-2 multispectral images with the assistance of Google Earth^{TM} images. Pure pixels were generally extracted from the center of a land cover patch to ensure their purity and were evenly distributed across each image to achieve high representativeness. For each of the eight training sites, 120 pixels were extracted for each land cover type, leading to 1080 pure pixels for each training site and 8640 for all sites.

3.2. Spectral Features of Water and Non-Water Types

Water indexes are typically mathematic combinations of several spectral features, aiming to enhance the contrast between water and non-water pixels [22]. Given the distinct separability of each feature for various land cover types, an optimal feature combination is required for an effective water index. Therefore, comprehensive analyses of the spectral characteristics of water and other land cover types are prerequisites for identifying the optimal feature combination to be applied in the formulation. Statistical distributions of pure pixel reflectance values of nine land cover types for blue, green, red, and NIR bands were obtained and displayed in Figure 2a–d. The results showed that the original bands generally demonstrated good performance in discriminating water from non-water types. However, considerable spectral overlap can be observed between water and dark shadows in all bands, making it difficult to extract water information while suppressing the shadow noise. To account for this issue, the TSUWI was proposed, which consisted of two subindexes of UWI and USI derived from different feature combinations. The UWI is formulated to effectively discriminate water and dark shadows from other non-water types, and the USI is formulated to remove the dark shadow pixels included in the extraction result of the UWI.

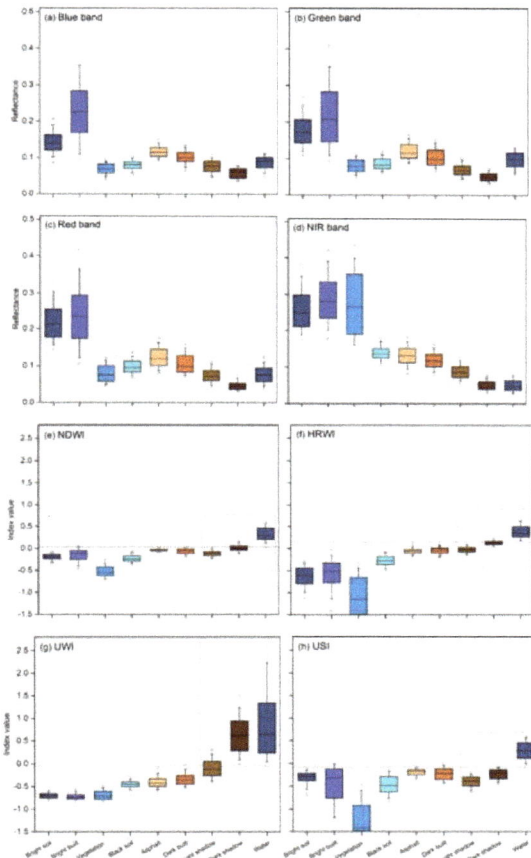

Figure 2. Distributions of reflectance and water index values (**a–h**) for the major urban land cover types, including bright soil, bright built, vegetation, black soil, asphalt, dark built, light shadow, dark shadow, and water. Horizontal lines in each box plot (boxes and whiskers) indicate the locations of the 10th, 25th, 50th, 75th, and 90th percentiles, and the circles indicate the 5th and 95th (blue dashed line) percentiles. The red dashed rectangles show the contrast between shadows and water in each water index. NIR: near-infrared; NDWI: Normalized Difference Water Index; HRWI: High Resolution Water Index; UWI: Urban Water Index; USI: Urban Shadow Index.

For the UWI, the optimal feature combination was identified from the image bands. Figure 2a–d showed that each band has a certain ability to separate water and dark shadows from other land cover types, and the red band achieves the best separability. However, the unstable reflectance values in the blue band may cause obvious variations in the optimal threshold values of the index [29], and no significant improvement in accuracy was observed after the blue band was introduced (discussed in Section 3.3.2). Therefore, the green, red and NIR bands were finally selected to formulate the UWI.

Band ratios were found to be capable of amplifying the minor differences between the spectral reflectance of water and dark shadows. Meanwhile, band ratios can also help stabilize the discrimination abilities of indexes by diminishing the undesired influence posed by topographic relief and light intensity change. Hence, six band ratios composed of either two of all four bands were calculated for the identification of the optimal feature combination for USI formulation, including NIR/B, NIR/G, NIR/R, B/G, B/R, and G/R, where G, R, B, and NIR refer to the reflectance values of the green, red, blue, and near infrared bands, respectively. Information redundancy exists among the six band ratios, and only three of them can cover all four bands. The three band ratios with maximum separability and minimum correlation were then chosen to formulate the USI. Scatter plots and M-statistical tests were used to qualitatively and quantitatively measure the separability of water and dark shadows in the band ratios. In the scatter plots shown in Figure 3, the area encompassed by the two dashed lines in each plot shows the locations of pure dark shadow pixels in the corresponding band ratio. Therefore, the more pure water pixels that fell out of this area, the better separability the band ratio has. The M-statistical test (Equation (1)) is defined by quantifying the histogram difference between two classes [33]. M values above 1.0 indicate fine separation, while M values below 1.0 indicate poor separation.

$$M = \frac{\mu_1 - \mu_2}{\sigma_1 + \sigma_2} \tag{1}$$

where $\mu_1 - \mu_2$ is the difference in the means of two classes, and $\sigma_1 + \sigma_2$ is the sum of their standard deviations. Considering that the USI is a linear combination of features, Pearson's Correlation Coefficient (PCC) analysis [34] was used to examine the linear correlation between two band ratios. The closer a correlation coefficient to 1 or −1 is, the more significant the linear relation is, indicating that one band ratio is more likely to be superseded by the other.

Separability results revealed that NIR/G, NIR/B, and NIR/R showed similar scattering distribution patterns (Figure 3), and NIR/G achieved the best performance at separating water and dark shadows with a high M value of 1.12 (Table 4). The PCC analysis further confirmed that there are high correlations among NIR/G, NIR/B, and NIR/R, with correlation coefficients greater than 0.91. Hence, NIR/G was chosen, while both NIR/B and NIR/R were discarded. The other three band ratios of B/G, G/R, and B/R have M values of 0.91, 0.57, and 0.24, respectively. B/G and B/R are highly correlated with a coefficient of 0.71, while G/R and B/G are only weakly correlated with a coefficient of −0.25 (Table 5). Consequently, B/G and G/R were selected as the other two band ratios used in the formulation of the USI. Any two of the selected band ratios (NIR/G, B/G and G/R) were significantly correlated (Table 5).

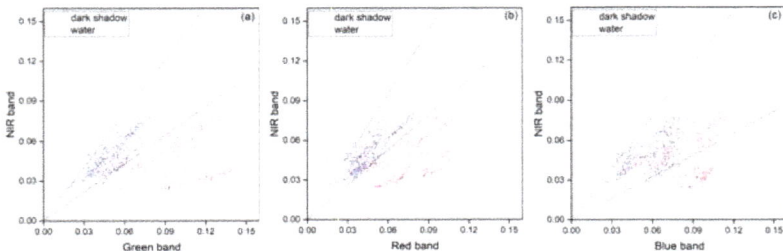

Figure 3. *Cont.*

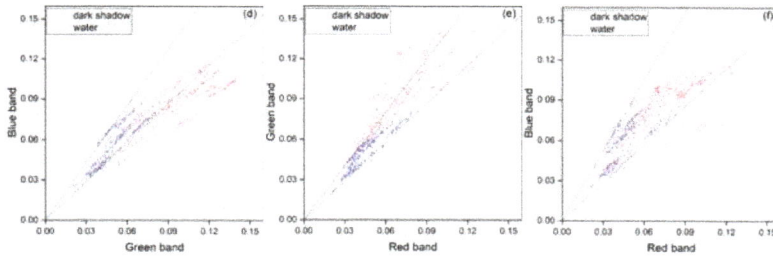

Figure 3. Separability of dark shadows and water in the six band ratios, including NIR/G, NIR/R, NIR/B, B/G, G/R, and B/R, denoted by the amounts of pure water pixels that fell out of the area encompassed by the two dashed lines in each plot. (**a**–**f**) display the scatter plots of surface reflectance of pure dark shadow and water pixels. The slopes of the two dashed lines in each plot correspond to the 5th and 95th percentiles of the corresponding band ratios of pure dark shadow pixels, respectively.

Table 4. Separability of water and dark shadows in six band ratios using the M-statistical test.

Class Pair	M Value					
	NIR/G	NIR/R	NIR/B	B/G	G/R	B/R
Water vs. Dark shadows	1.12	0.87	0.65	0.91	0.57	0.24

Table 5. Pearson's Correlation Coefficient (PCC) among the six band ratios.

Value	Band Ratio Pair						
	NIR/G vs. NIR/B	NIR/G vs. NIR/R	B/G vs. B/R	G/R vs. B/R	B/G vs. G/R	NIR/G vs. B/G	NIR/G vs. G/R
Pearson's r	0.91	0.94	0.71	0.50	−0.25	0.51	−0.58

3.3. Constructing the Two-Step Urban Water Index

The TSUWI was devised to effectively suppress non-water surfaces and extract urban water with improved accuracy. As discussed in Section 3.2, the spectral features used to eliminate water dark shadows differ from those used for other non-water types. Therefore, the TSUWI was designed to compose the two subindexes of the UWI and USI, the coefficients of which were obtained using linear SVM.

3.3.1. Linear Support Vector Machine

The coefficients of the water indexes imply the contribution of a corresponding feature to the separation of water and non-water pixels and become a significant issue for the design of a water index. The coefficients of conventional water indexes (e.g., NDWI, MNDWI, and AWEI) primarily resulted from reflectance pattern analysis of various land cover types, and therefore are characterized by certain subjectivity. In addition, because urban water bodies are typically sediment-rich and algae polluted and exhibit complicated optical features [35], it would become a great challenge for index designers to empirically determine the coefficients of an effective urban water index. In this paper, creating a new index is essentially a linear problem. Hence, the linear SVM was adopted to identify the optical coefficients for the new water indexes.

Linear SVM is a nonparametric statistical learning machine based on the structural risk minimization criterion [36]. By recovering an optical linear hyperplane in the feature space that maximizes the margin separation of two classes, it has been proven to be an advanced coefficient training model [29]. Given a set of labeled training data $(X, Y) = \{(x_i, y_i) \,|\, i = 1, \dots, N, y_i \in \{-1,1\}\}$, the margin of the positive class is represented by equation $w^T x + b \geq 1$, while the margin of the negative class is represented by equation $w^T x + b \leq -1$. That is to say, the minimum margin difference between these two classes is 2, which ensures that the classifier has stable discrimination ability. The linear SVM can be explicitly formulated by solving the following constrained optimization problem (Equations (2) and (3)) [37].

$$\underset{w,b}{\text{min}} \underset{\alpha_i}{\text{max}} \ \frac{1}{2}\| w \|^2 - \sum_{i=1}^{N} \alpha_i \left(y_i \left(w^T x_i + b \right) - 1 \right) \tag{2}$$

$$\text{subject to } 0 \leq \alpha \leq C \text{ and } \sum_{i=1}^{N} \alpha_i y_i = 0 \ \ \forall i \tag{3}$$

where $x_i \in R_d$ is the feature vector of training sample i, here referring to the optimal feature combination selected for index formulation. $y_i \in \{-1, 1\}$ is the corresponding class label. N is the total number of training samples. α_i is the Lagrangian multiplier ranging from 0 to a constant C. Weight vector w is a normal vector that is perpendicular to the hyperplane, and parameter b stands for the intercept term of the hyperplane.

The optical hyperplane is then represented by Equation (4).

$$w^T x + b = 0 \tag{4}$$

For a test pixel x, if the expression $w^T x + b$ output is greater than 0, it belongs to the positive class, and if the expression output is less than 0, it belongs to the negative class. Obviously, the expression $w^T x + b$ can be used as an index, and parameters $[w^T, b]$ are the coefficients. In addition to enhancing the separability of the positive and negative classes, the linear SVM also provided a default threshold of 0, which could be used as a reasonable starting threshold for binary classification.

3.3.2. Formulation of the Urban Water Index (UWI)

The UWI was formulated using the linear SVM to discriminate water and dark shadows from other land cover types. Pure pixels of all land cover types were used to train the linear SVM, where water and dark shadow pixels are labeled as 1, and the other pixels are labeled as −1. To help determine whether the blue band should be introduced into the new index, two linear SVM training experiments were conducted with and without the blue band. By comparing their classification abilities using pure pixels, it was found that the addition of the blue band led to a reduced accuracy of 94.14% compared with 94.27%. The feature combination composed of the green, red, and NIR bands was thus used as the input training vector. After training, the coefficients for the optimal hyperplane were obtained (Equation (5)).

$$P_{\text{UWI}} = 5.83 \times G - 6.57 \times R - 30.32 \times NIR + 2.25 \tag{5}$$

As shown in Figure 4a, the P_{UWI} values of water and dark shadows did not display great discrepancy with the values of other land cover types. To further enhance the separation ability, P_{UWI} was then divided by the expression $|5.83 \times G - 6.57 \times R - 30.32 \times NIR|$ to create the UWI. This division enlarged the difference that water and dark shadows had from other types. Providing insights into the histogram of pure pixel samples, it functioned by shifting water and dark shadow pixels towards larger positive values and shifting other land cover pixels towards smaller negative values, leading to a larger interval between them (Figure 4a,b). The modulus keeps the plus-minus sign unchanged, which means the water and dark shadow pixels in the UWI remain above 0 and other non-water pixels remain below 0. For ease of use, the common divisor 5.83 was removed in the final index, and the coefficients were rounded to one decimal digit, which did not cause a significant reduction in accuracy. The UWI formula is then represented by Equation (6).

$$UWI = \frac{G - 1.1 \times R - 5.2 \times NIR + 0.4}{|G - 1.1 \times R - 5.2 \times NIR|} \tag{6}$$

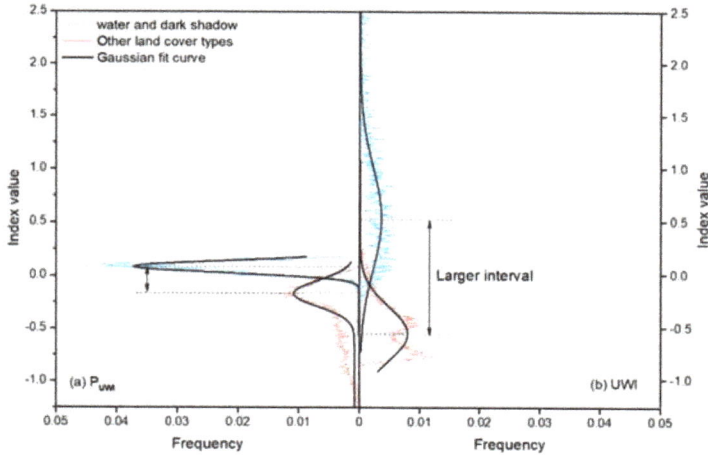

Figure 4. Histograms of P_{UWI} (a) and UWI (b) for pure pixels of positive classes (water and dark shadows) and negative classes (non-water types except dark shadows). The dashed gray lines indicate the peaks of the Gaussian fit curves.

3.3.3. Formulation of the Urban Shadow Index (USI)

The USI was designed to further improve the accuracy by removing the dark shadows that can be confused with water classes from the extraction result of the UWI. Herein, pure water and dark shadow pixels with NIR/G, B/G and G/R features were fed into the linear SVM to create the new USI. Pure water pixels were labeled as 1, and pure dark shadow pixels were labeled as -1. For ease of use, the coefficients of band ratios were rounded to two decimal digits, while one decimal digit was reserved in the constant term. The USI was finally formulated as shown in Equation (7).

$$USI = 0.25 \times \frac{G}{R} - 0.57 \times \frac{NIR}{G} - 0.83 \times \frac{B}{G} + 1.0 \tag{7}$$

3.3.4. The Two-Step Urban Water Index

The TSUWI was developed by combining the UWI and USI. The TSUWI extracts urban water by sequentially applying the UWI and USI to the image. The UWI was first applied to generate a temporary water mask. The USI was then used to eliminate dark and light shadow pixels included in the temporary water mask and obtain the final water extraction result. Therefore, the TSUWI can then be expressed as Equation (8).

$$TSUWI = (UWI > T_1) \wedge (USI > T_2) \tag{8}$$

Here, the TSUWI is a binary index with its possible values being 0 or 1. A value of 0 indicates non-water, while 1 indicates water. T_1 and T_2 denote the optimal thresholds of the UWI and USI, respectively. Zero could theoretically be used as their default value. However, due to the variation in scene brightness and contrast with time and space, the optimal thresholds should be determined in accordance with specific conditions.

The water extraction results of the TSUWI were generated by intersecting the threshold segmentation results from both the UWI and USI; thus, the commission error caused by one index could be corrected by the other. The UWI demonstrated remarkable performance in suppressing non-water land cover types, including bright built, bright soil, vegetation, black soil, dark built, and asphalt. But in areas with dark or light shadow covered surfaces, the UWI may misclassify such surfaces as water (Figure 2g). As a remedy for the UWI, although the USI showed limited ability to eliminate some non-water pixels, such as bright built, this index performed well in suppressing dark shadows and performed even better in suppressing light shadows (Figure 2h).

3.4. Assessment Methods

The assessment of the TSUWI method included an accuracy assessment and stability analysis. An accuracy assessment was used to measure how close the classification results were to the real world. The threshold stability analysis was used to investigate the stability of the optimal threshold close to the default threshold of 0 and the accuracy of water extraction near the optimum threshold.

3.4.1. Accuracy Assessment

To compare the accuracy of the proposed TSUWI with other methods, two well-known water indexes for visible and near-infrared imagery, NDWI and HRWI, were chosen in this study. The comparison between the TSUWI, NDWI, and HRWI was made at their optimum thresholds, which were captured by an iterative approach on the principle of balance of commission and omission errors [22]. Moreover, a nonlinear SVM with a Gaussian radial basis function was employed as a classic and commonly used supervised classifier [35], and its classification accuracy was also compared with that of the TSUWI. For the SVM classifier, the four multispectral bands of GF-2 imagery were chosen as the feature vector input, and the parameters of the SVM were determined by the performance with the highest accuracy. The training samples for each test site were taken from the pure pixel data of the nine land cover types. For the additional test sites in Fuzhou, Haerbin, Yinchuan, and Dongguan, pure pixels were acquired in the same way as other test sites (Section 3.2). After SVM classification, pixels belonging to non-water types were assigned to one category, and binary water results were then produced.

Classification accuracies of the TSUWI, NDWI, HRWI, and SVM, were assessed by calculating the KCs, commission error (CE) and omission error (OE) derived from the confusion matrix [38]. The confusion matrix was produced via a pixel-by-pixel comparison between the classification and reference images. As the reference image was the same for the different classification methods, dependence between their confusion matrixes can easily occurs. This dependence may result in too conservative inference about the superiority of one classification method over another [39]. McNemar's test was thus adopted to provide an assessment of the confidence in the accuracy difference between the TSUWI and the other three methods. The test was based on a chi-square statistic, computed as shown in Equation (9) [39].

$$\chi^2 = \frac{(|f_{12} - f_{21}| - 1)^2}{f_{12} + f_{21}} \tag{9}$$

where f_{12} and f_{21} denote the proportions of pixels that are correctly classified by one method but wrongly classified by the other.

3.4.2. Threshold Stability Assessment

Threshold stability analysis is an important paradigm in the context of index development and application. Because the NDWI and HRWI are similar to the UWI and USI and were formulated to discriminate water from non-water pixels by forcing water pixels above 0 and non-water pixels below 0, the NDWI and HRWI were also chosen to further compare the stability of the proposed TSUWI. For the three methods, a default value of 0 is, in theory, the optimum threshold that could extract water with the highest accuracy. However, due to the variation in scene brightness and contrast with time and space, the optimum threshold may not always lie at 0 but at a certain value near 0. As a result, a range of multiple thresholds of approximately 0 at regular intervals are iteratively tested to find the optimum threshold. To reduce the iteration times in adjusting the threshold, the threshold data for testing are expected to have a small range but a large interval. Water extraction methods are thus required to (1) stabilize the optimal threshold as close as possible to the 0 value and (2) maintain good performance near the optimum threshold. Therefore, the threshold stability comparison between the TSUWI, NDWI, and HRWI was made from these two perspectives. The former perspective was assessed by examining the variation in the optimal threshold values for the three methods across the twelve test sites, while the latter one was tested by comparing their accuracy variability in a range of thresholds near the optimum value. When testing the accuracy variability of the UWI and USI, the variation in the accuracy of one index was calculated by fixing the other index at its optimum threshold.

4. Results

4.1. Water Extraction Maps

The water extraction maps generated by the TSUWI, NDWI, HRWI, and SVM at the twelve test sites are presented in the Supplementary Material (Figure S1). Visual inspection of Figure S1 indicates that the TSUWI was effective in extracting surface water in the presence of complex urban surfaces. Compared to the NDWI and HRWI, the proposed TSUWI consistently performed better in suppressing shadows and other non-water surfaces, particularly at the test sites in Shanghai, Yinchuan, and Dongguan. In most cases, the NDWI and HRWI resulted in noisy results with a large number of misclassified pixels. The SVM resulted in classification outputs that were (visually) similar to the TSUWI at first sight. However, closer inspection revealed that the proposed TSUWI did improve the water extraction accuracy at most test sites compared to SVM.

4.2. Water Extraction Accuracy

The water classification accuracies of the TSUWI, NDWI, HRWI, and SVM methods at the twelve sites are presented in Table 6. Statistical analysis of Table 6 indicated that the TSUWI successfully achieved high accuracy of urban surface water mapping at all test sites, with a mean KC equal to 0.97 and a mean TE (the sum of the CE and OE) of 5.82%. In contrast, the other three methods consistently exhibited lower classification accuracy, with an exception at the test site in Aksu for SVM (TE = 6.89% for TSUWI, while TE = 6.22% for SVM). The two conventional indexes, NDWI and HRWI, exhibited similar performance and resulted in a lower classification accuracy than the other two methods, and their mean KC and mean TE were 0.90, 17.41% and 0.93, 13.21%, respectively. The SVM classifier fell between, with a mean KC of 0.95 and a mean TE of 8.81%. For the overall stability, it clearly appeared that the classification accuracy of the TSUWI at different test sites exhibited smaller variations compared to the other three methods (Figure A1). By comparing the TEs at each test site, it is found that at most test sites, the TE of TSUWI was less than 55% of that of NDWI or HRWI and 85% of that of the SVM classifier (Figure A2). In other words, the proposed TSUWI could generally decrease the classification error by more than 45% compared to NDWI or HRWI, and 15% for the SVM.

Table 6. Summary of classification accuracies of the three methods by test site. TSUWI: Two-Step Urban Water Index; NDWI: Normalized Difference Water Index; HRWI: High Resolution Water Index; SVM: Support Vector Machine.

Test Sites	Kappa Coefficient				Total Error (%)			
	TSUWI	NDWI	HRWI	SVM	TSUWI	NDWI	HRWI	SVM
Guangzhou	0.96	0.92	0.92	0.95	7.72	14.46	13.90	8.35
Aksu	0.97	0.96	0.96	0.97	6.89	7.58	7.07	6.22
Chengdu	0.94	0.84	0.84	0.83	11.33	30.15	29.75	29.70
Lhasa	0.96	0.94	0.92	0.94	7.97	12.16	14.84	11.05
Nanchang	0.98	0.95	0.96	0.98	3.11	9.45	7.41	3.28
Qingdao	1.00	0.99	0.99	0.99	0.34	0.65	0.69	0.75
Shanghai	0.99	0.96	0.94	0.99	1.24	5.38	7.80	1.48
Shigatse	0.96	0.90	0.87	0.93	7.10	19.43	24.51	13.81
Fuzhou	0.98	0.95	0.96	0.96	4.31	8.10	7.82	6.25
Haerbin	0.97	0.90	0.91	0.97	6.17	18.09	17.64	6.50
Yinchuan	0.96	0.57	0.93	0.94	7.97	63.79	13.25	11.04
Dongguan	0.97	0.89	0.92	0.96	5.65	19.69	13.78	7.33

Table 7 summarizes the significance of the accuracy difference at the twelve test sites by McNemar's chi-square test. Overall, significant accuracy improvement was achieved by the TSUWI compared to the NDWI, HRWI, and SVM (*p*-value < 0.001). Exceptions were found in the test site in Aksu for the HRWI and SVM. At this site, the superiority of the TSUWI over the HRWI was nonsignificant (*p*-value = 0.364), and the TSUWI performed significantly worse than the SVM because the TE of the TSUWI (6.89%) was greater than that of the SVM, and the *p*-value was below 0.001.

Table 7. Summary of McNemar's χ^2 test for accuracy difference between the TSUWI and the NDWI, HRWI and SVM.

Test Sites	TSUWI vs. NDWI		TSUWI vs. HRWI		TSUWI vs. SVM	
	χ^2	*p*-Value	χ^2	*p*-Value	χ^2	*p*-Value
Guangzhou	132,839.0	<0.001	115,749.6	<0.001	3154.6	<0.001
Aksu	77.3	<0.001	0.8	0.364	93.6	<0.001
Chengdu	17,469.1	<0.001	16,782.9	<0.001	29,634.9	<0.001
Lhasa	789.6	<0.001	1611.7	<0.001	565.7	<0.001
Nanchang	156,460.7	<0.001	90,288.0	<0.001	473.3	<0.001
Qingdao	27,076.7	<0.001	29,515.0	<0.001	37,705.9	<0.001
Shanghai	365,387.6	<0.001	614,546.5	<0.001	5523.5	<0.001
Shigatse	5466.0	<0.001	9882.5	<0.001	2461.9	<0.001
Fuzhou	93,062.7	<0.001	82,875.6	<0.001	33,324.0	<0.001
Haerbin	146,443.9	<0.001	148,493.8	<0.001	474.1	<0.001
Yinchuan	1,637,618.0	<0.001	41,856.2	<0.001	18,304.1	<0.001
Dongguan	242,347.7	<0.001	118,820.2	<0.001	10,548.3	<0.001

4.3. Threshold Stability Analysis

A comparison of the stability of the optimal thresholds of the UWI, USI, NDWI, and HRWI is shown in Figure 5. The optimal thresholds of the UWI and USI at different test sites presented similar ranges, which were from −0.38 to 0.15 and −0.38 to 0.11, respectively. Compared to the NDWI and HRWI, the optimal thresholds of these two new indexes have smaller ranges of approximately 0. Themaximum deviations of the optimal thresholds for the UWI and USI were both 0.38, whereas those for the NDWI and HRWI reached 0.56 and 0.85. It was concluded that the optimal thresholds of the UWI and USI at different test sites exhibited small variations from the default threshold of 0 compared to the NDWI and HRWI. Therefore, 0 could be used as the initial threshold in the iteration to find the optimum thresholds for both the UWI and USI.

Figure 6 shows the accuracies of the UWI, USI, NDWI, and HRWI in the range of [−0.1, 0.1] near the optimal thresholds. At all twelve sites, the UWI exhibited almost unnoticeable variations, whereas the variations in the USI variation were relatively more obvious. This result means that the accuracy stability of the TSUWI near the optimal threshold is mainly dependent on that of the USI. In most cases, the accuracy of the USI is much more stable and higher than that of the NDWI and HRWI. Therefore, the TSUWI can alleviate the manual iteration issue for the optimum threshold, which is often normal and serious in the application of water indexes [40]. Moreover, the UWI can maintain the best performance in the range [−0.1, 0.1], while the USI can maintain the best performance in the range [−0.01, 0.01]. In the application of the TSUWI, we thus recommend 0.2 as the iteration step size for the UWI and 0.02 for the USI.

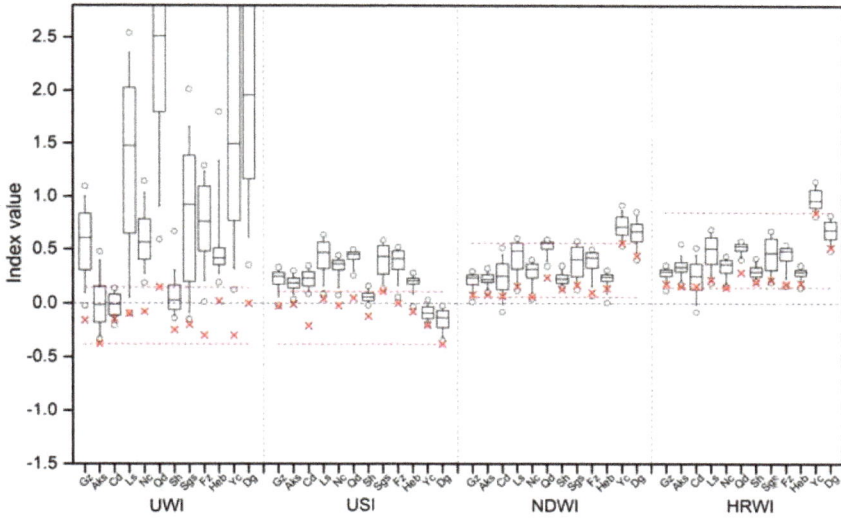

Figure 5. Threshold variability and distribution of index values of water pixels for the UWI, USI, NDWI, and HRWI. Dashed lines show the maximum and minimum of the optimal threshold at the twelve test sites, and the "x" symbol shows the optimal threshold for each site.

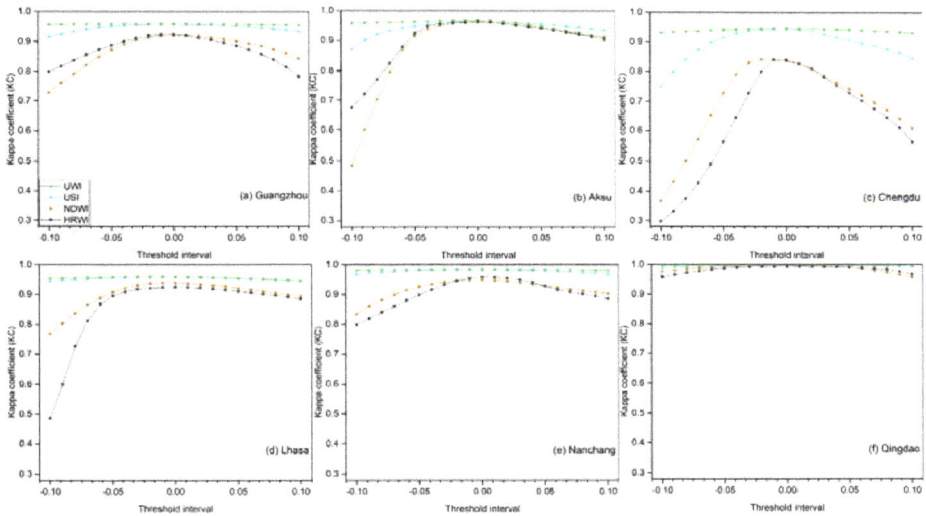

Figure 6. *Cont.*

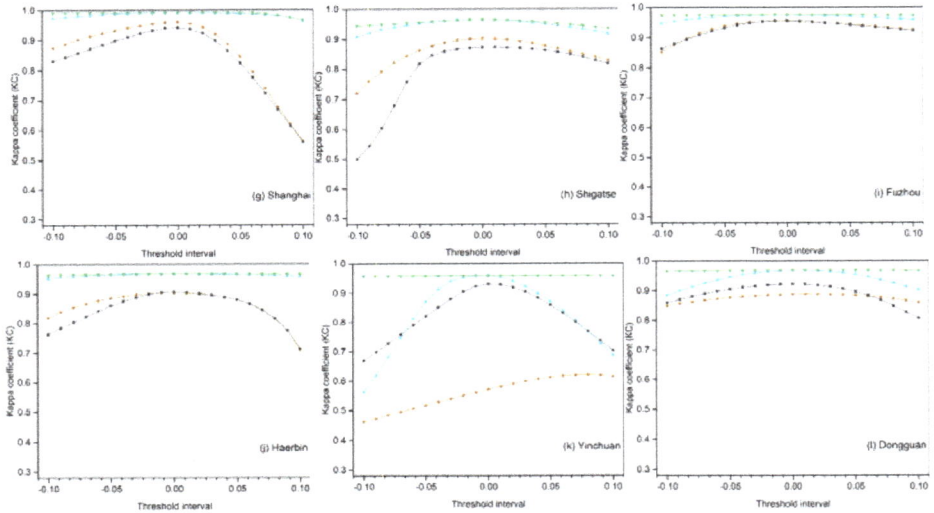

Figure 6. The accuracies of the UWI, USI, NDWI, and HRWI at twelve test sites (**a–i**) in a range of thresholds near the optimal threshold: (**a**) Guangzhou; (**b**) Aksu; (**c**) Chengdu; (**d**) Lhasa; (**e**) Nanchang; (**f**) Qingdao; (**g**) Shanghai; (**h**) Shigatse; (**i**) Fuzhou; (**j**) Haerbin; (**k**) Yinchuan; (**l**) Dongguan.

5. Discussion

5.1. Effects of Shadow Detection

The fact that shadows are widely distributed throughout urban areas and exhibit spectral patterns that are similar to those of water makes shadow removal a challenging problem in urban water extraction [25,41]. To address this issue, many researchers have contributed to previous research on the improvement of extraction accuracy by introducing additional shadow detection methods, such as object-oriented classification [35,42], shadow detection model based on SVM feature training [29], morphological shadow indexes [43,44] and invariant color model [45–47]. These methods may achieve expected results but are relatively difficult to apply and are time-consuming. Our new USI automatically suppresses shadow pixels through the arithmetic of bands. The circumvention of complex shadow detection procedures may simplify urban water mapping.

As shown in Section 3.3.4, the USI was preliminarily verified to have good separation through statistical analysis of pure pixels. To further confirm the role of this shadow detection index, we compared the accuracy results of the NDWI and HRWI, as well as their combination with the USI, and the proposed TSUWI at the twelve test sites (Table 8). Compared to the NDWI and HRWI, the combination of both with the USI achieved improved accuracy at each test site. For the NDWI and HRWI, the OE at most test sites was greater than the CE. The reason for this difference is that only the NDWI or HRWI cannot suppress the signal from shadows (Figure 2e,f), and the threshold has to be increased to achieve high accuracy at the cost of increasing the OE. By combining these indexes with the USI, the USI can successfully remove the noise from shadows (Figure 2h), and the NDWI or HRWI can then reduce the threshold to decrease the OE, thus resulting in improved accuracy. However, reducing the threshold of the NDWI (or HRWI) may simultaneously increase the number of misclassified pixels, such as dark built, asphalt and bright built (dark built and asphalt for HRWI) pixels, on which the USI also has limited effects (Figure 2h). Among the three combination methods with USI, the proposed TSUWI (UWI + USI) demonstrated the best performance with the highest accuracy in detecting urban water bodies at all test sites. Therefore, we recommend using the TSUWI method to extract urban water rather than the NDWI or the HRWI combined with the USI. However, the accuracies delivered by the combination of USI with NDWI and HRWI were quite similar to that of the TSUWI. This finding not only implies that the USI is much more important for the performance of the TSUWI, but also highlights the potential of USI to further improve the performance of other water indices.

Table 8. Summary of the accuracy using HRWI, NDWI, NDWI+USI, HRWI+USI and TSUWI (UWI+USI) at each optimal threshold.

Test Sites	NDWI			HRWI			NDWI + USI			HRWI + USI			TSUWI		
	Kappa	OE%	CE%	Kappa	OE%	CE%	Kappa	OE%	CE%	Kappa	OE%	CE%	Kappa	OE%	CE%
Guangzhou	0.92	10.22	4.24	0.92	10.68	3.22	0.93	9.08	3.46	0.95	5.84	3.48	0.96	5.37	2.35
Aksu	0.96	2.04	5.55	0.96	2.96	4.11	0.96	2.04	5.51	0.96	2.98	4.03	0.97	1.98	4.91
Chengdu	0.84	25.29	4.86	0.84	24.49	5.26	0.88	11.76	10.96	0.93	6.66	8.02	0.94	5.91	5.42
Lhasa	0.94	8.61	3.55	0.92	7.30	7.55	0.95	7.42	3.27	0.94	7.32	4.09	0.96	6.68	1.30
Nanchang	0.95	6.15	3.30	0.96	4.43	2.98	0.96	3.72	2.64	0.97	2.66	2.19	0.98	1.99	1.11
Qingdao	0.99	0.35	0.31	0.99	0.33	0.36	1.00	0.19	0.33	0.99	0.19	0.38	1.00	0.17	0.17
Shanghai	0.96	2.05	3.33	0.94	3.25	4.55	0.99	0.92	0.45	0.99	0.86	0.47	0.99	0.83	0.41
Shigatse	0.90	10.80	8.63	0.87	7.51	17.00	0.91	9.67	7.50	0.91	9.17	8.23	0.96	3.83	3.28
Fuzhou	0.95	6.64	1.46	0.96	6.05	1.77	0.96	5.93	1.65	0.97	3.69	2.01	0.98	3.05	1.25
Haerbin	0.90	13.60	4.49	0.91	9.67	7.98	0.91	11.39	4.49	0.92	9.55	4.67	0.97	4.20	1.97
Yinchuan	0.57	7.93	55.86	0.93	9.23	4.02	0.96	5.42	2.70	0.96	6.14	1.89	0.96	5.41	2.56
Dongguan	0.89	7.54	12.16	0.92	8.09	5.69	0.97	3.38	2.61	0.96	3.72	3.01	0.97	3.04	2.61

5.2. Advantages of the Proposed Method

The TSUWI proposed in this paper contributes to the efforts to improve the accuracy of urban water extraction for various environmental studies. Although a number of improved water mapping indexes [22,24] have been proposed, few of them were established based on pure pixels derived from various water body types in various environments with a sufficient number of study sites. This method is constructed by combining the UWI and USI. To create effective indexes, a linear SVM model and numerous pure pixels were used in this study. As an outstanding machine learning technique for training the coefficient index, the linear SVM will not only provide an inherent default threshold of zero but also automatically achieve the largest separation between positive and negative classes [29]. The pure pixels were selected from eight sites located in different regions across China, which were deliberately chosen to cover various water body types and urban surfaces. As expected, the TSUWI was shown to extract urban surface water with high accuracy and remain robust for different types of water bodies under various urban environments.

The lack of a stable threshold is a problem in many water indexes, which may make the decision of a cut-off threshold more time-consuming and easily lead to a subjective choice of threshold with decreased accuracy [22]. In addition to accuracy improvement, the two indexes in the TSUWI were also shown to have a relatively stable optimal threshold that is close to zero and maintain good performance in the range of neighborhood thresholds near the optimal value. In the determination of optimal thresholds, 0 could be used as the starting point for the iterations for both the UWI and USI; 0.2 is recommended as the iteration step size for the UWI, and 0.02 is recommended for the USI. Benefiting from this, the application of this method is simplified, and the likelihood of achieving the highest urban water accuracy is improved. However, our findings are based on the suggestion that radiometric calibration and atmospheric correction were carefully undertaken for the images from all test sites. If either of these corrections is ignored, the accuracy and optimal thresholds may be different from those observed in this study.

Although high-resolution images have been available for a few decades, simple yet efficient indexes to characterize urban water extent with adequate detail are still lacking. This deficiency mainly results from the limited bands and surface noise in these images, which are often major causes of misclassification in urban surface water mapping. Our new TSUWI fills this gap. The TSUWI is calculated by the simple arithmetic of four standard bands prevalent at high resolution images. Using a simple threshold segmentation approach, the TSUWI consistently provides accurate water results in various water conditions with regard to depth, turbidity, chemical composition, and surface appearance. The extracted urban surface water can be further used as basic information for various urban studies, such as water quality analysis, urban heat island effect, and urban surface water change under the context of urbanization.

5.3. Further Improvements

Although the proposed TSUWI achieved satisfactory results in this study, some issues remain, such as atmospheric composition, transferability of the proposed method to other image data, seasonal variation in the angle of the sun, and seasonal behavior of water bodies themselves. All of these factors have an impact on the performance of the TSUWI. The use of different atmospheric correction methods may also influence thresholds and accuracies, especially when there is heavy haze. Heavy haze has been a serious issue in Chinese urban areas during wintertime in recent years. Current atmospheric correction models may not necessarily work well when correcting atmospheric haze. Therefore, one may need to consider the importance and type of atmospheric correction applied in the image preprocessing stage when evaluating the accuracies of different water extraction methods. Because the TSUWI is designed based on the land cover reflectance using GF-2 images, it is theoretically free of the constraints in terms of satellite image type with similar spectral bands. However, due to the inevitable differences among different sensors, it is still necessary to test the TSUWI on image data from other sources. Seasonal variation in the angle of the sun leads to changes in the brightness of images, and may also influence the performance of TSUWI. In addition, the spectral properties of water bodies will vary with seasonal changes in precipitation, biodegradability, domestic animals, and aquatic plants. In our test cases, we did not consider the influence of seasonal variation in the angle of the sun as well as the seasonal behavior of water bodies themselves. Therefore, the robustness of the new method also needs to be tested during different seasons. These issues are worth a follow-up study and verification.

6. Conclusions

The main purpose of this study was to devise a method that improves the accuracy of urban water extraction by increasing the spectral separability between water and non-water surfaces in the presence of shadows, which are often major causes of low classification accuracy. Using GF-2 data, we proposed an urban water extraction method called the TSUWI, which is a combination of two new indexes (UWI and USI) and compared its accuracy and threshold stability with that of the NDWI, HRWI, and SVM classifiers. In twelve cities across China, the accuracy assessment results showed that this method exhibited good performance, with an average KC of 0.97 and an average TE of 5.82%. Compared with the NDWI, HRWI, and SVM, the TSUWI generally exhibited improved accuracy by decreasing the TEs by more than 45% for the NDWI or HRWI and 15% for the SVM. In addition, both the UWI and USI were shown to have stable thresholds that were close to 0 and maintained good performance near their optimum thresholds with images from different locations and times compared to the NDWI and HRWI. Therefore, the TSUWI is an alternative and improved method for urban water mapping using high-resolution imagery. Moreover, the USI can be used alone to combine with other water indices for the further improvement of their performance in more accurate water extraction.

Supplementary Materials: The following are available online at http://www.mdpi.com/2072-4292/10/11/1704/s1, Table S1: List of water indexes. B, G, R and NIR refer to the surface reflectance of the green, red, blue and near infrared bands. SWIR1 and SWIR2 donate the surface reflectance of two shortwave infrared bands (band 5 and band7) in the Landsat TM/ETM+ imagery; Figure S1: Water extraction results using the TSUWI, NDWI, HRWI and SVM at the twelve test sites.

Author Contributions: W.W. provided the conception, conducted the data analysis, and wrote the manuscript. W.W., Q.L. and Y.Z. developed the methodology. Y.Z. revised the manuscript. X.D. and H.W. provided valuable insights and edited the manuscript.

Funding: This research is funded by the Special Scientific Research Fund of Public Welfare Profession of China (Grant No. 201511010).

Acknowledgments: Great thanks to the anonymous reviewers whose comments and suggestions significantly improved the manuscript.

Conflicts of Interest: The authors declare no conflict of interest.

Appendix A

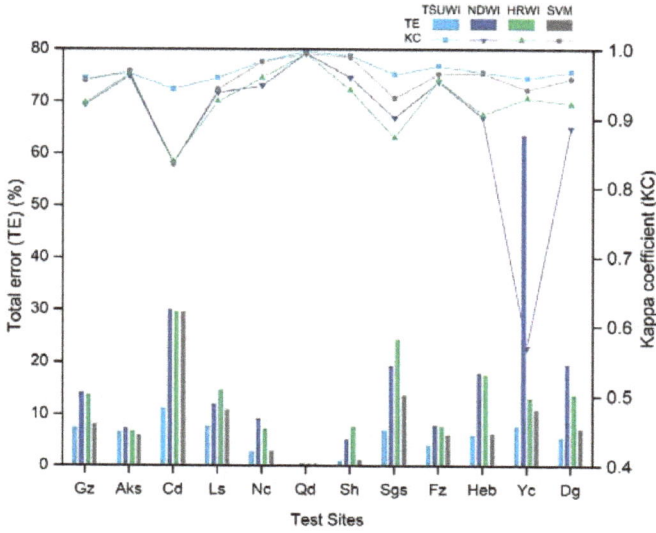

Figure A1. KCs and TEs obtained by the TSUWI, NDWI, HRWI and SVM for the twelve test sites. The twelve sites here were Guangzhou (Gz), Aksu (Aks), Chengdu (Cd), Lhasa (Ls), Nanchang (Nc), Qingdao (Qd), Shanghai (Sh), Shigatse (Sgs), Fuzhou (Fz), Haerbin (Heb), Yinchuan (Yc) and Dongguan (Dg).

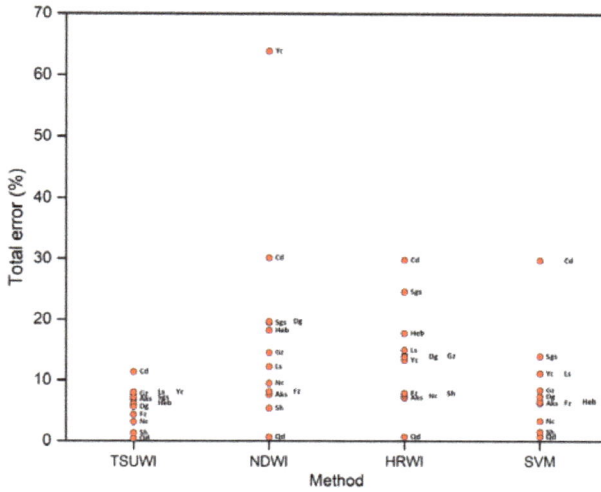

Figure A2. Total errors of the four methods for the twelve test sites.

References

1. Kandulu, J.M.; Connor, J.D.; Macdonald, D.H. Ecosystem services in urban water investment. *J. Environ. Manag.* **2014**, *145*, 43–53. [CrossRef] [PubMed]
2. Robitu, M.; Musy, M.; Inard, C.; Groleau, D. Modeling the influence of vegetation and water pond on urban microclimate. *Sol. Energy* **2006**, *80*, 435–447. [CrossRef]

3. Bond, N.R.; Lake, P.S.; Arthington, A.H. The impacts of drought on freshwater ecosystems: An Australian perspective. *Hydrobiologia* **2008**, *600*, 3–16. [CrossRef]
4. Yao, Y.; Zhang, Y.; Liu, J.; Shen, Z.; Liu, B. Model for evaluating urban water shortage risk: A case study in Beijing. *Int. J. Digit. Content Technol. Appl.* **2012**, *6*, 68–79.
5. Vermonden, K. Key Factors for Biodiversity of Urban Water Systems. Ph.D. Thesis, Radboud University, Nijmegen, The Netherlands, 2010.
6. Arnfield, A.J. Two decades of urban climate research: A review of turbulence, exchanges of energy and water, and the urban heat island. *Int. J. Climatol.* **2003**, *23*, 1–26. [CrossRef]
7. Dong-Hai, L.I.; Bin, A.I.; Xia, L.I. Urban water body alleviating heat island effect based on RS and GIS: A case study of Dongguan City. *Trop. Geogr.* **2008**, *28*, 414–418.
8. Jofre, J.; Blanch, A.R.; Lucena, F. Water-Borne Infectious Disease Outbreaks Associated with Water Scarcity and Rainfall Events. In *Water Scarcity in the Mediterranean: Perspectives Under Global Change*; Sabater, S., Barceló, D., Eds.; Springer: Berlin/Heidelberg, Germany, 2009; pp. 147–159.
9. Kalnay, E.; Cai, M. Impact of urbanization and land-use change on climate. *Nature* **2003**, *423*, 528–531. [CrossRef] [PubMed]
10. Mckinney, M.L. Urbanization as a major cause of biotic homogenization. *Biol. Conserv.* **2006**, *127*, 247–260. [CrossRef]
11. Zhang, D.L.; Shou, Y.X.; Dickerson, R.R. Upstream urbanization exacerbates urban heat island effects. *Geophys. Res. Lett.* **2009**, *36*, 88–113. [CrossRef]
12. Huang, S.L.; Chen, C.W. Urbanization and socioeconomic metabolism in Taipei. *J. Ind. Ecol.* **2010**, *13*, 75–93. [CrossRef]
13. Morss, R.E.; Wilhelmi, O.V.; Downton, M.W.; Gruntfest, E. Flood risk, uncertainty, and scientific information for decision making: Lessons from an interdisciplinary project. *Bull. Am. Meteorol. Soc.* **2005**, *86*, 1593–1601. [CrossRef]
14. Giardino, C.; Bresciani, M.; Villa, P.; Martinelli, A. Application of remote sensing in water resource management: The case study of Lake Trasimeno, Italy. *Water Resour. Manag.* **2010**, *24*, 3885–3899. [CrossRef]
15. Lira, J. Segmentation and morphology of open water bodies from multispectral images. *Int. J. Remote Sens.* **2006**, *27*, 4015–4038. [CrossRef]
16. Davranche, A.; Lefebvre, G.; Poulin, B. Wetland monitoring using classification trees and SPOT-5 seasonal time series. *Remote Sens. Environ.* **2012**, *114*, 552–562. [CrossRef]
17. Sethre, P.; Rundquist, B.; Todhunter, P. Remote detection of prairie pothole ponds in the devils lake basin, North Dakota. *Mapp. Sci. Remote Sens.* **2005**, *42*, 277–296. [CrossRef]
18. Asis, A.M.D.; Omasa, K.; Oki, K.; Shimizu, Y. Accuracy and applicability of linear spectral unmixing in delineating potential erosion areas in tropical watersheds. *Int. J. Remote Sens.* **2008**, *29*, 4151–4171. [CrossRef]
19. Xie, H.; Luo, X.; Xu, X.; Pan, H.; Tong, X. Automated subpixel surface water mapping from heterogeneous urban environments using Landsat 8 OLI imagery. *Remote Sens.* **2016**, *8*, 584. [CrossRef]
20. Bryant, R.G.; Rainey, M.P. Investigation of flood inundation on playas within the zone of Chotts, using a time-series of AVHRR. *Remote Sens. Environ.* **2002**, *82*, 360–375. [CrossRef]
21. Jain, S.K.; Singh, R.D.; Jain, M.K.; Lohani, A.K. Delineation of flood-prone areas using remote sensing techniques. *Water Resour. Manag.* **2005**, *19*, 333–347. [CrossRef]
22. Feyisa, G.L.; Meilby, H.; Fensholt, R.; Proud, S.R. Automated water extraction index: A new technique for surface water mapping using landsat imagery. *Remote Sens. Environ.* **2014**, *140*, 23–35. [CrossRef]
23. Mcfeeters, S.K. The use of the normalized difference water index (NDWI) in the delineation of open water features. *Int. J. Remote Sens.* **1996**, *17*, 1425–1432. [CrossRef]
24. Xu, H. Modification of normalised difference water index (NDWI) to enhance open water features in remotely sensed imagery. *Int. J. Remote Sens.* **2006**, *27*, 3025–3033. [CrossRef]
25. Tong, X. New hyperspectral difference water index for the extraction of urban water bodies by the use of airborne hyperspectral images. *Int. J. Remote Sens.* **2014**, *8*, 085098.
26. Verpoorter, C.; Kutser, T.; Tranvik, L. Automated mapping of water bodies using Landsat multispectral data. *Limnol. Oceanogr. Methods* **2015**, *10*, 1037–1050. [CrossRef]
27. Gessner, M.O.; Hinkelmann, R.; Nützmann, G.; Jekel, M.; Singer, G.; Lewandowski, J.; Nehls, T.; Barjenbruch, M. Urban water interfaces. *J. Hydrol.* **2014**, *514*, 226–232. [CrossRef]

28. Dare, P.M. Shadow analysis in high-resolution satellite imagery of urban areas. *Photogramm. Eng. Remote Sens.* **2005**, *71*, 169–177. [CrossRef]

29. Yao, F.; Wang, C.; Dong, D.; Luo, J.; Shen, Z.; Yang, K. High-resolution mapping of urban surface water using ZY-3 multi-spectral imagery. *Remote Sens.* **2015**, *7*, 12336–12355. [CrossRef]

30. Exelis. Exelis Visual Information Solutions. Available online: http://www.exelisvis.com (accessed on 15 November 2017).

31. ExelisHelp. Exelis Visual Information Solutions. Available online: http://www.exelisvis.com/Support/HelpArticles.aspx (accessed on 15 November 2017).

32. Tachikawa, T.; Kaku, M.; Iwasaki, A.; Gesch, D.B.; Oimoen, M.J.; Zhang, Z.; Danielson, J.; Krieger, T.; Curtis, B.; Haase, J. *ASTER Global Digital Elevation Model Version 2—Summary of Validation Results*; NASA: Pasadena, CA, USA, 2011.

33. Kaufman, Y.J.; Remer, L.A. Detection of forests using mid-IR reflectance: An application for aerosol studies. *Geosci. Remote Sens. IEEE Trans.* **1994**, *32*, 672–683. [CrossRef]

34. Neto, A.M.; Rittner, L.; Leite, N.; Zampieri, D.E. Pearson's correlation coefficient for discarding redundant information in real time autonomous navigation system. In Proceedings of the 16th IEEE International Conference on Control Applications, Singapore, 1–3 October 2007.

35. Yang, F.; Guo, J.; Tan, H.; Wang, J. Automated extraction of urban water bodies from Zymulti Log Pectral imagery. *Water* **2017**, *9*, 144. [CrossRef]

36. Sun, J.; Li, Q.; Lu, W.; Wang, Q. Image recognition of laser radar using linear SVM correlation filter. *Chin. Opt. Lett.* **2007**, *5*, 549–551.

37. Fu, Z.; Robles-Kelly, A.; Zhou, J. Mixing linear SVMs for nonlinear classification. *IEEE Trans. Neural Netw.* **2010**, *21*, 1963–1975. [PubMed]

38. Provost, F.; Kohavi, R. Guest editors' introduction: On applied research in machine learning. *Mach. Learn.* **1998**, *30*, 127–132. [CrossRef]

39. Leeuw, J.D.; Jia, H.; Yang, L.; Liu, X.; Schmidt, K.; Skidmore, A.K. Comparing accuracy assessments to infer superiority of image classification methods. *Int. J. Remote Sens.* **2006**, *27*, 223–232. [CrossRef]

40. Ji, L.; Zhang, L.; Wylie, B. Analysis of dynamic thresholds for the normalized difference water index. *Photogramm. Eng. Remote Sens.* **2009**, *75*, 1307–1317. [CrossRef]

41. Xie, C.; Huang, X.; Zeng, W.; Fang, X. A novel water index for urban high-resolution eight-band Worldview-2 imagery. *Int. J. Digit. Earth* **2016**, *9*, 925–941. [CrossRef]

42. Zhou, W.Q.; Huang, G.L.; Troy, A.; Cadenasso, M.L. Object-based land cover classification of shaded areas in high spatial resolution imagery of urban areas: A comparison study. *Remote Sens. Environ.* **2009**, *113*, 1769–1777. [CrossRef]

43. Chen, Y.; Wen, D.; Jing, L.; Shi, P. Shadow information recovery in urban areas from very high resolution satellite imagery. *Int. J. Remote Sens.* **2007**, *28*, 3249–3254. [CrossRef]

44. Huang, X.; Zhang, L. Morphological building/shadow index for building extraction from high-resolution imagery over urban areas. *IEEE J. Sel. Top. Appl. Earth Observ. Remote Sens.* **2012**, *5*, 161–172. [CrossRef]

45. Tsai, V.J.D. A comparative study on shadow compensation of color aerial images in invariant color models. *IEEE Trans. Geosci. Remote Sens.* **2006**, *44*, 1661–1671. [CrossRef]

46. Arévalo, V.; González, J.; Ambrosio, G. Shadow detection in colour high-resolution satellite images. *Int. J. Remote Sens.* **2008**, *29*, 1945–1963. [CrossRef]

47. Chung, K.L.; Lin, Y.R.; Huang, Y.H. Efficient shadow detection of color aerial images based on successive thresholding scheme. *IEEE Trans. Geosci. Remote Sens.* **2009**, *47*, 671–682. [CrossRef]

remote sensing

MDPI

Letter

Use of WorldView-2 Along-Track Stereo Imagery to Probe a Baltic Sea Algal Spiral

George Marmorino * and Wei Chen

Remote Sensing Division, Naval Research Laboratory, Washington, DC 20375, USA; wei.chen@nrl.navy.mil
* Correspondence: marmorino@nrl.navy.mil

Received: 26 March 2019; Accepted: 8 April 2019; Published: 10 April 2019

Abstract: The general topic here is the application of very high-resolution satellite imagery to the study of ocean phenomena having horizontal spatial scales of the order of 1 kilometer, which is the realm of the ocean submesoscale. The focus of the present study is the use of WorldView-2 along-track stereo imagery to probe a submesoscale feature in the Baltic Sea that consists of an apparent inward spiraling of surface aggregations of algae. In this case, a single pair of images is analyzed using an optical-flow velocity algorithm. Because such image data generally have a much lower dynamic range than in land applications, the impact of residual instrument noise (e.g., data striping) is more severe and requires attention; we use a simple scheme to reduce the impact of such noise. The results show that the spiral feature has at its core a cyclonic vortex, about 1 km in radius and having a vertical vorticity of about three times the Coriolis frequency. Analysis also reveals that an individual algal aggregation corresponds to a velocity front having both horizontal shear and convergence, while wind-accelerated clumps of surface algae can introduce fine-scale signatures into the velocity field. Overall, the analysis supports the interpretation of algal spirals as evidence of a submesoscale eddy and of algal aggregations as indicating areas of surface convergence.

Keywords: High-resolution satellite imagery; submesoscale; spiral eddy; cyanobacteria; surface convergence; western Baltic Sea

---

1. Introduction

High spatial resolution imagery from satellites capable of along-track stereo, or from satellites that follow each other closely in time on similar orbits, can be exploited for target motion given appropriate time lags between the image acquisitions [1–9]. In the ocean, targets take the form of spatial gradients and other features in ocean color and surface temperature, suspended material, surface films, and (as in this work) algae. Ocean currents can be deduced from time-lagged images by using various techniques such as maximum cross-correlation [2,4,8] and optical flow [4–9]. Particularly exciting is the possibility of using high-resolution time-lagged imagery to explore the realm of the ocean submesoscale, in which strong surface convergences and downwelling become associated with horizontal density fronts and cyclonic vortices e.g., References [10,11].

In this letter, a single pair of along-track stereo images from the WorldView-2 satellite (DigitalGlobe, Inc., Westminster, CO, USA; pixel sizes of ~1 m) is used to examine aspects of small-scale dynamical features as revealed through an algal bloom in the western Baltic Sea. A large-scale view of the study area is provided by a Sentinel-3 satellite image (Figure 1a). The numerous green filaments and spiral patterns in the imagery are caused by buoyant cyanobacteria (blue-green algae), which commonly form summer blooms, often toxic, on the surface of the Baltic Sea under relatively low wind conditions [12,13]. An understanding of and an ability to predict concentrations of cyanobacteria are of practical importance and interdisciplinary interest. The WorldView imagery captures a small algal spiral pattern (Figure 1b) that provides the focus of this study. High-resolution stereo views of submesoscale phenomena are rare in the Baltic Sea and other areas as well; and in the present case, may provide insight into how the various algal patterns arise and what might be learned from them. For instance, the relationship of the spiral pattern to ocean dynamics remains unclear. It is reasonable to suppose that a spiral pattern derives from the action of an underlying eddy, but to our knowledge this has never been confirmed directly. Assuming an eddy, then how does the spiral pattern arise? Is it from kinematic distortion (i.e., from a differential angular velocity of the fluid) of any initial distribution of algae material [14], or is it from the dynamics of the fluid, in which the cyanobacteria, like any other surface floating material, become concentrated along frontal convergence zones that are being swept into the eddy [10]?

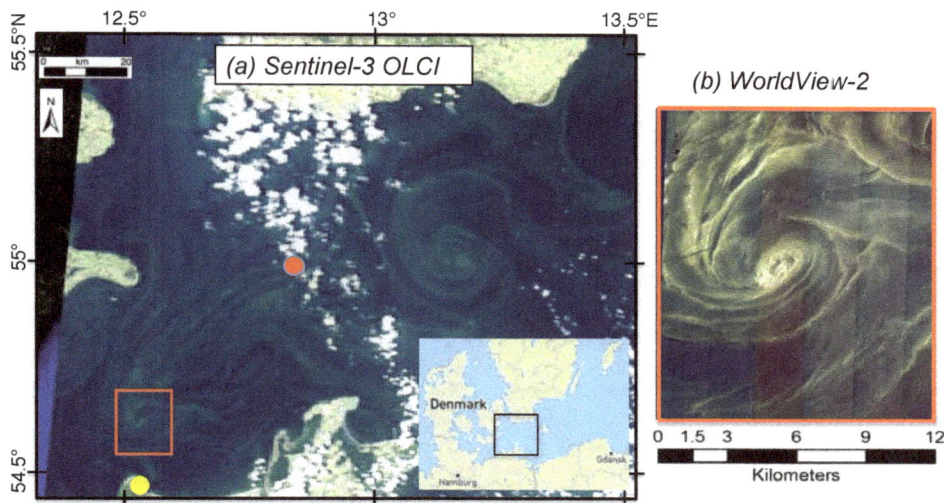

Figure 1. (**a**) Conditions in the Western Baltic Sea on 1 July 2018 as measured by the OLCI sensor aboard the Sentinel-3 satellite (https://s3view.oceandatalab.com/). Red-filled circle shows location of research platform FINO-2 (winds and in-water data); yellow-filled circle (just south of study site) indicates a land-based weather station. (**b**) WorldView-2 RGB image of the area of the red rectangle in panel a, showing the spiral pattern that is the focus of this study.

In order to address such questions, the WorldView stereo data are analyzed here using an optical-flow algorithm to deduce the velocity field. There are, however, a variety of noise sources—such as signal processing, data collection configuration, and environmental factors—that can contaminate the results. A scheme for ameliorating the effect of one processing artifact ("striping") is described and used in this work. As will be shown, the analysis does indeed reveal a vortex at the center of the algal spiral, which supports the interpretation of such algal spirals as evidence of a submesoscale eddy, and that an individual algal aggregation corresponds to a velocity front having both horizontal shear and convergence (and hence, downward transport).

2. Materials and Methods

2.1. Dataset

The WorldView-2 data were acquired on 1 July 2018, at 10:14:35 UTC (time t_1; shown in Figure 1b) and 10:15:52 UTC (time t_2); the time interval between data collections being $\Delta t = t_2 - t_1 = 77$ s. While only a segment of the imagery is examined in this study, browser versions of the full image strips can be accessed at https://discover.digitalglobe.com/; select "Area of Interest"; then "Search by image ID", where the identification numbers for the t_1 and t_2 images are 10300100803F8400 and

10300100819E2800. The t_1 and t_2 collections had similar off-nadir angles (25° and 31°), but necessarily different target azimuth angles (261° and 333°). Solar elevation and azimuth angles were 57° and 155°. After geo-referencing and resampling to a uniform UTM map grid, spatial resolution is $\Delta x = 2$ m for the eight color and near-infrared bands and $\Delta x = 0.5$ m for the panchromatic band. A further step of image-to-image registration using a simple translation and ground control points, results in rms displacement errors of about 1 pixel. High spatial resolution combined with relatively large image time difference yields a velocity detection threshold $\Delta x / \Delta t$ of 0.01 to 0.03 m s^{-1}.

Wind data are available from the German research platform FINO-2 (55.007°N, 13.154°E), and a close-by land station (54.443°N, 12.558°E). (See Figure 1a for locations; see Figure S1 for time series). During the 48-h period prior to acquisition of the WorldView imagery, the mean 10-m wind speed at FINO-2 was 3.3 m s^{-1} (s.d. = 1.3 m s^{-1}). For the hour immediately preceding image acquisition, the mean wind speed was 4.2 m s^{-1} at FINO-2 and 2.2 m s^{-1} at the land station, and the mean wind directions were 41.1° and 22.5°. In-water temperature and salinity measurements are also available from FINO-2, and profile and time series plots are shown in Figures S2 and S3. Stratification is dominated by temperature; the water column is weakly stratified in the upper 10 m, and becomes increasingly stratified toward the bottom. The stratification changes little over time, suggesting that horizontal spatial gradients are relatively weak.

2.2. Selection of Wavelength Bands to Analyze

An optimal signal would have a large dynamic range or image contrast and it would represent radiation backscattered from features at the sea surface, so that the derived velocity field could be ascribed to a fixed depth. The latter is an issue because the algae, in addition to forming accumulations at or very near the sea surface, will be dispersed by wind-induced shear and Langmuir circulation, and thus will have some vertical distribution. A dynamic range for an individual wavelength band can be defined as the rms signal in areas having real ocean structure minus that in areas where the signal is near the noise level. Values for the various WorldView bands were thus found to be as follows: highest (6.1 and 5.3 counts) for green and yellow wavelengths (bands 3 and 4); lowest (0.5 to 1.5 counts) for the shortest and the near-infrared wavelengths (bands 1, 2, 7, and 8); and intermediate (about 3 counts) for the red and red-edge wavelengths (bands 5 and 6). As the green wavelength band (band 3; 510 to 580 nm) had the best dynamic range, it was chosen for analysis. A segment of panchromatic data (dynamic range of 5.8 counts) will also be examined; although those data overall have a large percentage of signal near near-noise level, they do capture the finest-scale variations in the algal distributions.

As to the issue of what depth to associate with the derived velocity field, we can note that algal features were found to be highly coherent across bands 3 through 6. Band 6 (wavelengths of 706 to 746 nm) has a penetration depth of 0.75 m, based on the inverse attenuation coefficient for pure water; in the midst of an algal bloom, that depth is likely less. This suggests that, even though the algae may be mixed vertically downward several meters (see Section 3.2), the backscattered signal will be weighted toward the upper meter or so; hence, we assume the velocity field derived using band 3 represents the flow over the upper 1 to 2 meters of the water column.

2.3. Treatment of Noise Stripes

A push-broom instrument such as WorldView-2 uses a linear array of detectors arranged perpendicular to the flight direction of the spacecraft; different areas of the Earth's surface are then imaged as the spacecraft flies forward. As different electronic amplifiers are used to process sequential sub-arrays of detector elements, small residual errors inevitably arise across the array [15]. Such errors give rise to a pattern of prominent vertical stripes in imagery having a small dynamic range, which is often the case for imagery of the ocean. Such stripes are clearly visible, for example, in Figure 1b, which combines data from red, green, and blue wavelengths (bands 5, 3, and 2, respectively). Striping occurs in every wavelength band, including the panchromatic band, but to a varying degree; and, after geo-referencing data from any wavelength band, a particular stripe will appear in a different geographic location in the t_1 and t_2 images. The stripe—particularly the abrupt transition, or a jump from one stripe to another—can then be falsely interpreted as an ocean feature that has moved, and this will introduce noise into an image-derived velocity field. An initial de-striping of the data is thus an important step.

Approaches described in the literature [16,17] are based on deriving a polynomial to describe the detector variations across the entire image; this typically results in different corrective adjustments or offsets for every pixel across the push broom. Our approach differs in that it attempts to correct for only the effects of the most significant jumps, which often are regularly spaced across the image and readily identified. The approach consists of the following steps, all of which are done prior to geo-referencing the image data. First (step 1), the position of each jump and its height are determined through examination of an otherwise locally homogeneous part of the image. Second (step 2), the heights are applied uniformly and cumulatively within each stripe to create a one-dimensional array or row of offset values. (A minor complication arises because of a small overlap in the sub-array processing; as a consequence, a jump from one stripe to the next occurs over a range of pixels—about 7 pixels in this dataset.) Lastly (step 3), the offsets are used to de-stripe the data image.

We found it convenient to do both steps 1 and 3 within the image-processing environment of ImageJ [18]. For step 1, we use the ImageJ "Plot Profile" tool in areas of the image where the real ocean signal variation is weak compared to the jump. (To increase the precision to which the jump heights are determined, all data values are first multiplied by one hundred.) For step 3, the imported row of offsets is replicated sufficiently to make an image of the same size as the data image, and is then subtracted from it using ImageJ's "Image Calculator". Figure S4 shows the data both before and after the de-striping.

2.4. Method for Calculating Currents

Optical-flow methods estimate the velocity field by assuming conservation of a passive tracer; i.e., the total (Lagrangian) derivative of the tracer is equal to zero for a suitably short time interval. This condition is represented as an exact integral of the nonlinear tracer-conservation equation

$$I(\mathbf{r} + \mathbf{u}\,\Delta t, t_2) = I(\mathbf{r}, t_1) \tag{1}$$

where I is tracer intensity, Δt is the time difference between the two images, and $\mathbf{r}$ and $\mathbf{u}$ are the position and velocity vectors for a particular image pixel. For the present application, we use a nonlinear optical-flow method called the Global Optimal Solution [5–7]. In this approach, the image is partitioned into a number of square sub-domains (called tiles), n pixels on a side. The velocity field within each tile is modeled by a bilinear function. Velocity vectors at the tile nodes are derived through minimizing a cost function that is the sum of errors arising from the use of equation (1) at every pixel; an iterative technique is used that employs Gauss-Newton and Levenberg-Marquardt methods. Choice of tile size depends on the correlation length scale and noise level of the image data [7]. Our procedure at present is to try a range of tile sizes, then choose the one that is small enough to capture the features of interest without excessive noise. In this study, a value of $n = 300$ pixels with the band-8 data yielded a satisfactory result (see later Section 3.1); a value of $n = 50$ pixels was used with the panchromatic data.

A problem with any optical-flow method as applied to ocean imagery is sensitivity to intensity value changes that do not result from local surface currents; this is always a potential source of noise in the calculation [4]. An example would be changes in reflectance as the result of changes in viewing angles; another is contamination by non-ocean features such as clouds. To help mitigate such problems, two pre-processing steps are taken. The first is normalizing the images to have the same overall standard deviation and mean intensity levels. The second is creation of an image mask. In the present case, the elements of the mask include small clouds (and their shadows) that occur in the northwest corner of the image scene, an aircraft in the center-left part of the t_2 image, and an underway boat near the top edge. The mask of that boat includes the part of the boat's turbulent wake that formed between the t_1 and t_2 views, as this is a case where the tracer (the wake segment) is clearly not conserved between the two images. We did not, however, mask boat-generated internal waves that extend across the upper-middle part of the scene. (These are visible probably because of variations in reflected sky radiation resulting from internal wave-induced modulation of sea-surface roughness.) As internal waves propagate relative to the ambient fluid, they may contaminate estimates of the ambient water velocity.

After computing the velocity field, the vertical component of vorticity and horizontal divergence are calculated. These have their usual definitions: vorticity $\zeta = \partial v / \partial x - \partial u / \partial y$, and horizontal divergence $\delta = \partial u / \partial x + \partial v / \partial y$, where (u, v) are the (x, y) components of horizontal flow.

3. Results

3.1. Spiral Pattern (Band 3 Data)

Results from analysis of the spiral pattern using band 3 data are shown in Figure 2. To help distinguish any circulation pattern associated with the spiral, the velocity field (Figure 2a) was computed after first spatially translating the t_2 image to compensate for a 17-m southwestward drift of the spiral pattern between the t_1 and t_2 views. The dominant feature in the velocity field is an area of cyclonic (counter-clockwise) swirling flow that overlies what seems to be, based on the algae pattern, the visual center of the spiral pattern. This oval-shaped vortex is the central feature in the vorticity field (Figure 2b), though other features occur that may not all represent real dynamical features of the flow; see next section. In order to derive some properties of the vortex, we chose a contour level of 2.2×10^{-4} s^{-1} (e^{-1} times the maximum vorticity) to delineate a core region of the vortex. The objective here is to capture as much of the core vorticity as possible without including a possible errant signal, as represented by the lower-level contours that deviate from the overall oval shape. The area of the core was then used to estimate an approximate radius $R = 1.1$ km, and a mean vorticity $\zeta_{core} = 3.8 \times 10^{-4}$ s^{-1} (s.d.$=1.0 \times 10^{-4}$ s^{-1}); equivalently, $\zeta_{core} = 3.2 f$, where $f = 1.19 \times 10^{-4}$ s^{-1} is the local Coriolis frequency. As could be expected, the core radius and vorticity do vary with computational scale (i.e., tile size n): At lower resolution ($n = 400$), the radius is larger (1.37 km) and mean vorticity smaller (2.7×10^{-4} s^{-1}); at higher resolution ($n = 200$), the radius is smaller (0.91 km) and mean vorticity larger (4.1×10^{-4} s^{-1}). The divergence field (not shown), when averaged over the core area, yields values not significantly different from the background.

The character of the vortex core is of course just a part of a description of the eddy's hydrodynamics, as the spiraling arms of cyanobacteria accumulations extend over a much larger area. In synthetic aperture radar (SAR) imagery, at wind speeds of 0.2 to 5.6 m s^{-1}, the spiral arms would appear as relative dark streaks because of the wave-damping effect of algae-derived surface films. Sub-mesoscale eddies are thus manifested in SAR imagery as "black" spirals [19]. In the Baltic Sea, such black eddies have a mean diameter $D = 6.4$ km (s.d. $= 4.0$ km) [20], where D is measured between the most remote edges of the spiral pattern. One way spiral eddies can form is through ageostrophic baroclinic instability associated with a background horizontal density gradient; and theoretical studies [21] show such eddies to have cyclonic vorticity and a spiral diameter $D = 2 R_d$, where $R_d = (g \, \Delta\rho / \rho \, H)^{1/2} / f$ is the baroclinic deformation radius. Analysis of SAR spirals in the Baltic, Black, and Caspian seas shows that nearly all the spirals have a morphology consistent with cyclonic vorticity [19,20], a diameter proportional to R_d [20], and are statistically associated with lateral density gradients [20]. Our analysis explicitly reveals the cyclonic vorticity, and based on the available measurements of water stratification, we estimate a value of $R_d \sim 3.7$ km, which is close to a climatological value of 3.9 km near the study site [22]. These values of R_d can be compared with $D/2 \sim 3$ km, where $D \sim 6$ km is the distance between the farthest spiral arms in Figure 2. Our estimate of $\sim$1.1 km for the radius of the vortex core is smaller than both $D/2$ and R_d, thus making the vortex itself a submesoscale feature. Other studies [10,23] support the notion of a relatively small submesoscale vortex core and convergence bands at larger radius that spiral inwards toward the core.

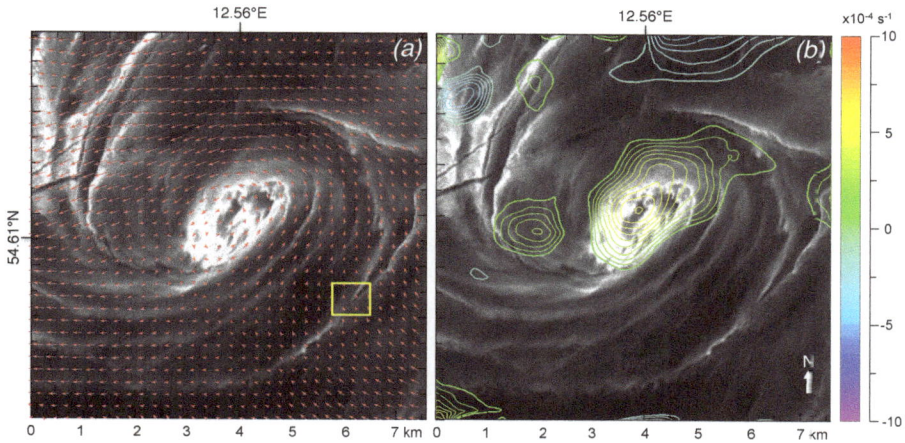

Figure 2. Results for area of the spiral pattern using data from band 3 (510 to 580 nm): (**a**) Velocity field; (**b**) Vorticity field. The largest vector in (**a**) has a magnitude of 0.21 m s^{-1}. Contours in (**b**) are shown at an interval of 5×10^{-5} s^{-1}, but only where the vorticity magnitude exceeds 1.5×10^{-4} s^{-1}. A subset area (yellow rectangle; **a**) is examined in the next figure.

3.2. Algal Aggregations (Panchromatic Data)

As an example of panchromatic data, we zoom into a representative area (Figure 3) within the spiral that shows two classes of algal features: windrows (the numerous bright streaks) and a long, individual algal aggregation that extends from the x axis toward the northeast, which we examine first.

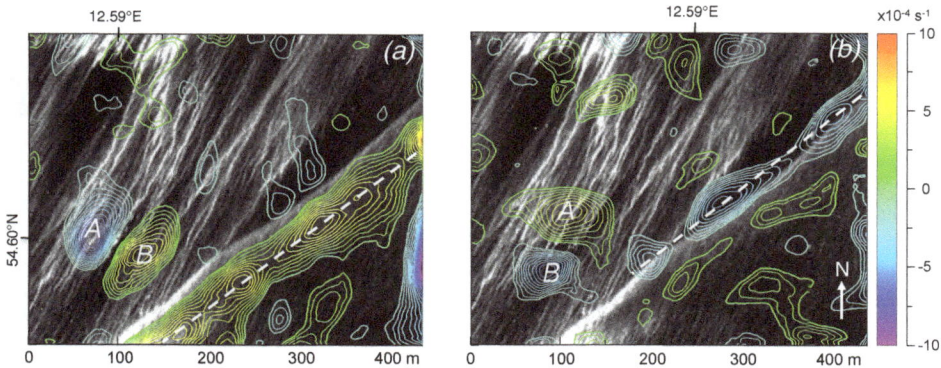

Figure 3. Example of algal aggregations in panchromatic data: (**a**) Vorticity; (**b**) Horizontal divergence. Contours are shown at an interval of 5×10^{-5} s^{-1}, but only where the vorticity or divergence magnitude exceeds 1.5×10^{-4} s^{-1}. Dashed lines and labeled features (A, B) are discussed in the text.

The dashed line in Figure 3a highlights a band of positive vorticity (horizontal shear) that lies to one side but parallels the aggregation. Spatially averaging in a 20-m wide strip along the dashed line yields a cyclonic vorticity of 4.81×10^{-4} s^{-1} (s.d. $= 1.07 \times 10^{-4}$ s^{-1}). The dashed line in Figure 3b connects areas of negative divergence that overlay the aggregation. The mean divergence is -2.9×10^{-4} s^{-1} (s.d. $= 1.42 \times 10^{-4}$ s^{-1}), and hence indicates surface convergence. These bands of vorticity and convergence indicate a velocity front. This supports an assumption made in the literature that such algal aggregations are frontal convergence zones and thus mark the locations of downwelling, i.e., that submesoscale convergence occurs within specific structures in the flow field [10]. Drifter measurements made in such structures show convergences of 2 to 6 f [10]; our value yields 2.4 f.

The windrows in Figure 3 are likely algae accumulations resulting from surface convergence created by Langmuir circulation cells [24,25]. Evidence for this is that the windrows are approximately aligned with the wind, i.e., they have an orientation of about 30°, which is within about 10° of the wind direction in the hour preceding image acquisition (Section 2.1). The spacing of the windrows in these data is in the range of 5 to 8 m, which would imply a mixing layer of 2.5 to 4 m depth. Amongst the windrows are a variety of relatively weak vorticity and divergence features, but an adjacent pair of negative and positive vorticity and divergence features (A and B in Figure 3a,b) stands out. These features can be accounted for as follows. Figure 4 shows an enlarged view of the area at times t_1 and t_2. The two views have been aligned using the windrow patterns; and, by referencing to a common set of grid lines, one can see that windrows patterns are indeed approximately stationary. On the other hand, numerous bright pixels near the middle of the scene, and associated with individual windrows, do move between views—by about 5 m, which corresponds to a relative speed of 0.06 m s^{-1}, or about 1% of the wind speed. Bright pixels such as in this figure have an enhanced response in the WorldView near-infrared bands, and that response is characteristic of surface cyanobacteria [26]. The bright pixels are thus assumed to correspond to small clumps of floating algae, and these clumps are likely to be affected by both the wind-drift layer and wind drag on parts of the clumps that extend above the sea surface. The enhanced speed of the clumps is detected by the optical-flow analysis and gives rise to the dipolar vortices (A, B) in Figure 3a, and areas of divergence (A) and convergence (B) in Figure 3b as the flow first accelerates upwind of the clumps and then slows downwind. These fine-scale (~100 m) flow variations might, if desired, be suppressed by using the near-infrared response of the algae clumps to mask them prior to the velocity computation.

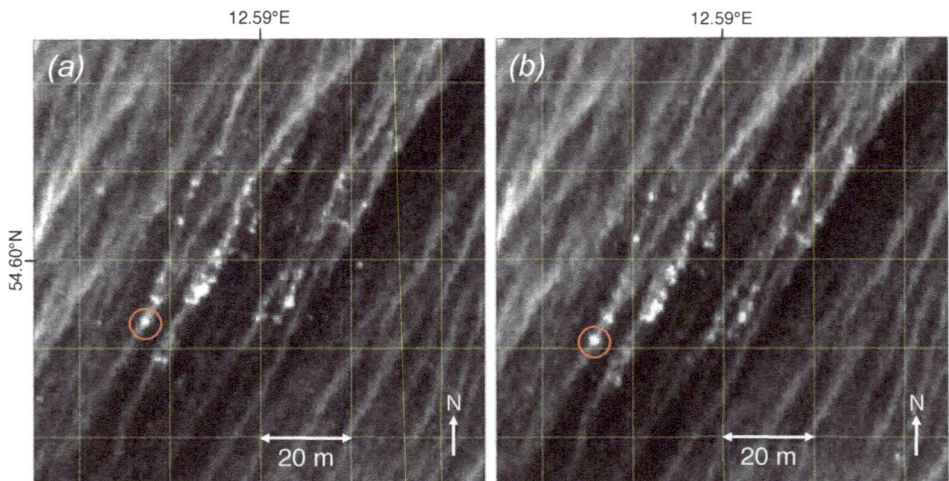

Figure 4. Enlargement of area near features A and B at times t_1 (**a**) and t_2 (**b**). The images shown here have been aligned using the windrow patterns. A common set of grid lines (spaced 20 m apart) helps illustrate how the brightest pixels (clumps of surface algae) move relative to the approximately stationary windrows. An example is circled in red and shows movement of about 5 m towards the south-southwest.

4. Conclusions

Very high-resolution, time-lagged satellite imagery has been used to examine aspects cf small-scale ocean dynamical features as revealed through a cyanobacteria bloom in the western Baltic Sea. In this case, a single pair of along-track stereo WorldView images was analyzed using an optical-flow algorithm to derive fields of velocity, vertical vorticity, and horizontal divergence. The results show (we believe for the first time) that an algal spiral pattern has at its center a cyclonic vortex, and that an individual algal aggregation corresponds to a velocity front having both horizontal shear and convergence (and hence, downward transport). While some sources of processing and environment noise have been identified, further attention is needed for identifying and quantifying noise in the derived flow field. Despite these shortcomings, the approach examined here is generally applicable, as previous studies using WorldView imagery analyzed with the same optical-flow algorithm have shown it provides a new way to quantify spatially complex phenomena and to assess the fidelity of high-resolution coastal circulation models [8,9].

Supplementary Materials: The following are available online at http://www.mdpi.com/2072-4292/11/7/865/s1.

Author Contributions: G.M. conceived the idea for this study and drafted the manuscript. W.C. analyzed the imagery using his Global Optimal Solution algorithm, and contributed to the writing of the final draft.

Funding: This work was supported by the Office of Naval Research under Naval Research Laboratory (NRL) Project 72-1C02-09.

Acknowledgments: Data from the FINO-2 platform were provided courtesy of the BMWi (Bundesministerium fuer Wirtschaft und Energie, Federal Ministry for Economic Affairs and Energy) and the PTJ (Projekttraeger Juelich, project executing organization). This is contribution NRL/JA/7230-19-0341.

Conflicts of Interest: The authors declare no conflict of interest.

References

1. Kääb, A.; Leprince, S. Motion detection using near-simultaneous satellite acquisitions. *Rem. Sens. Environ.* **2014**, *154*, 164–179. [CrossRef]
2. Qazi, W.A.; Emery, W.J.; Fox-Kemper, B. Computing ocean surface currents over the coastal California current system using 30-min-lag sequential SAR images. *IEEE Trans. Geosci. Rem. Sens.* **2014**, *52*, 7559–7580. [CrossRef]
3. Matthews, J.P.; Yoshikawa, Y. Synergistic surface current mapping by spaceborne stereo imaging and coastal HF radar. *Geophys. Res. Letts.* **2012**, *39*. [CrossRef]
4. Gade, M.; Seppke, B.; Dreschler-Fischer, L. Mesoscale surface current fields in the Baltic Sea derived from multi-sensor satellite data. *Int. J. Remote Sens.* **2012**, *33*, 3122–3146. [CrossRef]
5. Chen, W. A global optimal solution with higher order continuity for the estimation of surface velocity from infrared images. *IEEE TGRS* **2010**, *48*, 1931–1939.
6. Chen, W. Nonlinear inverse model for velocity estimation from an image sequence. *J. Geophys. Res.* **2011**, *116*, C06015. [CrossRef]
7. Chen, W.; Mied, R.P. River velocities from sequential multispectral remote sensing images. *Water Resour. Res.* **2013**, *49*, 3093–3103. [CrossRef]
8. Delandmeter, P.; Lambrechts, J.; Marmorino, G.O.; Legat, V.; Wolanski, E.; Remacle, J.F.; Chen, W.; Deleersnijder, E. Submesoscale tidal eddies in the wake of coral islands and reefs: Satellite data and numerical modelling. *Ocean Dyn.* **2017**, *67*, 897–913. [CrossRef]
9. Marmorino, G.; Chen, W.; Mied, R.P. Submesoscale Tidal-Inlet Dipoles Resolved Using Stereo WorldView Imagery. *IEEE Geosci. Rem. Sens. Lett.* **2017**, *14*, 1705–1709. [CrossRef]
10. D'Asaro, E.A.; Shcherbina, A.Y.; Klymak, J.M.; Molemaker, J.; Novelli, G.; Guigand, C.M.; Haza, A.C.; Haus, B.K.; Ryan, E.H.; Jacobs, G.A.; et al. Ocean convergence and the dispersion of flotsam. *Proc. Natl. Acad. Sci. USA* **2018**, *115*, 1162–1167. [CrossRef]
11. Taylor, J.R. Accumulation and subduction of buoyant material at submesoscale fronts. *J. Phys. Oceanogr.* **2018**, *48*, 1233–1241. [CrossRef]

12. Huisman, J.; Codd, G.A.; Paerl, H.W.; Ibelings, B.W.; Verspagen, J.M.; Visser, P.M. Cyanobacterial blooms. *Nature Rev. Microbiol.* **2018**, *16*, 471. [CrossRef]

13. Kahru, M. Using satellites to monitor large-scale environmental change: A case study of cyanobacteria blooms in the Baltic Sea. In *Monitoring Algal Blooms: New Techniques for Detecting Large-Scale Environmental Change*; Springer: Berlin, Germany, 1997; pp. 43–61.

14. Flohr, P.; Vassilicos, J.C. Accelerated scalar dissipation in a vortex. *J. Fluid Mech.* **1997**, *348*, 295–317. [CrossRef]

15. Updike, T.; Comp, C. *Radiometric Use of WorldView-2 Imagery*; Tech. Note; DigitalGlobe Inc.: Longmont, CO, USA, 2010.

16. Corsini, G.; Diani, M.; Walzel, T. Striping removal in MOS-B data. *IEEE Trans. Geosci. Rem. Sens.* **2000**, *38*, 1439–1446. [CrossRef]

17. Lyon, P.E. An automated de-striping algorithm for Ocean Colour Monitor imagery. *Int. J. Rem. Sens.* **2009**, *30*, 1493–1502. [CrossRef]

18. Schneider, C.A.; Rasband, W.S.; Eliceiri, K.W. NIH Image to ImageJ: 25 years of image analysis. *Nat. Methods* **2012**, *9*, 671. [CrossRef]

19. Karimova, S.; Gade, M. Improved statistics of sub-mesoscale eddies in the Baltic Sea retrieved from SAR imagery. *Int. J. Rem. Sens.* **2016**, *37*, 2394–2414. [CrossRef]

20. Karimova, S. Spiral eddies in the Baltic, Black and Caspian seas as seen by satellite radar data. *Adv. Space Res.* **2012**, *50*, 1107–1124. [CrossRef]

21. Eldevik, T.; Dysthe, K.B. Spiral eddies. *J. Phys. Oceanogr.* **2002**, *32*, 851–869. [CrossRef]

22. Fennel, W.; Seifert, T.; Kayser, B. Rossby radii and phase speeds in the Baltic Sea. *Cont. Shelf Res.* **1991**, *11*, 23–36. [CrossRef]

23. Marmorino, G.; Smith, G.; North, R.; Baschek, B. Application of airborne infrared remote sensing to the study of ocean submesoscale eddies. *Front. Mech. Eng.* **2018**, *4*. [CrossRef]

24. Thorpe, S.A. Langmuir circulation. *Annu. Rev. Fluid Mech.* **2004**, *36*, 55–79. [CrossRef]

25. Szekielda, K.H.; Marmorino, G.O.; Maness, S.J.; Donato, T.F.; Bowles, J.H.; Miller, W.D.; Rhea, W.J. Airborne hyperspectral imaging of cyanobacteria accumulations in the Potomac River. *J. Appl. Rem. Sens.* **2007**, *1*, 013544.

26. McKinna, L.I. Three decades of ocean-color remote-sensing Trichodesmium spp. in the World's oceans: A review. *Progr. Oceanogr.* **2015**, *131*, 177–199. [CrossRef]

remote sensing

MDPI

Article

Impact of the Acquisition Geometry of Very High-Resolution Pléiades Imagery on the Accuracy of Canopy Height Models over Forested Alpine Regions

Livia Piermattei [1,*], Mauro Marty [2], Wilfried Karel [1], Camillo Ressl [1], Markus Hollaus [1], Christian Ginzler [2] and Norbert Pfeifer [1]

[1] Department of Geodesy and Geoinformation, TU Wien, 1040 Wien Austria;
Wilfried.Karel@geo.tuwien.ac.at (W.K.); Camillo.Ressl@geo.tuwien.ac.at (C.R.);
Markus.Hollaus@geo.tuwien.ac.at (M.H.); Norbert.Pfeifer@geo.tuwien.ac.at (N.P.)

[2] Swiss Federal Institute for Forest, Snow and Landscape Research, WSL, 8903 Birmensdorf, Switzerland;
mauro.marty@wsl.ch (M.M.); christian.ginzler@wsl.ch (C.G.)

* Correspondence: livia.piermattei@geo.tuwien.ac.at; Tel.: +43-(1)58-8011-2257

Received: 11 August 2018; Accepted: 21 September 2018; Published: 25 September 2018

Abstract: This work focuses on the accuracy estimation of canopy height models (CHMs) derived from image matching of Pléiades stereo imagery over forested mountain areas. To determine the height above ground and hence canopy height in forest areas, we use normalised digital surface models (nDSMs), computed as the differences between external high-resolution digital terrain models (DTMs) and digital surface models (DSMs) from Pléiades image matching. With the overall goal of testing the operational feasibility of Pléiades images for forest monitoring over mountain areas, two questions guide this work whose answers can help in identifying the optimal acquisition planning to derive CHMs. Specifically, we want to assess (1) the benefit of using tri-stereo images instead of stereo pairs, and (2) the impact of different viewing angles and topography. To answer the first question, we acquired new Pléiades data over a study site in Canton Ticino (Switzerland), and we compare the accuracies of CHMs from Pléiades tri-stereo and from each stereo pair combination. We perform the investigation on different viewing angles over a study area near Ljubljana (Slovenia), where three stereo pairs were acquired at one-day offsets. We focus the analyses on open stable and on tree covered areas. To evaluate the accuracy of Pléiades CHMs, we use CHMs from aerial image matching and airborne laser scanning as reference for the Ticino and Ljubljana study areas, respectively. For the two study areas, the statistics of the nDSMs in stable areas show median values close to the expected value of zero. The smallest standard deviation based on the median of absolute differences (σ_{MAD}) was 0.80 m for the forward-backward image pair in Ticino and 0.29 m in Ljubljana for the stereo images with the smallest absolute across-track angle ($-5.3°$). The differences between the highest accuracy Pléiades CHMs and their reference CHMs show a median of 0.02 m in Ticino with a σ_{MAD} of 1.90 m and in Ljubljana a median of 0.32 m with a σ_{MAD} of 3.79 m. The discrepancies between these results are most likely attributed to differences in forest structure, particularly tree height, density, and forest gaps. Furthermore, it should be taken into account that temporal vegetational changes between the Pléiades and reference data acquisitions introduce additional, spurious CHM differences. Overall, for narrow forward–backward angle of convergence (12°) and based on the used software and workflow to generate the nDSMs from Pléiades images, the results show that the differences between tri-stereo and stereo matching are rather small in terms of accuracy and completeness of the CHM/nDSMs. Therefore, a small angle of convergence does not constitute a major limiting factor. More relevant is the impact of a large across-track angle (19°), which considerably reduces the quality of Pléiades CHMs/nDSMs.

Keywords: very high-resolution Pléiades imagery; canopy height model; acquisition geometry; forested mountain; accuracy assessment

1. Introduction

Mountain forests provide a wide range of ecosystem services in terms of protective, productive, social and economic functions. Therefore, the timely information on the state and the change of land-use and forest cover, productivity, and structure is crucial for different stakeholders from local to regional scales. To quantify the provided forest services, detailed forest information is required with high spatial and temporal resolutions. Among the forest metrics, canopy height, which describes the top of the vegetated canopy, is the basis for deriving other parameters such as forest gaps, crown coverage, canopy density, volume, and biomass. For deriving canopy height at wide spatial coverage i.e., at the landscape scale, remote sensing observations with fine spatial resolutions are required. Current remote sensing systems that fulfil this requirement include airborne laser scanning (ALS) and multi-view aerial or very high resolution (VHR) satellite imagery [1]. Since the last decade, airborne laser scanning (ALS) has been the primary data source for three-dimensional (3D) information on forest vertical structure [2–4]. The main advantage of ALS for forest applications is the capability to penetrate through the vegetation and thus to obtain the top height, the forest vertical structure, and the bare-earth, with the latter being needed to define the terrain height. By contrast, passive optical sensors can provide only the topmost surface of forest canopies, where at least two images (a so-called image stereo pair) share a common area in the scene [5]. Both aerial and satellite images have been used to record forest change for more than 30 years. In the last decade, thanks to great technological improvements, the gap between aerial and VHR satellite imagery has become smaller in terms of image resolution (up to 30 cm ground sample distance (GSD)). However, in comparison with airborne remote sensing, VHR satellite imagery has the benefits of worldwide availability without any access restrictions, large area coverage and high temporal resolutions of only a few days.

Forest mapping over large spatial extents with VHR satellites started with IKONOS in 1999 [6,7]. Since then, several VHR satellites have been launched such as QuickBird, GeoEye-1, WorldView-1, -2, and -3, Ziyuan-3A, and the Pléiades satellites. WorldView multispectral stereo imagery was largely used to analyse forest structure [8,9], the size of tree crowns [10] and for the 3D modelling of forest canopies [1,11,12].

Among the available VHR satellite systems, we consider Pléiades imagery over mountain regions for deriving DSMs. This first European VHR satellite system is comprised of two identical satellites, Pléiades-1A (PHR1A) and Pléiades-1B (PHR1B), which were launched in December 2011 and in December 2012, respectively. Both satellites fly at an altitude of 694 km in sun-synchronous orbits with 98.2° inclination and an offset of 180° from each other, which provides a daily revisit capability [13]. An outstanding feature of the Pléiades system is the great agility of its sensors that allows for optimised acquisitions of areas of interest, with stereo angles varying from ~6° to ~28° [14]. The time difference between along-track images is in the range of a few seconds only, which guarantees almost constant sun illumination conditions, limited changes in the scene and similar cloud coverages in all of them [15]. Moreover, the sensor is designed to acquire panchromatic and multispectral images, which are delivered at nominal resolutions of 0.5 m and 2 m, respectively, in stereo (forward-, backward-view) and tri-stereo (forward-, nadir-, and backward-view) modes along-track.

To explore the full potential of Pléiades images over forest mountain areas, an investigation of the relationship between DSM accuracy and imaging geometry, such as tri-stereo imagery, convergence angle and across-track angle, is essential for an optimal image acquisition planning. Therefore, we focus on answering the question of which combination and acquisition setting of tri-stereo or stereo pairs of Pléiades images can produce the highest DSM accuracy over Alpine forest regions.

Related Work and Research Questions

Despite the advantages of the Pléiades system, only a limited number of studies have investigated the versatility of the Pléiades system through the controllable viewing angle and stereo and tri-stereo views for forestry applications. The generation of height models with Pléiades triplets with large and short base length image combinations has been investigated in different urban areas by References [16–19]. In Alpine areas, the capability of Pléiades tri-stereo to deliver reliable DSMs in complex terrain was demonstrated by References [20–22]. Additionally, the benefit of using tri-stereo images was tested to estimate the lava flow volume [12], and the height changes induced by earthquakes [23]. Among the few researchers that have used the Pléiades satellites for forestry purposes, the Pléiades image texture for forest structure mapping and forest classification has been investigated previously in [24,25]. Recently, the potential for use Pléiades images has been investigated for estimating forest attributes for 10 m plots in a boreal forest [26], for deriving forest biomass by combining spectral and geometric information [27,28], for predicting forest inventory attributes in New Zealand's planted forests [29] and for modelling tree diversity [30].

In contrast to these previous works on the use of Pléiades satellite images over forest areas, our study is, to the best of our knowledge, the first to explore the accuracy of DSMs/CHMs derived from Pléiades imagery over large mountain regions. Specifically, our work aims at answering the following two questions, both regarding the accuracy of derived DSMs: (1) what is the benefit of using tri-stereo images versus stereo pairs, and (2) what is the impact of different viewing angles and topography? According to this twofold goal, the investigation is carried out in two accordingly selected study areas located in Alpine forest regions. In order to derive the height above ground (nDSM), we subtract external high resolution DTMs from Pléiades image matching DSMs. We focus the analyses of the nDSMs/CHMs on open stable areas, and on tree covered areas. For evaluating the accuracy of Pléiades CHMs, we use CHMs from aerial image matching and ALS as reference for the Ticino and Ljubljana study areas, respectively.

2. Test Sites and Pléiades Image Data Sets

We tasked a new tri-stereo Pléiades data acquisition over Canton Ticino, Switzerland (site "Ticino") to investigate the potential of triplet scenes and the impact of tree height and slope on CHM accuracy in forest mountain areas. The area was chosen for its topographic characteristics, with elevations between 220 m and 2265 m a.s.l. and an average slope of 37°. The tri-stereo data were acquired with platform PHR1A within 20 s on 3 September 2017 around 10:00 in North-South direction, covering ~125 km^2 in total (Figure 1). To study the impact of different along- and across-track angles, we used three Pléiades stereo pairs near Ljubljana, Slovenia (site "Ljubljana") available in the supplier's archive, which were acquired one day apart from each other with platform PHR1A on 27 and 29 July 2013, and with platform PHR1B on 28 July 2013. This study area is rather flat with elevations between 346 m and 1900 m a.s.l., and an average slope of 8.6°. It comprises about 400 km^2 that largely consist of agricultural land and managed forest.

The optical satellite images for both study areas were delivered as four bands (blue, green, red, near infrared), pan-sharpened with spatial resolutions (i.e., mean GSDs) between 0.71 m and 0.78 m, depending on the viewing angle. The viewing angles and consequently the convergence angles and baseline to height ratios are different for each stereo pair. For the Ticino dataset, according to our request, we received one quasi-nadir image (viewing angle close to the vertical), and one backward-and one forward-view with symmetric along-track angles, and a small convergence angle of about 12°. The archive images over Slovenia were collected in stereo mode with symmetric and asymmetric along-track angles and rather large convergence angles. In the across-track direction, the mean angles are −5.3°, 6.8°, and 19.6°. The acquisition properties of the satellite images for each study area are given in Table 1. Figure 1 shows the Pléiades satellite positions over Ticino and Ljubljana.

Figure 1. Study areas and imaging geometries of the Pléiades data sets (Google Earth preview of the footprints and the satellite's position).

Table 1. Acquisition properties for the satellite images over the two study areas.

Study Area	Acquisition Date	View	Acqu. Time	GSD (m)		Incidence Angles (°)		
				Along Track	Along Track	Across Track	Angle of Convergence
Ticino	3 September 2017	Forward	10:29:03	0.71	6.05	−0.42	5 (FN)
		Nadir	10:29:13	0.70	0.89	0.78	6 (NB)
		Backward	10:29:22	0.70	−5.37	2.49	12 (FB)
Ljubljana	27 July 2013	Forward	10:10:09	0.73	−12.12	−1.98	24
		Backward	10:10:51	0.74	12.36	−8.59	
	28 July 2013	Forward	10:03:43	0.71	−7.39	9.36	22
		Backward	10:04:21	0.74	14.78	3.95	
	29 July 2013	Forward	09:55:04	0.77	−10.20	22.31	27
		Backward	09:55:50	0.78	16.91	16.90	

3. Data Processing and Analyses

3.1. Reference Data

For Ticino, an aerial image matching DSM [31] was used as reference data. The aerial images were acquired in spring 2015 (two years before the Pléiades images) and the corresponding DSM was provided by the Swiss National Forest Inventory as a raster with 1 m resolution. The digital terrain model swissAlti3D from the Federal Office of Topography (swisstopo) with a resolution of 2 m was upsampled to 1 m and was used to derive the Pléiades and Aerial nDSM and to improve the absolute geolocation of the Pléiades DSMs (see Section 3.2.2). The reference data in Slovenia was based on ALS data collected in 2015 (i.e., two years after the Pléiades images), having a mean density of 14 points/m^2. From the classified ALS point cloud, the DTM and the nDSM were derived with a resolution of 1 m using OPALS [32]. For both study areas, an existing orthophoto of 0.2 m spatial resolution was used to pick the ground control points (GCPs) and check points (CPs) over stable areas, whose height coordinates were extracted from the corresponding DTM. The GCPs were used in the image orientation phase to optimize the Pléiades projection parameters provided along with the image data, whereas the purpose of the CPs was to validate the accuracy of the image orientations and the DSMs. The GCPs and CPs were homogeneously distributed in planimetry and height across the area. The accuracies of the measured GCP and CP object coordinates were in the order of 20 cm in planimetry and 25 cm in the vertical direction.

3.2. Pléiades Image Processing and DSM Generation

3.2.1. Pléiades Image Processing

The reconstruction of 3D points from VHR satellite imagery requires at least two overlapping images, and it is performed by applying photogrammetric techniques and dense image matching algorithms. The transformation between image and object space is given in terms of the Rational

Polynomial Coefficients (RPCs) model [33]. The RPCs are provided for each image by the satellite vendor. For the Pléiades pan-sharpened imagery the RPCs are reported to have a geo-location accuracy of 8.5 m CE90 (circular error at 90% confidence) for nadir view in the nadir direction, which corresponds to a standard deviation of 4 m [34]. Therefore, the declared absolute accuracy is not sufficient if sub-meter accuracy is required, but this can be achieved with the use of GCPs. Thus, in total, we employed 18 GCPs for each study area and 7 and 12 CPs for Ticino and Ljubljana, respectively. Additionally, in order to quantify the geolocation accuracy of the original RPCs, the tri-stereo images over Ticino were also evaluated without GCPs.

The Pléiades images were processed using Trimble/Inpho software. To generate a 3D point cloud from stereo imagery the main procedure is (i) to import the images and the RPC information, (ii) to (optionally) refine the orientation of the images based on tie points and GCPs, and (iii) to apply dense image matching to determine the dense point cloud. The general workflow from the images to the final DSM is illustrated in Figure 2.

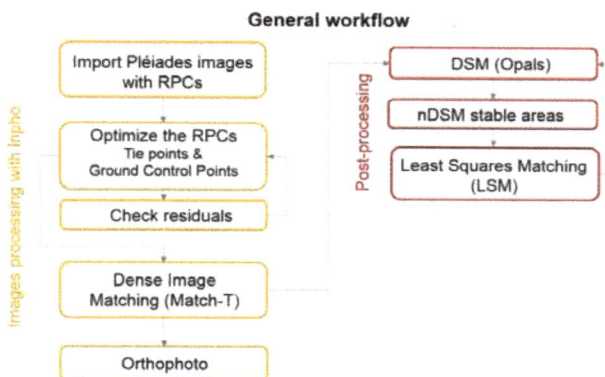

Figure 2. General workflow to generate the nDSMs from Pléiades images.

For each study area, the image orientation refinement was performed jointly for all available images in order to minimize the influence of the image orientation quality on the quality of the DSMs derived for particular combinations of images. Therefore, tie points were extracted automatically for all images, and they were employed together with the GCPs to refine the initial RPC coefficients. During bundle block adjustment, the residuals of tie points, GCPs, and CPs were computed by the software. Consequently, tie points with image residuals larger than 2 pixels were considered as blunders and removed, and the RPCs refined again. For dense image matching, Match-T was used. This is a module of the Trimble/Inpho software, which adopts a feature-based on the higher and a cost-based strategy on the lower pyramid levels. The cost-based matching is a version similar to the semi-global matching algorithm [35], which computes an object point for every pixel. For the Ticino study area, after a simultaneous orientation refinement of the tri-stereo images, dense matching was performed independently for the tri-stereo data (forward, nadir, backward, FNB) and for each stereo pair i.e., forward-nadir (FN), nadir-backward (NB), and forward-backward (FB). A similar approach was adopted for the Ljubljana study area. Thus, the six images were jointly oriented based on tie points and GCPs, and dense matching was performed for the forward-backward pair of each day of acquisition. For both study areas, four band orthophotos were generated to derive normalised difference vegetation index (NDVI) maps.

3.2.2. Pléiades DSM

The 3D points generated by dense image matching were turned into regular rasters of height values (i.e., DSMs) with 1 m resolution using the moving (tilted) plane interpolation with a search radius of 3 m. We have found this grid interpolation to be the optimal compromise between the preservation of detail and the reconstruction of void areas over vegetation. This interpolation approach was used for generating the DSMs for all combinations of images and for both study areas. We obtained the nDSMs by subtracting the ALS DTMs from the Pléiades DSMs. Because the reconstructed area of each image combination was slightly different due to different image footprints, the analyses were performed on common regions of interest, which were about 103 km^2 and 344 km^2 for Ticino and Ljubljana, respectively (see yellow rectangles in Figure 3). Since for both study areas, systematic errors were visible between the Pléiades and reference nDSMs, we applied least squares matching (LSM) to reduce them (compare [36]). LSM estimated the full 3D affine transformation parameters of Pléiades DSMs that minimize the errors with respect to their reference DTMs over common stable areas. These stable areas were identified based on several features. Mainly, the absolute values of nDSM cells of the reference data needed to be less than 2 m for the aerial image data (Ticino), and less than 0.5 m for the ALS data (Ljubljana). Subsequently, Pléiades nDSM cells with values greater than 60 m were identified as clouded areas and were removed from the mask of stable areas. Additionally, NDVI map cells exceeding the thresholds of 0.1 for Ticino and −0.1 for Ljubljana were classified as rivers and lakes, and were removed from their mask. The final stable areas consisted of approximately 28.3% of the scene for Ticino and 38.7% for Ljubljana. Subsequently, the 3D point cloud of each Pléiades image combination was transformed according to the estimated LSM parameters and re-interpolated to generate the final DSM. For comparison purposes and due to the low quality of the reference ALS DTM of the Ticino study area, LSM was also applied to the aerial reference DSM. Having applied the estimated transformation parameters, the aerial reference DSM points were interpolated using the same method as for the Pléiades DSMs.

3.3. CHM Generation

To focus the analysis on forested areas and thus to derive the CHMs, a tree mask across the entire area of interest was generated. The tree mask of the Ticino area was derived by applying a lower threshold of 0.6 to the NDVI map. Furthermore, in order to exclude meadows and cloud cover, cells of the Pléiades and aerial nDSMs smaller than 2 m or greater than 60 m were removed from the tree mask. These nDSM thresholds were also used for masking the trees in the Ljubljana area. However, because several forest areas had been harvested during the long time lag between the acquisitions of the Pléiades images and the reference ALS data, the NDVI map could not be used. Instead, the tree mask was further restricted to areas with ALS points classified as vegetation. According to the final masks, the area covered by trees was approximately 57% in Ticino and 43% in Ljubljana. For evaluating the accuracy of Pléiades CHMs, the aerial image matching and ALS CHMs were used as reference for the Ticino and Ljubljana study areas, respectively.

3.4. Accuracy Assessment

For both study areas, the quality of the photogrammetric Pléiades image processing and the derived products was assessed by considering three aspects. Firstly, we evaluated the quality of the image orientation by means of the image residual errors of the tie points, GCPs, and independent CPs. The ground coordinates of the GCPs and CPs were used to calculate the horizontal and vertical RMSE of the residuals of measured and transformed coordinates. Secondly, the vertical accuracies of the Pléiades DSMs were assessed in more detail for the entire scene and for each image combination, both before and after applying LSM, and separately for stable areas and forested areas. The vertical accuracy over stable areas was quantified by considering (a) the vertical RMSEs between the measured and predicted object coordinates of the GCPs and CPs for each generated DSM and (b) the Pléiades nDSMs, having an expected value of zero. In forest areas, the reference CHM was subtracted from the Pleiades CHM to calculate the height differences (ΔH). Subsequently, the error distribution of these ΔH was analysed, and for the vertical accuracy assessment we derived measures such as mean, standard deviation (σ), median and a robust standard deviation based on the median of absolute differences (σ_{MAD}). For both study areas, potential factors controlling the quality of the Pléiades CHMs like the tree height and terrain slope were evaluated with respect to vertical accuracy. Thirdly, a detailed analysis of the Pléiades nDSMs was performed on small selected forest areas of 500 m by 500 m (blue squares in Figure 3). Profiles within these areas were analysed to provide a detailed view of the structure of the produced Pléiades nDSMs in comparison to the reference data set Specifically, in Ticino, one area that exhibits steep terrain and sparse forest coverage was chosen to investigate the performance of the tri-stereo dense matching in comparison to each stereo pair. The analysis of the impact of different viewing angles on the reconstruction of the forest canopy height was performed on two selected areas in Ljubljana, where the first (area 1) is characterised by homogenous forest height, adult trees and high topography variation, and the second (area 2) by coarse forest cover with several gaps and a relatively flat terrain.

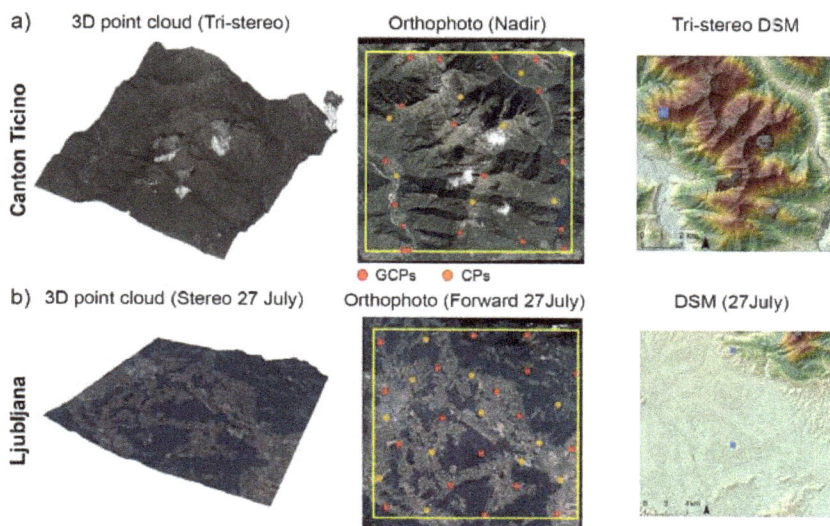

Figure 3. For (**a**) the Ticino and (**b**) the Ljubljana study areas, left: the 3D point cloud. Centre: the orthophoto generated from the 4-bands Pléiades images, visualised as true colour RGB, and overlaid with the GCPs (red circles) and CPs (orange circles). The yellow rectangles represent the common regions of interest for all scene combinations. Right: the colour coded, reconstructed DSMs. The blue squares are the 500 × 500 m selected areas.

4. Results

For each study area, the image orientation refinement was performed jointly for all available images. In Ticino, the bundle adjustment was performed with all three images in a single block. Automatic tie point extraction identified ~580 points. The RMSE of the GCPs is in the range of a decimetre both in the horizontal and vertical directions. At the CPs, a similar accuracy is achieved in planimetry, whereas with 1.04 m, the RMSE results were much larger in the vertical direction (Table A1). In Ljubljana, the RPCs of the six images were improved simultaneously in one single block using 18 GCPs and automatically extracted tie points. The number of the automatically extracted tie points ranges between 690 and 806. The standard deviation of the tie point residuals ranges between 0.39 and 0.52 pixel. In the vertical direction, the RMSE of the ground coordinates is 0.80 m at the CPs, whereas at the GCPs it is 0.17 m. In planimetry, the accuracy at the GCPs and CPs is almost the same (Table A2). For details of adjustment results, see Appendix A for Ticino and Ljubljana, respectively.

Dense image matching was successful on forest areas (Figure 3). With ~1362 million points, FNB over Ticino provided a larger point cloud than each of the three stereo pairs, having ~700 (FN),

~696 (NB), and ~657 (FB) million points. For each of the Ljubljana stereo pairs, around 2000 million points were matched, where the lowest number of points was generated for the stereo pair with the widest angle of convergence.

Over the entire scene, the interpolation ensured an almost complete reconstruction of the void areas since those areas were small enough to be reliably filled by the used grid interpolation with 3 m search radius. In all Pléiades DSMs, less than 1.5% of pixels result as void, being located within or close to the clouds or, for Ticino, on the lake surface. Despite this, the image triplet reduced the missing height values by up to 0.7 percentage points when compared to the standard stereo FB DSM. Concerning Ljubljana, the image pair with the widest angle of convergence resulted in a notably larger amount of missing data than the other pairs (Table 2).

Table 2. For both study areas, the performance of image matching and the percentage of cloud coverage and empty cells (no data in %) within the regions of interest. The cloud coverage was defined by selecting the nDSM cells with absolute values greater than 60 m.

	Ticino (Switzerland)				Ljubljana (Slovenia)				
Scene Comb.	Matching Time (h)	LAS File (GB)	Cloud Coverage (%)	No Data (%)	Image Acquis.	Matching Time (h)	LAS File (GB)	Cloud Coverage (%)	No Data (%)
FNB	16	34	2.1	0.8	27 July	25	54	1.0	0.1
NB	10	17	2.1	1.2	28 July	26	53	0.2	0.1
FN	10	18	2.3	1.3	29 July	28	49	0.0	0.4
FB	9	17	1.8	1.5					

4.1. Impact of the Tri-Stereo Acquisition on the Quality of the Derived Products in Ticino

4.1.1. Accuracy Assessment of the VHR nDSMs Over Stable Areas and of the VHR-CHMs

The GCPs and CPs were employed to assess the vertical accuracy of the DSM before and after LSM for each image combination (Table 3). The total RMSE of the DSMs before LSM at the GCPs and CPs ranged between 0.79 m (FB) and 1.25 m (FN). When the GCPs were not used within the bundle adjustment, the vertical accuracy of the tri-stereo DSM resulted with an offset of approximately 37 m at the GCP and CP locations. Nevertheless, this vertical offset was completely removed by application of LSM, resulting in the same accuracy achieved with GCPs. The LSM transformation slightly reduced the total RMSEs for each image combination, ranging between 0.68 m (FB) and 1.10 m (FN).

Table 3. RMSE in the Z-direction of the GCPs and CPs for each generated DSM. (* No GCPs within the bundle adjustment).

	$RMSE_Z$ DSM (m)								
	Before LSM					After LSM			
	FNB *	FNB	FN	NB	FB	FNB *	FNB	FN	NB	FB
18 GCPs	36.99	0.69	0.99	0.88	0.60	0.72	0.52	0.85	0.85	0.46
7 CPs	37.41	1.21	1.67	0.90	1.09	1.09	1.20	1.48	0.90	1.00
Tot (25)	37.12	0.89	1.25	0.89	0.79	0.90	0.75	1.10	0.89	0.68

The spatial distribution of the normalised elevation (nDSM) for the tri-stereo and the aerial images are shown in Figure 4 before and after LSM.

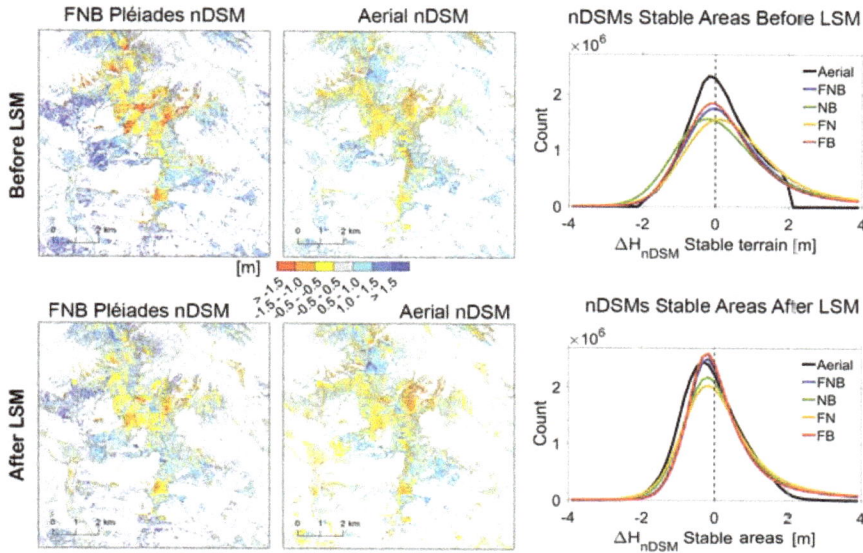

Figure 4. nDSM statistics for stable areas, with results before (**top**) and after (**bottom**) LSM. (**Left**) colour coding of the Pléiades (FNB) and aerial nDSMs (non-stable areas shown in white). (**Right**) distribution of nDSM heights for the aerial images, and for the Pléiades tri-stereo and all stereo pairs.

According to visual analysis, both positive and negative biases are visible on the stable areas for both Pléiades and aerial nDSMs before and after the LSM transformation. Before LSM, the vertical error distributions of the Pléiades nDSMs feature medians between 0.14 m and 0.43 m, with a maximum σ_{MAD} of 1.30 m (Table 4).

Table 4. Height differences (ΔH) before and after LSM of Pléiades and aerial nDSMs over stable areas.

	ΔH nDSM for Stable Areas (m)									
	Before LSM					**After LSM**				
	Mean	*Std*	*Median*	σ_{MAD}	*±1 m (%)*	*Mean*	*Std*	*Median*	σ_{MAD}	*±1 m (%)*
Aerial	0.11	0.84	0.05	0.85	74.8	−0.07	0.84	−0.15	0.80	77.6
Pléiades FNB	0.76	2.26	0.26	1.17	60.7	0.55	2.06	−0.05	0.85	72.3
NB	0.65	2.43	0.14	1.30	56.6	0.56	2.21	0.07	0.96	68.6
FN	0.88	2.33	0.43	1.29	56.2	0.54	2.23	0.08	1.02	66.1
FB	0.70	2.43	0.19	1.12	62.4	0.54	2.27	0.02	0.82	73.4

The highest accuracy was achieved by the standard stereo FB, although after LSM the accuracy is practically the same for all image combinations and comparable with the results from aerial image matching. LSM improved the accuracy of the Pléiades DSMs on the stable areas, which yielded a median close to zero for each scene combination, and a σ_{MAD} below 1 m. Only the FN stereo pair exceeds this value slightly. Moreover, the ratio of Pléiades cells with an absolute accuracy of better than 1 m increases by around 10 percentage points due to application of LSM. Nevertheless, both Pléiades and aerial nDSMs after LSM show a negative shift in the frequency distribution histograms.

The distributions of ΔH between the Pléiades and aerial CHMs before and after LSM are reported in Figure 5 by means of a map of the differences, histograms, and boxplots.

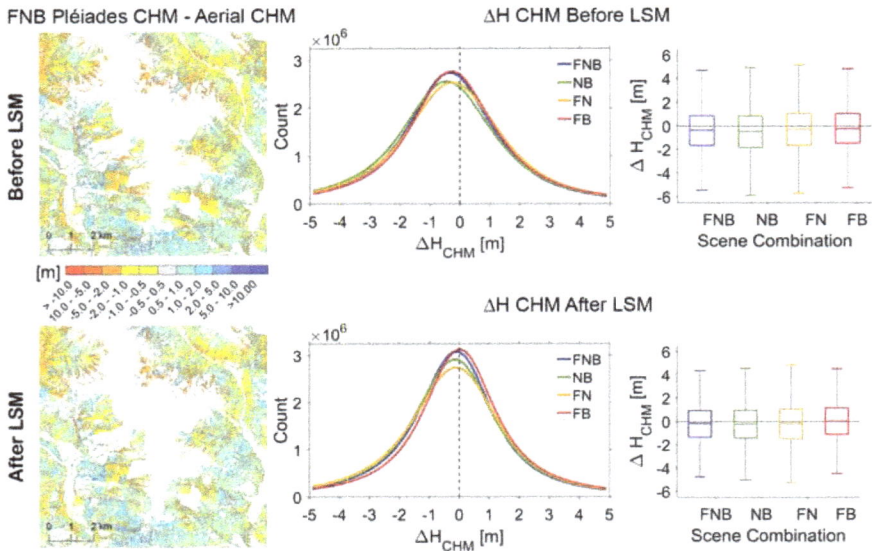

Figure 5. Height differences (ΔH) between Pléiades and aerial CHMs before (**top**) and after (**bottom**) LSM. (**Left**) spatial distribution for FNB as colour coding (areas outside forest mask shown in white). Centre: distributions of ΔH for tri-stereo and all stereo pairs as histograms. (**Right**) the same distributions shown as box plots.

For all image combinations, the histograms reveal unimodal symmetric distributions of the height differences, similar to normal distributions, and they feature negative shifts. The two stereo pairs involving the nadir image provided the lowest accuracy, whereas the tri-stereo and FB combinations resulted in similar error distributions. After LSM, the histograms show a slightly lower dispersion around zero, but the ratio of cells that fall in the range of ±1 m increase only by approximately 5 percentage points. The statistics of each scene combination are reported in Table 5.

Table 5. Height differences (ΔH) before and after LSM between the Pléiades CHMs for each stereo combination and the reference aerial CHM.

	ΔH Pléiades—Aerial CHMs (m)									
	Before LSM					**After LSM**				
	Mean	*Std*	*Median*	σ_{MAD}	*±1 m (%)*	*Mean*	*Std*	*Median*	σ_{MAD}	*±1 m (%)*
Pléiades FNB	−0.47	3.24	−0.22	2.04	40.5	−0.28	3.05	−0.17	1.88	45.2
NB	−0.56	3.35	−0.36	2.13	37.9	−0.30	3.12	−0.20	1.94	43.4
FN	−0.37	3.31	−0.46	2.12	38.7	−0.28	3.19	−0.18	2.00	41.5
FB	−0.21	3.32	−0.23	2.05	41.2	−0.04	3.16	0.02	1.90	45.9

Overall, the tri-stereo and nadir stereo Pléiades DSMs underestimated the canopy height, with medians of around 20 cm, whereas the FB combination shows a median of almost zero. The dispersion in terms of σ_{MAD} is about 2 m for each scene combination. The correlation between the Pléiades CHM errors and the canopy height itself, as given by the reference CHM, was calculated. In order to remove the spatial variation of the canopy height, a standard deviation moving window of 20 m was applied over the aerial CHM and subsequently pixels with a standard deviation greater than 5 m were excluded from the calculation. The time gap of two years can justify the positive ΔH of Pléiades CHMs for young trees with heights below 10 m (Figure 6a), although young trees occupied only about 8% of the entire area. For this tree height class, the median values of ΔH range from 0.29 m to 1.22 m, whereas negative values between −0.42 m and −0.17 m are identified for tree heights greater than 10 m. No significant correlation was identified between ΔH and forest roughness, quantified by the RMSE of the height values to a fitting a plane. Contrary, grouping the data by slope classes of 10° width, a rapid increase in the variance of ΔH was observed for the steep classes with slopes larger than 60°, which, however, represents a small proportion of the investigated area (14%) (Figure 6b).

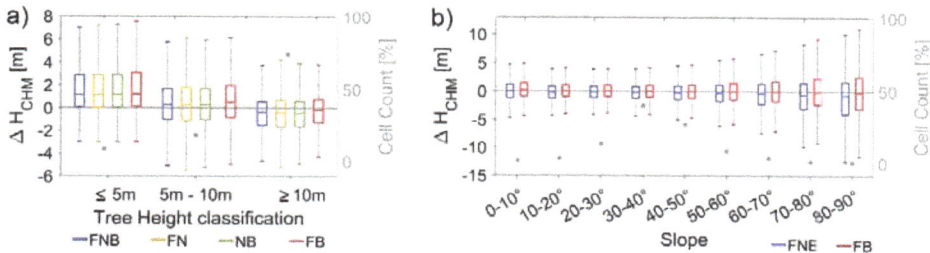

Figure 6. Box plots of the distributions of ΔH after LSM for different classes of (**a**) reference canopy height, and (**b**) terrain slope. Respective cell counts are plotted in grey and refer to the right axes.

4.1.2. Accuracy Assessment of the VHR nDSMs for a Selected Area

The profiles in the insets in Figure 7 show that with Inpho Match-T, there is no benefit of using three images i.e., FNB, because the resulting point cloud is simply the union of those of the two nadir stereo pairs.

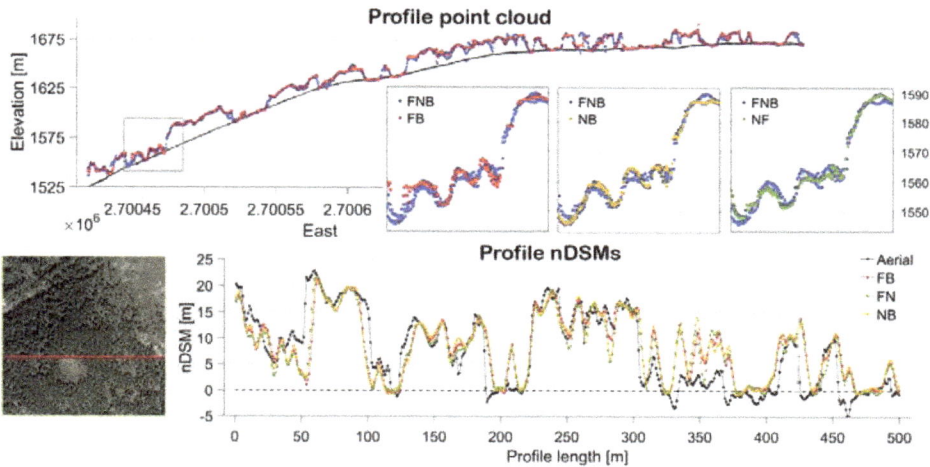

Figure 7. Profiles of 1 m width of the tri-stereo and stereo Pléiades point clouds (**top**) and of the Pléiades nDSMs for each stereo pair in comparison to the reference aerial nDSM (**bottom**). The red line in the orthophoto indicates the position of the profiles.

However, the questions which remain are: (1) if a better DSM than FNB can be derived by computing the three stereo point clouds independently (i.e., for FN, NB, FB after LSM), and then interpolate them into a tri-stereo DSM (FB-NB-FN) using the method described above, and (2) if selecting the locally best of the three stereo DSMs (MinAbsΔH) improves upon the DSMs computed so far. As the locally best DSM, we used the one with the minimum absolute error. To answer the above questions, we derived the histogram of the absolute errors for each of the corresponding CHMs, shown in Figure 8a. The generation of the DSM considering simultaneously the point clouds from the three stereo pairs (FB-NB-FN) does not provide significantly better results than the FB stereo pair. The optimum selection provided a significant improvement. However, no clear systematic pattern can be identified from the spatial distribution of the stereo pairs with the locally minimum absolute error (Figure 8b). Moreover, examining the influences of the canopy height and forest roughness on the absolute error, no significant relationships were identified. Nevertheless, for deriving an automatic approach to optimally select the best height of these three stereo results, a reference data surveyed at the same time should be considered.

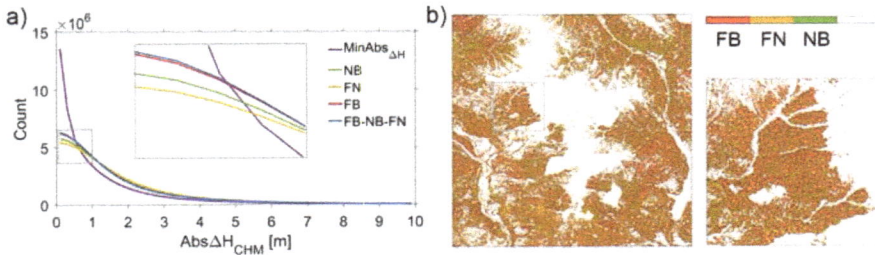

Figure 8. (a) Histograms of the absolute height errors (AbsΔH) for each Pléiades CHM (MinAbsΔH i.e., optimal selection from stereo DSMs, NB, FN, FB, and FB-NB-FN), and (b) the spatial distribution of the stereo pairs with locally minimum absolute differences (MinAbsΔH).

4.2. Impact of the Viewing Angles on the Quality of the Derived Product in the Area of Ljubljana

4.2.1. Accuracy Assessment of the VHR nDSM over Stable Areas and the VHR-CHMs

The vertical accuracy of the Pléiades DSMs in terms of the total RMSE at the GCPs and CPs is around 0.70 m (Table 6). The application of the LSM transformation led to a lower RMSE in the Z-direction, which ranges between 0.14 m and 0.27 m. The worst result is shown by the DSM derived from the stereo pair with the largest angle of convergence and across-track angle. The vertical accuracy of Pléiades DSMs over the entire scene was assessed by calculating the distribution of ΔH using the reference ALS dataset i.e., the ALS DTM for the stable areas, and the ALS CHM for the forested areas. Figure 9 allows for a visual comparison of the Pléiades nDSMs for each stereo scene before and after LSM, together with the ALS nDSM, and the orthophoto. Compared to the ALS nDSM, the spatial distribution of the Pléiades nDSMs over stable areas appears clustered before LSM and more homogeneous after LSM. The spatial pattern observed in the Pléiades nDSM suggests a tilt of around ±1 m in north-east/south-west direction between the Pléiades DSM and the ALS DTM, which, however, was significantly reduced by applying a full LSM transformation.

Table 6. RMSE in Z-direction of the GCPs and CPs for each generated DSM.

	RMSE$_Z$ DSM (m)					
	Before LSM			After LSM (m)		
	27 July	*28 July*	*29 July*	*27 July*	*28 July*	*29 July*
18 GCPs	0.65	0.64	0.77	0.14	0.22	0.25
12 CPs	0.74	0.63	0.88	0.15	0.17	0.30
Tot (30)	0.69	0.64	0.82	0.14	0.20	0.27

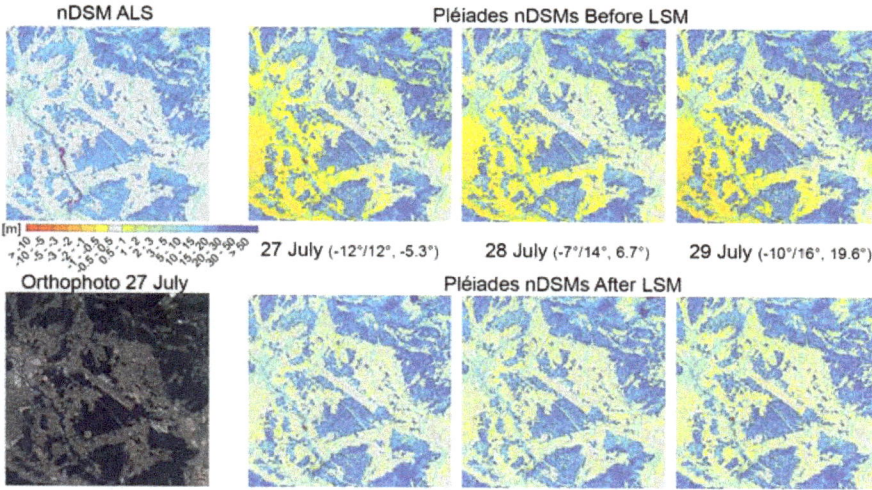

Figure 9. The spatial distribution of the nDSMs of ALS (**left**, **top**) and each Pléiades stereo scene before (**right**, **top**) and after (**righ**, **bottom**) LSM. At the (**left**, **bottom**), the orthophoto derived from the forward Pléiades image of 27 July is shown.

According to the histograms of Pléiades nDSMs in stable areas and ΔH (in forest areas) (Figure 10), the improvement of the accuracy of the CHMs due to LSM is not as significant as in the stable areas, where a considerable reduction of dispersion around zero is visible.

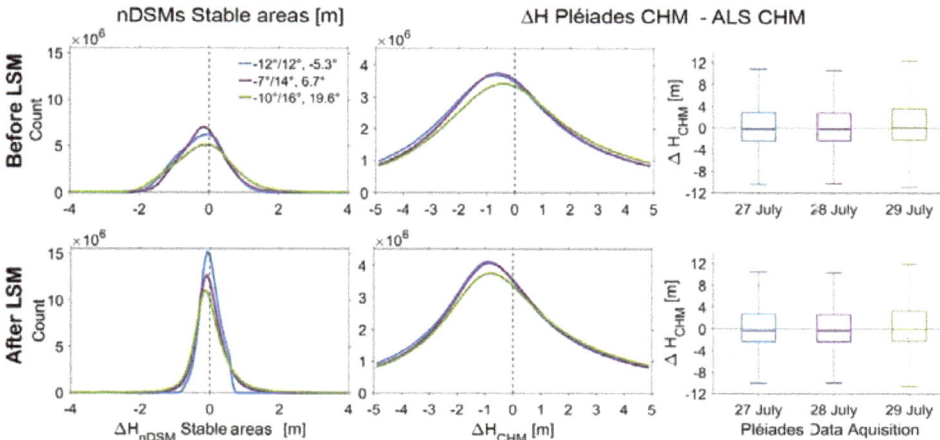

Figure 10. Distribution of nDSM heights in stable areas (**left**) and CHM errors (ΔH) before (**right**, **top**) and after (**right**, **bottom**) LSM for each stereo pair.

The statistics and the box plots of each stereo pair confirm this observation. Overall, after LSM, nDSM heights in stable areas have a median of almost zero, with a σ_{MAD} below 0.50 m. whereas CHM errors show a considerably higher dispersion, with a σ_{MAD} around 4 m (Table 7).

Table 7. nDSM heights in stable areas and CHM errors (ΔH) before and after LSM.

nDSM Stable Areas	Before LSM (m)				After LSM (m)			
	Mean	Std	Median	σ_{MAD}	Mean	Std	Median	σ_{MAD}
27 July (−12°/12°, −5.3°)	0.23	0.63	−0.23	0.48	0.00	0.29	−0.01	0.29
28 July (−7°/14°, 6.7°)	−0.05	0.64	−0.19	0.60	0.14	0.34	−0.03	0.36
29 July (−10°/16°, 19.6°)	−0.05	1.35	−0.14	0.71	0.11	1.23	−0.04	0.42
ΔH								
27 July (−12°/12°, −5.3°)	0.66	6.11	−0.21	3.85	0.67	6.08	−0.32	3.79
28 July (−7°/14°, 6.7°)	0.57	6.29	−0.17	3.89	0.54	6.28	−0.29	3.85
29 July (−10°/16°, 19.6°)	1.15	6.54	0.20	4.14	1.03	6.49	−0.04	4.07

In order to investigate the correlation between the accuracy of the Pléiades CHMs and the reference canopy height, the latter was divided into three height classes, and CHM cells with high variation, i.e., larger than 5.0 m standard deviation within 20 m moving window were excluded (Figure 11a). For all stereo pairs, the CHM errors for tree heights greater than 10 m displays negative median values ranging from −0.58 m to −0.82 m. This can be explained by the tree growth during the two years gap between the Pléiades and ALS data capture. Conversely, the accuracy of the Pléiades CHMs for tree heights below 10 m shows overall positive differences and a considerably higher dispersion for the tree heights smaller than 5 m (Figure 11b). Those trees occupied about 10% of the entire forest area and they are mainly distributed within managed forests areas, as shown in the spatial distribution of the ALS CHM grouped by the three height classes. For this study area, the effect of slope on the accuracy of the Pléiades CHMs cannot be assessed since less than 1% of it reveals a slope greater than 50°.

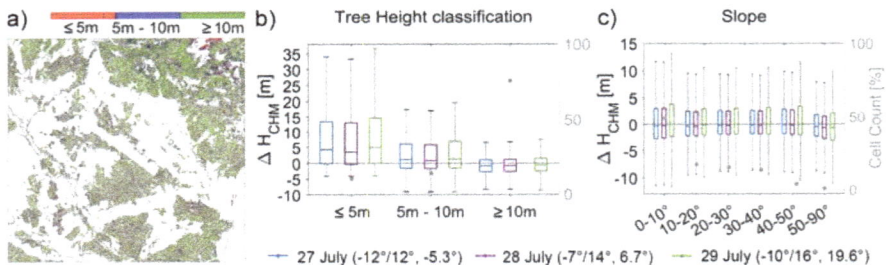

Figure 11. (**a**) The spatial distribution of the three reference CHM tree height classes, disregarding areas with large height variation. Distributions of the CHM error (ΔH) for each stereo pair after LSM grouped by (**b**) tree height class and (**c**) terrain slope.

4.2.2. Accuracy Assessment of the VHR CHMs for the Selected Areas

The quality of the Pléiades CHMs varies according to forest density and forest height. A detailed investigation on the Pléiades CHMs for the two selected areas shows that Pléiades CHMs provided comparable results to the ALS data for a homogenous forest canopy (Figure 12a). Conversely, distinct canopy height patterns that can be seen in the ALS canopy height map cannot be discerned in the Pléiades map where the tree crowns are significantly wider and less defined (Figure 12b). It is worth noting that those differences in the canopy gap characteristics can be attributed to forest management activities, but most likely to the severe freezing rain event that hit Slovenian forests in 2014, damaging 40% forest areas throughout the country [37]. Indeed, the orthophoto acquired simultaneously to ALS data (2015) shows larger canopy gaps than those visible in the Pleiades orthophoto (2013) (Figure 12c).

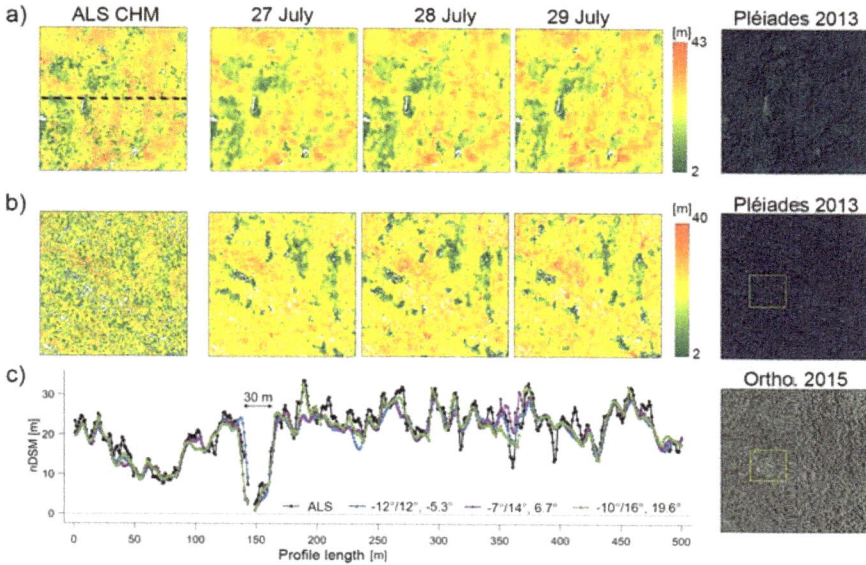

Figure 12. For the selected (**a**) area 1 and (**b**) area 2, the reference CHM (left), the Pléiades CHMs for each stereo pair (centre), and the orthophoto generated from the four bands Pléiades image (right). The dashed black line in ((**a**), left) indicates the position of the profiles of 1 m width of reference and Pléiades CHMs shown on the left of (**c**). On the right of (**c**), the aerial orthophoto for area 2 that was acquired simultaneously to the reference data.

These qualitative results are confirmed by the distribution of Pleiades CHM height errors, where a clear correlation exists with low forest heights (Figure 13b, area 2). Moreover, the trend also demonstrates that young forest (height < 5 m), which is typically underestimated by image matching [23], is completely missing in the Pleiades results, validating the differences in forest structure at the points in time of Pléiades and ALS data capture. The larger across-track angle shows a wider error dispersion for both areas i.e., forest types.

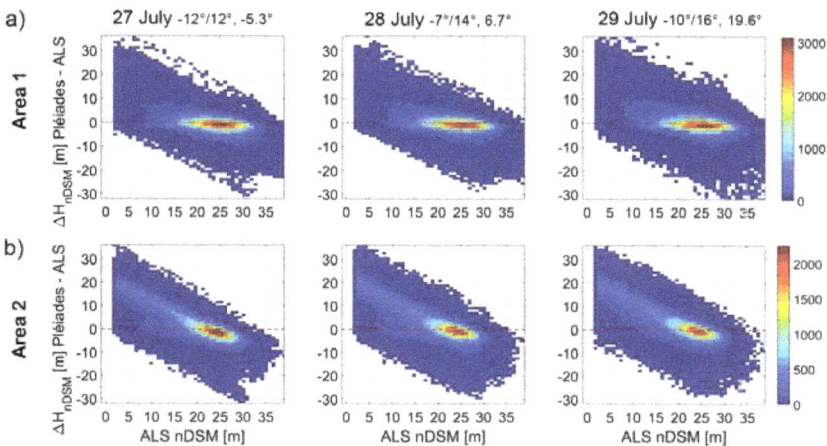

Figure 13. 2D histograms of Pléiades CHM errors and reference CHM for the selected (**a**) area 1 and (**b**) area 2, and each stereo pair (left, centre, right).

5. Discussion

This work focuses on analysing the accuracy of Pléiades DSMs over forest mountain regions in order to identify the optimal acquisition planning in terms of stereo or tri-stereo data and incidence angle along- and across-track. Tri-stereo (FNB) images were requested in 2017 over Ticino forest area in Switzerland. The angle of convergence of the forward-backward (FB) images was about 12°, with an average across track-angle of −5°. The three stereo pairs with different across- and along-track angles were acquired in 2013 over the same area in Ljubljana (Slovenia), one day apart from each other. Those images had an angle of convergence of about 24°, 22°, and 27°, respectively, with about −5°, +6° and +20° in across-track direction.

In this study, we focus on the reconstruction of the surface height from pan-sharpened Pléiades images. As reported by Reference [26], if only tree heights are of interest, there is a limited reimbursement of also acquiring and processing spectral and textural data, because multispectral and colour information typically does not contribute to the matching performance [38]. Therefore, in this work, the fourth band was only exploited to generate the NDVI map which was used to derive the tree mask for the Ticino study area and for removing the lake and river surfaces from the stable area mask. The surface covered by clouds was excluded from the analyses in post-processing based on the nDSM.

Processing in Inpho Match-T is highly automated. It requires only limited information to be entered by users, consisting of manually observing GCPs in the images, checking the residuals of tie points to remove blunders, and setting some parameters for the dense point cloud computation e.g., the type of filtering and the spatial resolution (in our work, one point per pixel). The GCPs were employed to achieve sub-meter accuracy. A high number of GCPs improves the accuracy, but no significant improvement can be reached by increasing the number of GCPs from 10 to 40 [39,40]. For both study areas we used 18 GCPs to refine the RPC in combination with automatically extracted tie points. For the tie points, we obtained a precision of better than 0.3 pixels for the tri-stereo images in Ticino and around 0.5 pixels for the three stereo images in Ljubljana. For both study areas, there is a good agreement between the horizontal RMSE of the adjusted coordinates of the GCPs and the CPs on the ground, but high discrepancies can be found in the vertical direction, where the RMSE of the CPs is about 1.0 m (Ticino) and 0.80 m (Ljubljana) in comparison to one decimetre of the GCPs. Because the GCPs were used within the bundle adjustment, only the CPs residuals represent external accuracy. Indeed, this result is validated by the vertical RMSE of the DSM at the GCPs and CPs, which is in total about 0.90 m and 0.70 m for Ticino and Ljubljana, respectively. For the area of Ticino, the vertical DSM RMSE of the CPs is almost twice as large as the one of the GCPs, which suggests a sub-optimal distribution of the GCPs. However, the large forest coverage and the steep terrain limited the selection of GCPs and CPs within this area.

To generate DSMs from dense point clouds, a moving plane interpolation was chosen that consider all the points within a 3 m search radius. This approach was found to be the optimal compromise between minimizing the number of void pixels, preservation of detail, and noise filtering. Considering that the vertical accuracy of photogrammetric DSMs largely depends on the target land cover [11], we assessed the global accuracy of Pléiades DSMs separately for stable areas and for forest areas (CHM) by comparison with the reference data. In stable areas, the Pléiades DSMs elevation errors showed a clustered bias for both study areas. Consequently, the application of an affine transformation estimated by LSM reduced them. However, to derive globally optimal transformation parameters using LSM, the common stable areas in the master and slave surfaces should be homogeneously distributed over the entire scene, which can be hard to achieve in forest mountain areas. Despite this limitation, we demonstrated that for both study areas, LSM improved the geolocation accuracy by removing the clustered error, especially over the flat area of Ljubljana. Moreover, in Ticino, LSM removed the 35 m geolocation error resulting from the original RPCs delivered with the imageries. This accuracy corresponds to the results reported by Reference [39], who estimated an RMSE of the absolute height of the Pléiades DSMs between 35.6 m and 41.9 m when using the original RPCs. Our results indicate that sub-meter geolocation accuracy can be achieved without the time-consuming measurement of

GCPs by applying an LSM transformation, if a high-resolution DTM and stable areas are available. In Ticino, some clustered errors were still present after LSM, likely due to the low quality of the reference DTM, which showed abrupt terrain discontinuities. This is confirmed by similar distributions of the nDSMs in stable areas for Pléiades (σ_{MAD} = 0.82 (FB)) and aerial image matching (σ_{MAD} = 0.80), where the same DTM was used for normalisation. The application of LSM in Ljubljana significantly removed the tilt effect in the flat stable areas: For the images of 29 July, 74% of the cells fell in the range ±1 m before LSM, whereas the percentage increased to 92% after LSM. The improvement due to LSM of the Pléiades DSM accuracies in forest areas is not as significant as in stable areas. If we quantify this improvement as the percentage of Pléiades DSM cells with height errors within the range of ±1 m, then LSM increases the accuracy by 5 and 4 percentage points for Ticino and Ljubljana, respectively. However, note that in both study areas there is a considerable time gap between the acquisition of the reference data and the Pléiades image.

After LSM, the vertical error distribution of the Pléiades CHMs in Ticino and Ljubljana show good agreement in terms of their median, ranging between 0.02 m (FB) and −0.20 m (NB) for Ticino and between −0.04 m and −0.32 m for Ljubljana. In contrast, their dispersions differ significantly, with a σ_{MAD} of about 4 m for Ljubljana, as opposed to 2 m for Ticino. This larger dispersion of the error distribution can be attributed to the different forest types and managements of the two study areas. In Ticino, the forest areas were more homogeneous, containing mainly adult trees (87%, >10 m) in broadleaf forest. The height of lower trees was generally overestimated by Pléiades image matching, contrary to the conclusion of Reference [26]. However, in our study, this overestimation can be attributed to tree growth between the time of aerial (2015) and Pléiades (2017) images acquisitions. However, in Ljubljana, only 76% of the canopy cover revealed heights greater than 10 m, and it consists of managed forests, several vegetated urban areas, and single trees in flat areas. Tree growth between the Pléiades (2013) and ALS (2015) data acquisitions can explain the negative differences between Pléiades and ALS CHMs for tree heights greater than 10 m. Younger trees were significantly overestimated by Pleiades CHMs in comparison to ALS. However, note that between these two data acquisitions, the Slovenian forests were hit by a strong ice storm, which caused severe damage. Consequently, many of the canopy gaps formed during this event and the younger trees due to forest regeneration were not yet present at the time of the Pléiades image. This was confirmed by visual comparison of a Pléiades orthophoto and an aerial one acquired at the time of ALS data acquisition. An additional aspect to consider in the comparison of the results for the two study areas is the different kinds of data used to compute the reference DSMs: ALS in Ljubljana versus aerial images in Ticino. Hence, two image sensors (with GSDs of 0.50 m for the aerial and of 0.70 m for the Pleiades images) were compared, which have similar issues concerning gap detection, because the same point has to be visible in at least two images. Hence, accurate CHM reconstruction in mountain areas remains difficult due to strong elevation contrast between trees and the surrounding ground, which results in occlusions. Therefore, we confirm that the ability to accurately measure points between trees heavily depends on the GSD, the base length of stereo images, dominating tree heights and the density of the forest [41]. The analysis of the height profiles confirms that the Pléiades DSMs follow well the aerial DSM (Figure 7) and the ALS DSM (Figure 12c) for homogenous canopy cover and adult trees, but overestimates the height between single trees close to each other and within canopy gaps. This result matches the expectation that stereo-photogrammetry reconstructions yield a relatively smooth surface in which height discontinuity between trees and their surrounding are represented by gradual changes [39,42]. The median of Pléiades CHM errors is not influenced by slope, but a rapid increase of its dispersion was observed for steep areas in Ticino (>60°).

When analysing the quality of Pléiades DSMs regarding the acquisition mode in Ticino, i.e., tri-stereo vs. stereo, we observed in a profile (Figure 7) that the nadir image increases completeness, reducing the data void left out by FB matching in steep forest areas, because of fewer occlusions, and larger image similarity. Anyway, the small angle of convergence (12°) of the FB stereo pair resulted in small unreconstructed areas only that were mostly filled by DSM interpolation. The study shows a limitation of the used software in the dense matching of from tristereo images: The resulting FNB dense point cloud is simply a subset of the two-nadir stereo (FN and NB) point clouds, without any contribution of the FB stereo pair. FB provided the highest accuracy in comparison to the tri-stereo and the two nadir stereo pairs, albeit differences are rather small. Among the two nadir stereo pairs, FN had the smaller angle of convergence, and it provided the worst results, as expected. The reached accuracy of the height model based on all three images (i.e., FN-NB-FB) is slightly better than for FB, as also observed by Reference [42]. Selecting the locally best of the three stereo results (i.e., with minimum local height error) yields an improved model. This selection is straightforward when having reference data at hand that was acquired at the same time as the satellite imagery. Without reference data, however, according selection criteria still remain an open question.

Incidence angles, both across- and along-track, affect DSM accuracy, although our results showed only moderate differences, especially within forest areas. When comparing the accuracies of the study areas, we observed that the larger angle of convergence (>24° of Ljubljana versus 12° in Ticino) provided a σ_{MAD} in stable areas that was two times smaller. However, this might also be attributed to the more mountainous topography in Ticino and the more homogeneous distribution of the GCPs in Ljubljana. Nevertheless, the median values of the nDSMs in stable areas is close to zero for both study areas, which suggests that a narrow angle of convergence doesn't constitute a major limiting factor for the quality of the Pléiades DSMs [39]. By contrast, based on the assumption that a wider angle of convergence (>15°) would help to enhance the accuracy of the measured heights [15,43], our results suggest that a wider across-track angle (20°) has a negative impact on the vertical accuracy of the DSM. Furthermore, the error dispersion is more affected by steep terrain than by small across-track angles (±6°).

6. Conclusions

Pléiades satellite images compared to other VHR sensors have the main advantages of a great agility, daily revisit capability, smaller time intervals between the image acquisition and the possibility to acquire a nadir-looking view. Considering these characteristics and if a digital terrain model (DTM) is available, the system offers a great potential for providing a high spatial resolution canopy height model (CHM), which can be used for supporting forest inventory and monitoring programs at the regional and national level. Therefore, to take this system into consideration, an important challenge is to understand the accuracy of the derived products. Specifically, since the acquisition mode (Tri-stereo/stereo) and the incidence angles can be planned for the new Pléiades images acquisitions, this work wants to answer the following two questions: (1) what is the benefit of using tri-stereo images and (2) what is the impact of different incidence angles along- and across-track on the image matching performance and on the accuracy of the DSM. In order to derive the height above the ground (i.e., the nDSM), available DTM was subtracted from the DSMs. The image orientation implied the use of GCPs and tie points to refine the RPC. However, in order to remove systematic errors on the generated DSMs, an affine transformation (LSM) was successfully applied to the dense point cloud. We demonstrated that by applying an LSM transformation sub-meter geolocation accuracy without the time-consuming GCPs measurements could be achieved.

Our results suggest that the differences between tri-stereo and stereo matching are rather small and the stereo forward-backward canopy height showed slightly higher accuracies than the tristereo results and the two nadir stereo pairs. In terms of completeness, the nadir image can minimize the issues of stereo matching in steep forest areas, but the adopted interpolation method and the narrow angle of convergence of the forward-backward pair yielded to small unreconstructed areas. Both incidence angles, across- and along-track are important parameters for determining DSM accuracy of a stereo pair, although our results do not show dramatic differences. However, a large across-track angle (19°) reduces the quality of Pléiades CHMs/nDSMs.

Author Contributions: Conceptualization, L.P., M.H., M.M. and C.G.; methodology, L.P., M.H., M.M., and C.R.; software, L.P. and C.R.; validation, L.P.; formal analysis, L.P.; investigation, L.P., resources, N.P. and C.G.; data curation, L.P., W.K. and C.R.; writing—original draft preparation, L.P.; writing—review and editing, W.K., C.R., M.M, N.P., C.G.; visualization, L.P.; supervision, M.H.; project administration, M.H.; funding acquisition, N.P.

Funding: This research was financed by the Austrian Space Applications Programme (ASAP) through the project PleiAlps (FFG project number 859774) and the Swiss Space Office (SSO). ASAP is a programme of the Austrian Ministry for Transport, Innovation and Technology. The Pléiades imageries were delivered by Airbus Defence and Space.

Conflicts of Interest: The authors declare no conflict of interest.

Appendix A

Figures A1 and A2, respectively for Ticino and Ljubljana, illustrate the image residual vectors for the GCPs, CPs and tie points at the positions of their image points, and the 2D scatter plot of the image residuals of the tie points for each image. Tables A1 and A2, respectively for Ticino and Ljubljana, report the σ of the tie points image residuals and the RMSE of each of the adjusted ground coordinates for the GCPs and CPs after the bundle adjustment.

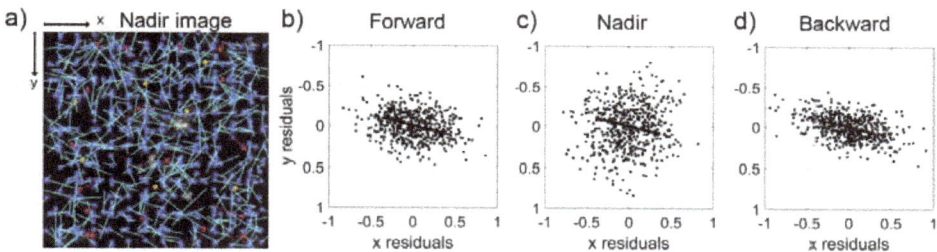

Figure A1. For Ticino study area (**a**) the image residual vectors (scale 5000) for GCPs (red circle), CPs (yellow circle) and tie points (blue circle) at their respective image positions after the RPC correction. (**b–d**) The 2D scatter plot of the image residuals of the tie points of each image in pixel units.

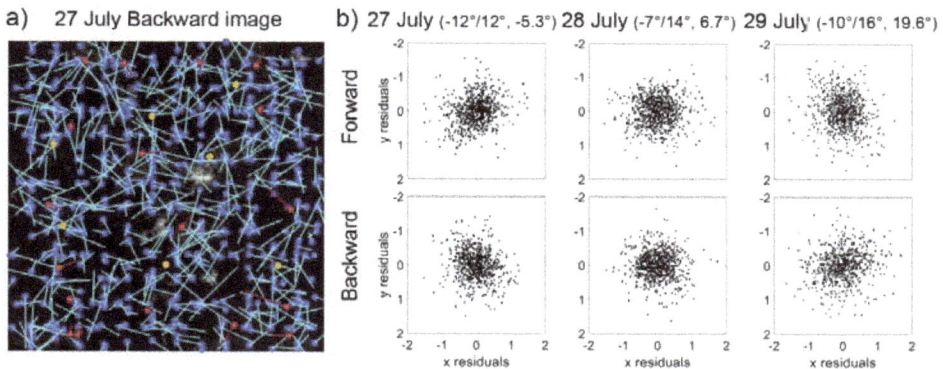

Figure A2. For Ljubljana study area (**a**) Image residual vectors (scale 5000) for GCPs (red circle), CPs (yellow circle) and tie points (blue circle) at their respective image positions after the RPC correction. (**b**) The 2D distribution of image residuals of tie points of each image.

Table A1. For Ticino study area, standard deviation (σ) of the tie points in image space after bundle adjustment and the RMSE for each object coordinate of the GCPs and CPs after the bundle adjustment.

	σ after RPC Refinement (pix)						RMSE (m)		
	FW		N		BW		Adjusted Coordinates		
	x	*y*	*x*	*y*	*x*	*y*	X	Y	Z
Tie points	0.27	0.16	0.27	0.26	0.29	0.14	18 GCPs	0.34	0.23	0.09
							7 CPs	0.18	0.35	1.04

Table A2. For Ljubljana study area, standard deviation (σ) of the tie points in image space after bundle adjustment and the RMSE for each object coordinate of the GCPs and CPs after the bundle adjustment.

	σ after RPC Refinement (pix)												RMSE (m)		
	27 July				28 July				29 July				Adjusted Coordinates		
	FW		BW		FW		BW		FW		BW		X	Y	Z
	x	*y*	*x*	*y*	*x*	*y*	*x*	*y*	*x*	*y*	*x*	*y*			
Tie points	0.43	0.43	0.39	0.44	0.46	0.42	0.44	0.40	0.44	0.52	0.52	0.49	18 GCPs	0.37	0.37	0.17
													12 CPs	0.56	0.54	0.80

References

1. Immitzer, M.; Stepper, C.; Böck, S.; Straub, C.; Atzberger, C. Use of WorldView-2 stereo imagery and National Forest Inventory data for wall-to-wall mapping of growing stock. *For. Ecol. Manag.* **2016**, *359*, 232–246. [CrossRef]
2. Lim, K.; Treitz, P.; Wulder, M.; St-Onge, B.; Flood, M. LiDAR remote sensing of forest structure. *Prog. Phys. Geogr.* **2003**, *27*, 88–106. [CrossRef]
3. Latifi, H.; Nothdurft, A.; Koch, B. Non-parametric prediction and mapping of standing timber volume and biomass in a temperate forest: Applications of multiple optical/LiDAR-derived predictors. *Forestry* **2010**, *83*, 395–407. [CrossRef]
4. Vastaranta, M.; Wulder, M.A.; White, J.C.; Pekkarinen, A.; Tuominen, S.; Ginzler, C.; Hyyppä, H. Airborne laser scanning and digital stereo imagery measures of forest structure: Comparative results and implications to forest mapping and inventory update. *Can. J. Remote Sens.* **2013**, *39*, 382–395. [CrossRef]
5. Leberl, F.; Irschara, A.; Pock, T.; Meixner, P.; Gruber, M.; Scholz, S.; Wiechert, A. Point clouds. *Photogramm. Eng. Remote Sens.* **2010**, *76*, 1123–1134. [CrossRef]
6. Kayitakire, F.; Hamel, C.; Defourny, P. Retrieving forest structure variables based on image texture analysis and Ikonos-2 imagery. *Remote Sens. Environ.* **2006**, *102*, 390–401. [CrossRef]
7. Uddin, K.; Gilani, H.; Murthy, M.S.R.; Kotru, R.; Qamer, F.M. Forest condition monitoring using very-high-resolution satellite imagery in a remote mountain watershed in Nepal. *Mt. Res. Dev.* **2015**, *35*, 264–277. [CrossRef]
8. Shamsoddini, A.; Trinder, J.C.; Turner, R. Pine plantation structure mapping using WorldView-2 multispectral image. *Int. J. Remote Sens.* **2013**, *34*, 3986–4007. [CrossRef]
9. Persson, H.J.; Perko, R. Assessment of boreal forest height from WorldView-2 satellite stereo images. *Remote Sens. Lett.* **2016**, *7*, 1150–1159. [CrossRef]
10. Song, C.; Dickinson, M.B.; Su, L.; Zhang, S.; Yaussey, D. Estimating Average Tree Crown Size Using Spatial Information from Ikonos and QuickBird Images: Across-sensor and Across-Site Comparisons. *Remote Sens. Environ.* **2010**, *114*, 1099–1107. [CrossRef]
11. Hobi, M.L.; Ginzler, C. Accuracy assessment of digital surface models based on WorldView-2 and ADS80 stereo remote sensing data. *Sensors* **2012**, *12*, 6347–6368. [CrossRef] [PubMed]
12. Hobi, M.L.; Ginzler, C.; Commarmot, B.; Bugmann, H. Gap pattern of the largest primeval beech forest of Europe revealed by remote sensing. *Ecosphere* **2015**, *6*, 1–15. [CrossRef]
13. Bagnardi, M.; González, P.J.; Hooper, A. High-resolution digital elevation model from tri-stereo Pleiades-1 satellite imagery for lava flow volume estimates at Fogo Volcano. *Geophys. Res. Lett.* **2016**, *43*, 6267–6275. [CrossRef]

14. De Lussy, F.; Greslou, D.; Dechoz, C.; Amberg, V.; Delvit, J.M.; Lebegue, L.; Blanchet, G.; Fourest, S. Pleiades HR in flight geometrical calibration: Location and mapping of the focal plane, ISPRS International Archives of the Photogrammetry. *Remote Sens. Spat. Inf. Sci.* **2012**, *39*, 519–523.

15. Poli, D.; Caravaggi, I. 3D modeling of large urban areas with stereo VHR satellite imagery: Lessons learned. *Nat. Hazards* **2013**, *68*, 53–78. [CrossRef]

16. Perko, R.; Raggam, H.; Gutjahr, K.; Schardt, M. Assessment of the mapping potential of Pléiades stereo and triplet data. *ISPRS Ann. Photogramm. Remote Sens. Spat. Inf. Sci.* **2014**, *2*, 103. [CrossRef]

17. Jacobsen, K.; Topan, H. DEM generation with short base length Pleiades triplet. *Int. Arch. Photogramm. Remote Sens. Spat. Inf. Sci.-ISPRS Arch.* **2015**, *40*, 81–86. [CrossRef]

18. Poli, D.; Remondino, F.; Angiuli, E.; Agugiaro, G. Radiometric and geometric evaluation of GeoEye-1, WorldView-2 and Pléiades-1A stereo images for 3D information extraction. *ISPRS J. Photogramm. Remote Sens.* **2015**, *100*, 35–47. [CrossRef]

19. Panagiotakis, E.; Chrysoulakis, N.; Charalampopoulou, V.; Poursanidis, D. Validation of Pleiades Tri-Stereo DSM in Urban Areas. *ISPRS Int. J. Geo-Inf.* **2018**, *7*, 118. [CrossRef]

20. Eisank, C.; Rieg, L.; Klug, C.; Kleindienst, H.; Sailer, R. Semi-Global Matching of Pléiades tri-stereo imagery to generate detailed digital topography for high-alpine regions. *J. Geogr. Inf. Sci.* **2015**, *2015*, 168–177. [CrossRef]

21. Himmelreich, L.C.; Ladner, M.; Heller, A. Pléiades Tri-Stereo-Bilder im Hochgebirge–eine Parameterstudie mit PCI Geomatics. *Agit–J. für Angewandte Geoinformatik* **2017**, *3*, 153–162. [CrossRef]

22. Himmelreich, L. DHM Ableitungen aus Pléiades Tri-Stereo Satellitenbildern im Hochgebirge. Digitale Höhenmodelle Verschiedener Softwareprodukte im Vergleich zu ALS Daten. Master Thesis, University of Innsbruck, Innsbruck, Austria, July 2017.

23. Zhou, Y.; Parsons, B.; Elliott, J.R.; Barisin, I.; Walker, R.T. Assessing the ability of Pleiades stereo imagery to determine height changes in earthquakes: A case study for the El Mayor-Cucapah epicentral area. *J. Geophys. Res. Solid Earth* **2015**, *120*, 8793–8808. [CrossRef]

24. Beguet, B.; Guyon, D.; Boukir, S.; Chehata, N. Automated retrieval of forest structure variables based on multi-scale texture analysis of VHR satellite imagery. *ISPRS J. Photogramm. Remote Sens.* **2014**, *96*, 164–178. [CrossRef]

25. Trisakti, B. Vegetation type classification and vegetation cover percentage estimation in urban green zone using pleiades imagery. In *IOP Conference Series: Earth and Environmental Science*; IOP Publishing: Bristol, UK, 2017; Volume 54.

26. Persson, H. Estimation of Boreal Forest Attributes from Very High Resolution Pléiades Data. *Remote Sens.* **2016**, *8*, 736. [CrossRef]

27. Maack, J.; Kattenborn, T.; Fassnacht, F.E.; Enßle, F.; Hernández, J.; Corvalán, P.; Koch, B. Modeling forest biomass using Very-High-Resolution data—Combining textural, spectral and photogrammetric predictors derived from spaceborne stereo images. *Eur. J. Remote Sens.* **2015**, *48*, 245–261. [CrossRef]

28. Abdollahnejad, A.; Panagiotidis, D.; Surový, P. Estimation and Extrapolation of Tree Parameters Using Spectral Correlation between UAV and Pléiades Data. *Forests* **2018**, *9*, 85. [CrossRef]

29. Akbari, H.; Kalbi, S. Determining Pleiades satellite data capability for tree diversity modeling. *iForest-Biogeosci. For.* **2016**, *10*, 348. [CrossRef]

30. Pearse, G.D.; Dash, J.P.; Persson, H.J.; Watt, M.S. Comparison of high-density LiDAR and satellite photogrammetry for forest inventory. *ISPRS J. Photogramm. Remote Sens.* **2018**, *142*, 257–267. [CrossRef]

31. Ginzler, C. Vegetation Height Model NFI. *Natl. For. Inventory (NFI)* **2018**. [CrossRef]

32. Pfeifer, N.; Mandlburger, G.; Otepka, J.; Karel, W. OPALS—A framework for Airborne Laser Scanning data analysis. *Comput. Environ. Urban Syst.* **2014**, *45*, 125–136. [CrossRef]

33. Fraser, C.S.; Dial, G.; Grodecki, J. Sensor orientation via RPCs. *ISPRS J. Photogramm. Remote Sens.* **2006**, *60*, 182–194. [CrossRef]

34. Astrium. *Pléiades Imagery User Guide V 2.0.*; Astrium: Toulouse, France, 2012.

35. Heuchel, T.; Köstli, A.; Lemaire, C.; Wild, D. Towards a next level of quality DSM/DTM extraction with Match-T. *Proc. Photogramm. Week* **2011**, *11*, 197–202.

36. Ressl, C.; Mandlburger, G.; Pfeifer, N. Investigating Adjustment of Airborne Laser Scanning Strips without Usage of GNSS/IMU Trajectory Data. In Proceedings of the ISPRS Workshop Laserscanning 09, Paris, France, 1–2 September 2009; Volume XXXVIII, pp. 195–200.

37. Nagel, T.A.; Firm, D.; Rozenbergar, D.; Kobal, M. Patterns and drivers of ice storm damage in temperate forests of Central Europe. *Eur. J. For. Res.* **2016**, *135*, 519–530. [CrossRef]

38. Bleyer, M.; Chambon, S. Does color really help in dense stereo matching. In Proceedings of the International Symposium 3D Data Processing, Visualization and Transmission, 3DPVT 2010, Paris, France, 17–20 May 2010; pp. 1–8.

39. Stumpf, A.; Malet, J.P.; Allemand, P.; Ulrich, P. Surface reconstruction and landslide displacement measurements with Pléiades satellite images. *ISPRS J. Photogramm. Remote Sens.* **2014**, *95*, 1–12. [CrossRef]

40. Topan, H.; Cam, A.; Özendi, M.; Oruç, M.; Jacobsen, K.; Taşkanat, T. Pléiades project: Assessment of georeferencing accuracy, image quality, pansharpening performance and DSM/DTM quality. *Int. Arch. Photogramm. Remote Sens. Spat. Inf. Sci.-ISPRS Arch.* **2016**, *41*, 503–510. [CrossRef]

41. Zarco-Tejada, P.J.; Diaz-Varela, R.; Angileri, V.; Loudjani, P. Tree height quantification using very high resolution imagery acquired from an unmanned aerial vehicle (UAV) and automatic 3D photo-reconstruction methods. *Eur. J. Agron.* **2014**, *55*, 89–99. [CrossRef]

42. White, J.C.; Tompalski, P.; Coops, N.C.; Wulder, M.A. Comparison of airborne laser scanning and digital stereo imagery for characterizing forest canopy gaps in coastal temperate rainforests. *Remote Sens. Environ.* **2018**, *208*, 1–14. [CrossRef]

43. Bernard, M.; Decluseau, D.; Gabet, L.; Nonin, P. 3D capabilities of Pleiades satellite. International archives of the photogrammetry. *Remote Sens. Spat. Inf. Sci.* **2012**, *39*, 553–557.

remote sensing

MDPI

Article

Long-Term Satellite Monitoring of the Slumgullion Landslide Using Space-Borne Synthetic Aperture Radar Sub-Pixel Offset Tracking

Donato Amitrano [1,*], Raffaella Guida [1], Domenico Dell'Aglio [2,3], Gerardo Di Martino [2], Diego Di Martire [4], Antonio Iodice [2], Mario Costantini [5], Fabio Malvarosa [5] and Federico Minati [5]

[1] Surrey Space Centre, University of Surrey, Guildford GU2 7XH, UK; r.guida@surrey.ac.uk
[2] Department of Electrical Engineering and Information Technology, University of Napoli Federico II, Via Claudio 21, 80125 Napoli, Italy; domenicoantoniogiuseppe.dellaglio@unina.it (D.D.); gerardo.dimartino@unina.it (G.D.M.); iodice@unina.it (A.I.)
[3] Benecon S.C.aR.L., Via Santa Maria di Costantinopoli 104, 80138 Napoli, Italy
[4] Department of Earth, Environmental and Resources Sciences, University of Napoli Federico II, Via Cintia 21, 80126 Napoli, Italy; diego.dimartire@unina.it
[5] e-geos., Via Tiburtina 965, 00156 Roma, Italy; mario.costantini@e-geos.it (M.C.); fabio.malvarosa@e-geos.it (F.M.); federico.minati@e-geos.it (F.M.)
* Correspondence: d.amitrano@surrey.ac.uk

Received: 29 December 2018; Accepted: 8 February 2019; Published: 12 February 2019

Abstract: Kinematic characterization of a landslide at large, small, and detailed scale is today still rare and challenging, especially for long periods, due to the difficulty in implementing demanding ground surveys with adequate spatiotemporal coverage. In this work, the suitability of space-borne synthetic aperture radar sub-pixel offset tracking for the long-term monitoring of the Slumgullion landslide in Colorado (US) is investigated. This landslide is classified as a debris slide and has so far been monitored through ground surveys and, more recently, airborne remote sensing, while satellite images are scarcely exploited. The peculiarity of this landslide is that it is subject to displacements of several meters per year. Therefore, it cannot be monitored with traditional synthetic aperture radar differential interferometry, as this technique has limitations related to the loss of interferometric coherence and to the maximum observable displacement gradient/rate. In order to overcome these limitations, space-borne synthetic aperture radar sub-pixel offset tracking is applied to pairs of images acquired with a time span of one year between August 2011 and August 2013. The obtained results are compared with those available in the literature, both at landslide scale, retrieved through field surveys, and at point scale, using airborne synthetic aperture radar imaging and GPS. The comparison showed full congruence with the past literature. A consistency check covering the full observation period is also implemented to confirm the reliability of the technique, which results in a cheap and effective methodology for the long-term monitoring of large landslide-induced movements.

Keywords: synthetic aperture radar; landslide monitoring; sub-pixel offset tracking; Slumgullion landslide; natural hazards; large displacements

1. Introduction

The kinematic estimation of landslide deformation is an important task for investigating the mechanisms affecting their evolution. An improved understanding of changes in landslide behavior as a function of predisposing factors (such as slope, aspect, land use, lithology, etc.) and triggering factors (such as intense or prolonged rainfall, seismicity, human actions, etc.) is crucial for risk mitigation, and thus brings significant social, environmental, and financial benefits [1].

Historically, landslides have been investigated with sparse point measurements acquired through demanding on-field campaigns using traditional instruments (inclinometers, piezometers topographic leveling) and/or GPS. However, recent years have been characterized by a rising interest in and exploitation of satellite technologies for the monitoring of active ground deformation, including landslide motion and hillslope creep, thus allowing for the overcoming of the insufficient spatiotemporal coverage achievable through traditional monitoring methods. In this sense, Differential Synthetic Aperture Radar Interferometry (DInSAR) techniques have largely demonstrated their effectiveness in measuring surface motion and deformation at different scales [2–6]. The spread of this methodology is mainly related to Permanent Scatterers (PS) [7] and Small BAseline Subset (SBAS) [3] approaches, and their combination (known as SqueeSAR) [8–10]. However, these methods are subject to some restrictions, such as the presence of vegetation (causing temporal decorrelation even within very short time intervals) and the 1-D line of sight (LOS) measurement sensitivity, which is also constrained by the adopted wavelength relatively to the maximum measurable displacement rate [11,12].

Today, a new class of methods can provide information complementary to that derived from DInSAR by working with the amplitude channel. These techniques, based on Sub-Pixel Offset Tracking (SPOT) [11,13], allow the measurement of displacements in the South–North and East–West directions without any limitation on the observable rate. This means that a pair of SAR images can be used to detect movements of several meters with a good degree of approximation [13]. SPOT methods are generally less precise than conventional DInSAR methods [11], however they are less sensitive to atmospheric effects and are not strictly only applicable to highly coherent targets, i.e., they can measure displacements even in densely vegetated areas [14].

SPOT methods have been successfully applied to estimate movements of glaciers and terrain [15–17] due to natural phenomena (landslides) [11,13] and/or human activities (subsidence induced by underground excavation) [18,19]. In this work, the suitability of this methodology for the long-term monitoring of the Slumgullion landslide (Colorado, US) is investigated. This landslide has been extensively studied in the past due to the fast/slow related displacements [20–23]. Previous investigations have mainly been implemented with field sensors. Starting from the early 2000s, synthetic aperture radar (SAR) airborne remote sensing was also employed. The first campaign was implemented by the Brigham Young University using an X-band sensor, with the purpose of correlating the measured movements to the soil water content [24]. More recently, some studies exploiting airborne L-band remote sensing have been presented [25–27]. However, so far, the use of satellite technologies in the literature was limited, despite they can provide a cost-effective and continuous update of the landslide state. In fact, kinematic characterization at landslide scale is today still rare, especially for long periods, due to the difficulty in implementing demanding ground surveys with adequate spatiotemporal coverage [22].

To this end, state-of-the-art SAR SPOT was applied to three X-band COSMO-SkyMed spotlight images with about one-meter spatial resolution to monitor the Slumgullion landslide over a time frame of two years, from August 2011 to August 2013. The landslide is investigated at both large and small scale, and, for the first time, the displacements retrieved using very high-resolution space-borne images are validated against ground data provided in the past literature.

The work is organized as follows. The case study and available data are presented in Section 2. Experimental results are presented and validated in Section 3, and discussed in Section 4. Conclusions are drawn at the end of the work, in Section 5.

2. Materials and Methods

2.1. Study area and available data

The study area concerns the Slumgullion landslide, US. It is depicted in Figure 1, with a LiDAR Digital Elevation Model (DEM) with a 0.5-meter spatial resolution superimposed and developed by the US National Center for Airborne Laser Mapping (NCALM). The landslide is located in the San Juan Mountains in southwestern Colorado, near the town of Lake City. It has been moving for about 350 years, with a maximum measured velocity of six meters per year [21]. According to [28], it has been classified as a debris flow which involves deeply weathered tertiary volcanic rocks. It extends for about seven kilometers from the Cannibal Plateau to Lake San Cristobal, with a mean width and depth of 300 meters and 14 meters, respectively [20].

Inside the landslide, an area of about one square kilometer (see black curve in Figure 1) is still active, and is characterized by a low ground-surface inclination (about eight degrees). Based on its average displacement rate (between 0.1–2.0 cm/day as reported in [2]), it can be considered a very slow landslide according to [28].

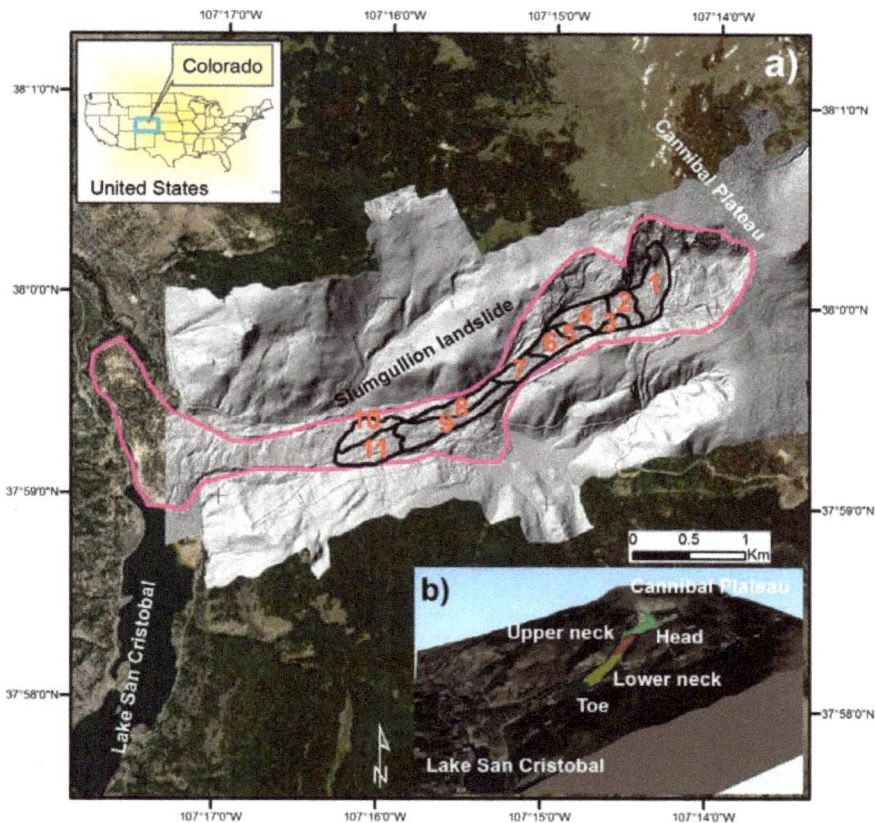

Figure 1. Study area. The Slumgullion landslide is represented by the area within the purple line. In box (a), the LIDAR Digital Elevation Model (DEM) by the US National Center for Airborne Laser Mapping (NCALM) has been superimposed onto the Google Earth view of the landslide.

As observed in [29,30], the Slumgullion landslide is composed of multiple kinematic elements, each of them moving like a rigid block sliding along faults. Accordingly, the landslide has been segregated into 11 primary kinematic elements (see labels in Figure 1) that can be assumed as temporally consistent [22].

Between 1985 and 1990, velocity of elements from 1 to 4, constituting the flattest part of the landslide, increased nearly linearly in the downslope direction, as reported in [18,20,21]. Displacement rate abruptly increased from element 4 to 5 and increased almost linearly up to element 7. The latter was the fastest and steepest element of the landslide. Then, it widens and flattens, slowing downslope to elements 8 and 9. Finally, the landslide moved along the oldest ground surface forming elements 10 and 11, whose speed was much slower than that of elements 8 and 9. As indicated in [31], these regions can be grouped into head (Region 1 to 4), upper neck (Region 5 to 7), lower neck (Region 8 and 9), and toe (Region 10 and 11) (see box (b) in Figure 1).

Satellite monitoring has been implemented by exploiting three COSMO-SkyMed spotlight images with about one-meter spatial resolution. The dataset covers a time frame of two years, from August 2011 to August 2013. Images were acquired with one-year frequency. They were combined in two co-registered pairs, the first covering the period August 2011–August 2012 and the second covering the period August 2012–August 2013. A third pair, covering the time frame August 2011–August 2013, was also considered to implement a consistency check of the obtained results (see Section 3).

2.2. Implemented Sub-Pixel Offset Tracking (SPOT) technique

The flow diagram of the implemented methodology is depicted in Figure 2. In particular, in the upper part, the general processing chain, from the input to the final displacement vector field, is reported. In the lower part, an exploded view of the SPOT processing block is displayed.

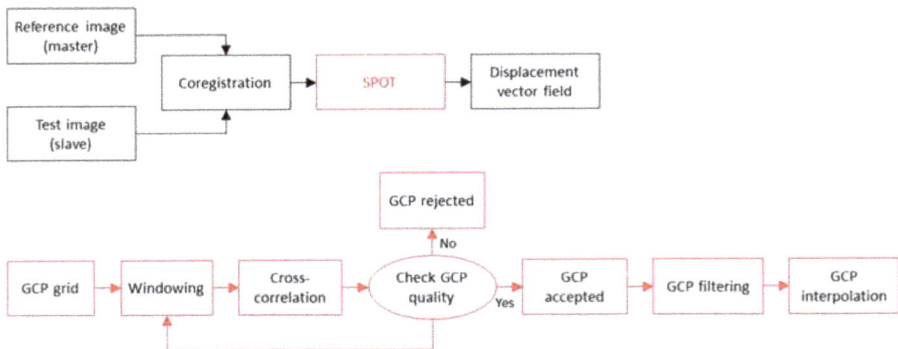

Figure 2. Flow diagram of the implemented methodology. The upper part of the picture shows the general processing chain, while in the lower part the flux represents an exploded view of the Sub-Pixel Offset Tracking (SPOT) processing block.

General processing starts from standard co-registration [32]. If precise orbit data are available, an only orbit-based co-registration can be implemented. Otherwise, full co-registration (i.e., including cross-correlation and coherence refinement steps) is suggested, provided that the scene is much larger than the area of interest. In this case, Ground Control Points (GCPs) located on the landslide will be automatically discarded since they typically exhibit very low cross-correlation and interferometric coherence values. The first acquired image is assumed to be the master image, i.e., the reference for the estimation of displacements. The image in which displacements are evaluated is the slave image.

The SPOT algorithm exploits cross-correlation measures on several windows extracted from the co-registered image pair to estimate the shift between the master patch and the slave patch. Windows are extracted around grid points usually regularly distributed across the images (see first two blocks of the lower diagram in Figure 2).

The cross-correlation between two null-mean patches M and S, taken, respectively, on the master and slave image, is computed as follows:

$$C = \frac{IFFT\{FFT\{M\} \times FFT\{S\}^*\}}{\sqrt{\langle M^2 \rangle \times \langle S^2 \rangle}}, \ C_{ij} \in [0,1] \tag{1}$$

where *FFT* and *IFFT* indicate the fast Fourier transformation and the inverse fast Fourier transformation, respectively, the apex $*$ the complex conjugation operation, and the symbol $\langle \cdot \rangle$ the mean operator. In this equation M and S are oversampled by a factor f (which must be a power of two in order to optimize the FFT calculation) to take into account sub-pixel movements, being the minimum detectable displacement (in pixel units) equal to $1/f$.

In Equation (1), C is a matrix. Its maximum identifies the amount of shift to be applied to the slave patch to have it superimposed onto the master patch. The higher (and sharper) the peak identifying the matrix maximum, the more reliable is the estimated shift. Note that C is a circular matrix, and therefore the maximum detectable shift is equal to $\pm d/2$, where d is the patch dimension (usually square).

In order to identify reliable shifts, two quality parameters are considered, i.e., the peak value c_{max} of the cross-correlation matrix and the ratio $q = c_{max}/\langle C \rangle$. For both of them, a pre-determined user-defined threshold is used to exclude invalid GCPs.

Accepted GCPs are then subjected to a filtering procedure to minimize noisy displacement patterns. To this end, shifts are classified based on their sign; specifically, a three-class classification map (positive shift, negative shift, and null shift) is generated. A connected component labeling algorithm [33] is used to segment this classification map. Small regions surrounded by a homogeneous background of shifts with the same sign are assumed to be noisy patterns and are discarded. Finally, filtered GCPs are interpolated into the final displacement map.

3. Results

An example of the results obtained with the SPOT technique is reported in Figure 3. It refers to the time frame August 2011–August 2012. As expected from the past literature, the highest velocity values have been recorded in the central part of the landslide (upper and lower necks). The arrows in Figure 3 represent the direction of the estimated vector field retrieved from the North–South and East–West component. All the results presented here have been obtained after re-projection of SAR images into a cartographic system using the NCALM LIDAR DEM.

The result for the time frame August 2012–August 2013 is very similar to that for the time frame August 2011–August 2012, and therefore is omitted for brevity. In both experiments, the size of the correlation window was set to 64 pixels, the cross-correlation threshold to 0.1, the threshold on the ratio q between the peak and the mean of the correlation matrix to 4, and the oversampling factor to 4. This means that the minimum retrievable displacement is on the order of 17 centimeters in the azimuth direction and 10 centimeters in the range direction. In Table 1, the parameters set to run the experiments are summarized. The third one (Run 3) concerns a consistency check of the retrieved displacements fields.

In Figure 4, the maps of quality parameters relative to the pair August 2011–August 2012 are shown. Figure 4a represents the map of the maximum correlation coefficient and Figure 4b is the q map, e.g., the ratio between the peak and the mean of the cross-correlation matrix. The noisy displacement patterns displayed in Figure 3 correspond to areas characterized by a low value of the considered quality parameters for the GCPs selection.

In Table 2, quantitative data about the estimated displacements rate are reported. These values were obtained by averaging the estimated velocities in the kinematic regions indicated in [22]. Data concerning the standard deviation of the measurements are also provided.

Two different window dimensions (64 and 128 pixels) were experimented, and produced similar results. The last columns of the table are reserved for the consistency check, which was conceived as follows. A is the result for the period 2011–2012, B that for the period 2012–2013, and C that for the period 2011–2013; reliable results should return A + B − C = 0. This is not strictly achieved, however the deviation from zero of the aforementioned equation is generally small. The distribution of the consistency check is Gaussian with a mean of about 1 cm/year and standard deviation of 42 cm/year in the 64-pixel window case. In the 128-pixel window case, the mean and standard deviation of the distribution are on the order of 2 cm/year and 42 cm/year, respectively.

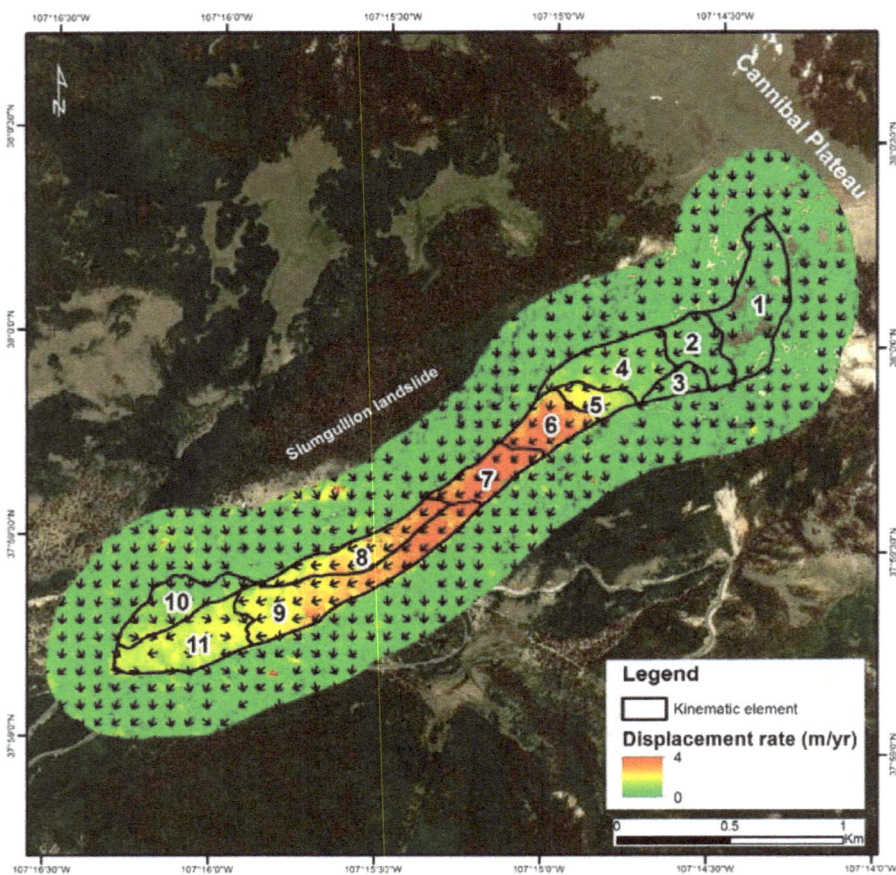

Figure 3. Slumgullion landslide displacement rate, in meters per year, for the time frame August 2011–August 2012. The superimposed arrows represent the direction of the estimated vector field.

Table 1. Summary of the adopted Sub-Pixel Offset Tracking (SPOT) parameters.

	Run 1	**Run 2**	**Run 3**
Parameter	Value	Value	Value
Master image	2011/08/03	2012/08/06	2011/08/03
Slave image	2012/08/06	2013/08/08	2013/08/08
Time span	369 d	367 d	736 d
Pixel spacing azimuth	0.70 m	0.70 m	0.70 m
Pixel spacing range	0.38 m	0.38 m	0.38 m
x-grid spacing	10 pixels	10 pixels	10 pixels
y-grid spacing	10 pixels	10 pixels	10 pixels
Oversampling factor	4	4	4
Cross-correlation threshold	0.1	0.1	0.1
q threshold	4	4	4
Max displacement	8 m	8 m	16 m
Computational time	≈ 45 min	≈ 45 min	≈ 45 min

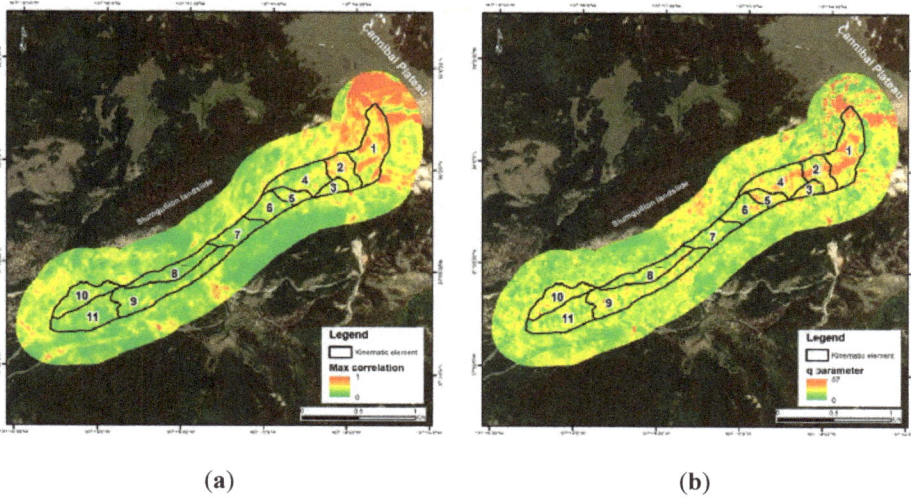

(a)   (b)

Figure 4. Ground Control Point (GCP) quality parameter maps for the satellite pair August 2011–August 2012. (**a**) Maximum of the cross-correlation matrix. (**b**) Ratio between the peak and the mean of the cross-correlation matrix. In both pictures, black contours indicate the kinematic regions as defined in [22].

Table 2. Summary of the obtained results. Data are reported in meters per year and have been averaged using the regions indicated in [22]. V–displacement velocity; CC–consistency check.

Region id	\<V\> 2011–2012		std(V) 2011–2012		\<V\> 2012–2013		std(V) 2012–2013		|CC|		std(CC)	
	w = 64	w = 128	w = 64	w = 128	w = 64	w = 128	w = 64	w = 128	w = 64	w = 128	w = 64	w = 128
Landslide	1.03	1.03	0.79	0.80	0.81	0.86	0.72	0.73	0.01	0.02	0.42	0.42
1	0.11	0.06	0.09	0.07	0.13	0.14	0.08	0.06	0.07	0.04	0.14	0.10
2	0.26	0.24	0.13	0.11	0.16	0.19	0.09	0.09	0.00	0.00	0.15	0.09
3	0.33	0.31	0.13	0.16	0.24	0.27	0.10	0.10	0.03	0.01	0.19	0.17
4	0.54	0.56	0.18	0.34	0.38	0.45	0.16	0.15	0.08	0.03	0.19	0.13
5	1.11	1.11	0.29	0.25	0.91	0.98	0.29	0.34	0.12	0.07	0.23	0.28
6	2.00	2.05	0.49	0.43	1.80	1.89	0.42	0.37	0.09	0.24	0.63	0.74
7	2.40	2.44	0.67	0.58	2.14	2.25	0.55	0.44	0.07	0.05	0.89	1.04
8	1.56	1.54	0.47	0.46	1.24	1.29	0.45	0.42	0.05	0.05	0.50	0.54
9	1.63	1.64	0.47	0.47	1.34	1.39	0.59	0.51	0.09	0.05	0.52	0.36
10	0.58	0.54	0.23	0.19	0.36	0.30	0.16	0.10	0.09	0.06	0.30	0.15
11	1.00	1.01	0.17	0.11	0.52	0.59	0.18	0.14	0.05	0.00	0.26	0.15

3.1. Validation

The Slumgullion landslide has been widely investigated in the past literature. Most of the studies rely on field surveys [1,20–23,29,30,34]. Recently, some papers exploiting remote sensing data have been presented [2,25–27,31]. In this context, only Reference [2] made use of space-borne SAR data. However, this work was focused on spotlight DInSAR methods covering one year of observations, with small insights in long-term displacement monitoring and limited validation of the presented results against literature data. All other works reviewed rely on airborne images acquired by the NASA/JPL L-band UAVSAR with 0.6- and 1.9-m spatial resolution in the azimuth and slant range directions, respectively [35].

The validation of the obtained results is implemented both at landslide scale and at point scale. A perfectly consistent validation set (i.e., ground measurements acquired over the same time span covered by the SAR images used in this study) is not available, especially concerning the landslide scale. In this case, the most referenced data are relevant to the period 1985–1990 and to the year 2010. They are reported in Table 3 for the ease of the reader.

Data concerning the period 1985–1990 were produced in [29,30] using photogrammetry and field surveys. Data relevant to the year 2010 were collected in [22] through Ground-Based SAR Interferometry (GBInSAR) measurements. The latter study highlighted that, by 2010, the landslide's velocity halved compared to its values in the 1985–1990 period, and that the landslide head was affected by the largest decrease in velocity. The authors ascribed this behavior to both geomorphological and climatic factors. At the climatic level, they suggested that the average increase in temperature and decrease in precipitation could have induced an overall slowing of the landslide. Reference [1] showed that the movement of the Slumgullion landslide is strongly correlated to the soil moisture, and that its decreasing trend reflected the general slowing of the landslide.

Table 3. Available literature data, in meters per year, as reported in [22,29,30]. Vf is the displacement velocity measured with field instruments between 1985 and 1990. VGB is the displacement velocity estimated through Ground-Based Synthetic Aperture Radar Interferometry (GBInSAR) in 2010.

Region id	<Vf> 1985–1990	std(Vf) 1985–1990	<VGB> 2010	std(VGB) 2010
Landslide	2.48	1.38	1.16	1.35
1	0.73	0.40	0.14	0.91
2	1.20	0.25	0.32	1.27
3	1.42	0.14	0.36	1.31
4	1.60	0.51	0.36	0.98
5	2.44	0.29	1.05	1.53
6	3.86	0.87	1.67	2.51
7	5.25	0.73	2.84	3.35
8	3.57	0.40	1.64	3.13
9	3.65	0.87	1.93	3.06
10	1.56	0.18	0.91	6.49
11	1.97	0.36	1.13	2.37

From a geomorphological point of view, the observed thinning of the landslide head [36] caused its stability to increase. The slowing of the head should have favored the overall slowing of the landslide by decreasing downslope-directed transfer of shear stresses [22].

The displacement rates retrieved in this study are congruent with those reported in Table 3 in the column relevant to the GBInSAR survey implemented in 2010 [22]. The direction of the estimated vector field (see arrows in Figure 3) mainly follows the landslide slope profile and is qualitatively quite similar to that presented in the past literature (see as an example [31]).

The obtained results showed that the effect of the variation of the correlation window is negligible from the viewpoint of the estimated displacements, being for all the regions below the theoretical sensitivity of the method, which, as previously stated, is given by $1/f$, where f is the applied oversampling factor. On the other hand, defining a smaller correlation window (e.g., 32 pixels) makes the frequency-domain cross-correlation less reliable, and this increases the standard deviation of the estimated displacement field (not reported here for brevity), which results in noisiness and physical inconsistency. Therefore, it is suggested to operate with the 64-pixel window. This allows a lower computational time compared to the 128-pixel window (for these experiments, the computational time was about 2.1 h per run, compared to about 45 min for each 64-pixel window run), as well as a higher level of detail and a better preservation of the landslide edges.

For the 64-pixel window, the registered values of the standard deviation range from 0.09 m/year in Region 1 (pair 2012–2013) to 0.67 m/year in Region 7 (pair 2011–2012). They are similar to those indicated in [29,30].

It is remarkable that the noisier displacement patterns, see as an example that north of landslide Region 8, are characterized by very low values of the quality parameters considered. This means that the peak of the correlation matrix is not sharp (i.e., it is not well-defined compared with the background), thus leading to an unreliable estimate of the displacements. In the landslide area, even though the peak of the correlation matrix c_{max} is not very pronounced (as expected based on the characteristics of the phenomenon under investigation), its ratio q with respect to the mean is quite high. The average values registered for the pair 2011–2012 for these quantities within the defined kinematic regions are, respectively: Region 1–0.33, 12.34; Region 2–0.39, 12.65; Region 3–0.32, 12.14; Region 4–0.30, 12.33; Region 5–0.30, 12.38; Region 6–0.28, 10.21; Region 7–0.24, 9.70; Region 8–0.23, 9.92; Region 9–0.25, 11.17; Region 10–0.28, 11.42; and Region 11–0.24, 10.27.

The consistency check involves the three pairs: August 2011–August 2012; August 2012–August 2013; and August 2011–August 2013. A consistent result should pose that the sum of the displacements is zero. As stated above, this is not strictly achieved, however the resulting distribution has a mean very close to zero either at landslide scale or within the 11 kinematic regions. The registered deviations from zero (using the 64-pixel window) range (in absolute value) from less than 1 cm/year (region 2) to 12 cm/year (region 5). Similar results were obtained using the 128-pixel window (see Table 2).

In the following, the obtained results will be discussed at a finer scale exploiting data concerning 19 measurement points (MPs) installed on the landslide by the US Geological Survey (USGS) [34]. Reference data for comparison were extracted from Reference [26], in which the kinematics of the landslide was analyzed in the time frame August 2011–April 2012 using airborne L-band remote sensing and was compared with GPS data collected at the USGS MPs. Note that these data were reproduced through graph digitalization, and it is therefore possible that they exhibit (negligible) differences with respect to their original version.

In Figure 5, the comparison between the results presented here (time frame August 2011–August 2012) and those from [26] is shown. The position of the MPs with respect to one of the available SAR acquisitions is also reported (note that MP1 and MP2 are not shown in this graphic since their position falls outside the image cut considered). Airborne SAR-derived data are rendered with blue bars and GPS data with gray bars. Outcomes from this study (orange bars) were produced by taking the maximum displacement in a window of 100×100 meters around each MP position (only for MP1, which is expected to be immobile, the mean displacement was considered). This choice was made in order to compensate geocoding errors (some of the MPs are at the landslide borders and/or at the transition between different kinematic zones) and SPOT maps inhomogeneity/noise.

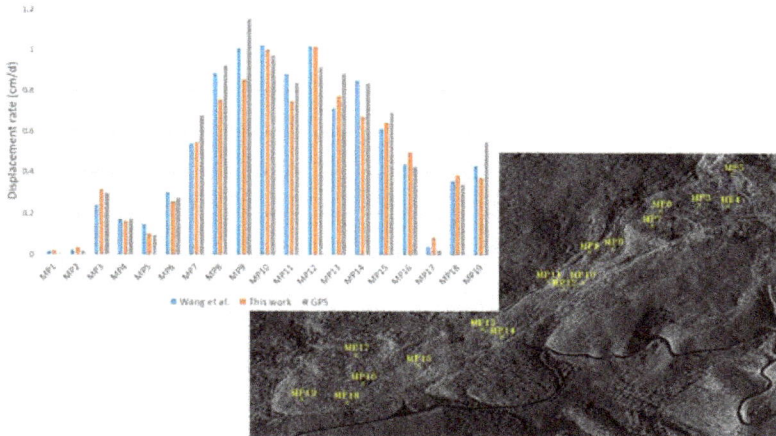

Figure 5. Comparison between space-borne COSMO-SkyMed displacement rates (orange), airborne UAVSAR displacement rates (blue) and GPS displacement rates (gray) for the 19 USGS measurement points. UAVSAR and GPS displacement rates have been extracted from [26].

The three datasets qualitatively show a good agreement. Disagreements are concentrated, as expected, in the neck sector, where the landslide is faster. In this area, estimates from this work exhibit the highest discrepancy when compared to airborne SAR and GPS data, in particular in MP8, MP9, and MP14. For those points, the landslide displacement rate is underestimated when compared to GPS data of about 0.17 cm/day, about 0.3 cm/day, and about 0.16 cm/day, respectively. If the measurements extracted from airborne SAR images are considered, the underestimation in correspondence of the aforementioned points is on the order of 0.13 cm/day, 0.15 cm/day, and 0.17 cm/day.

Assuming that points MP1 to MP7 belong to the head, that MP8 to MP14 belong to the neck, and that MP15 to MP19 belong to the toe, the registered Root-Mean-Square Error (RMSE) of the displacement rates here estimated with respect to GPS measurements is about 0.05 cm/day for the head, 0.15 cm/day for the neck, and 0.09 cm/day for the toe.

4. Discussion

SPOT methods allow for the estimation of two-dimensional displacements by using only the intensity channel. They are less sensitive to atmospheric effects and land cover types (i.e. they can be applied also in presence of vegetation) if compared with DInSAR. This is due to the lack of the phase information, which also allows for avoiding unwrapping procedures, often leading to the failure of DInSAR because of the low density of valid pixels [37].

With a lack of any facilities (i.e., corner reflectors) installed in situ, the accuracy of the estimated displacements depends on the resolution of input images, the magnitude of the real displacements, the acquisition geometry, and the correlation coefficient between the scatterers [38]. These parameters are clearly interconnected, since higher-resolution images can provide a better description of the scene features, thus maximizing their correlation at fine scale. On the other hand, when displacements are significant, the hypothesis of rigid shift is weakened, i.e., the correlation tends to be lower, as does the accuracy of the estimated offsets, as testified by the comparison with the GPS data provided in Section 3.1, which gave the worst results in correspondence with areas characterized by faster movements.

Spatial decorrelation can also be due to orbital issues. Differences in the acquisition geometry can significantly change speckle patterns, thus affecting the amplitude and the sharpness of the correlation peak [39]. Therefore, it is crucial that images ingested in the SPOT algorithm are acquired with similar orbit, i.e., with contained normal baselines. Their values for the examined image pairs were on the order of 30 m for the couple 2011–2012 and 460 m for the couples 2012–2013 and 2011–2013, which values are significantly below the X-band critical baseline (i.e., the baseline value causing complete image decorrelation), which is more than 4 km [40].

The implemented SPOT method, validated against the available literature data both at landslide and at point scale, was able to successfully reconstruct the kinematics of the Slumgullion landslide using very high-resolution space-borne SAR remote sensing. At landslide scale, the estimated average displacement rates in relevant kinematic were consistent with the closest available data, which were acquired a couple of years before the first considered observation period. The obtained displacement maps resulted quite homogeneous in those regions, as testified by the low values of the standard deviation reported in Table 2. As expected, its peaks were registered where the landslide is faster, i.e., where both the correlation value and the q parameter exhibited the lowest value (see Figure 4).

At the point scale, the performance of the implemented SPOT was fully comparable to that achieved by dedicated airborne campaigns, whose results showed a slightly better agreement with reference GPS data. However, this was expected, since UAVSAR L-band images exploited in [26], having higher penetration depth, are less sensitive to land cover (e.g., vegetation) compared to the X-band COSMO-SkyMed data used for this study. This means that the correlation between different images is expected to be higher in the first case [41] and, in turn, the displacement estimation is more precise.

However, the acquisition of space-borne images is surely cheaper than that of airborne ones and, since differences in performance are small, it can be argued that the first solution is more beneficial, being also independent from weather conditions, especially when monitoring is implemented with high temporal frequency.

5. Conclusions

Satellite technologies have widely demonstrated their potential in environmental monitoring and forecasting, response, and recovery of natural hazards. In this paper, the suitability of space-borne SAR observations for fast/slow landslide monitoring was explored. This application cannot be addressed using classic differential SAR interferometry due to the magnitude of the displacement rate and scarce preservation of interferometric coherence when movements are on the order of meters. Therefore, sub-pixel offset tracking methodology was exploited. This technique, using only the amplitude channel, is less sensitive to atmospheric and scene decorrelation factors and does not have particular restrictions on the observable displacement rates. Thus, it can be applied to fast/slow landslides such as the Slumgullion landslide (Colorado, US) which represented the objective of this study.

This landslide has been extensively investigated in the past through field surveys. Recently, airborne remote sensing has been exploited for monitoring displacements, however, as far as we know, there have been no results obtained using space-borne images validated against ground data. Therefore, state-of-the-art sub-pixel offset tracking was applied to pairs of SAR images acquired with about one-year temporal baseline in spotlight modality with sub-meter resolution by the COSMO-SkyMed constellation. These were combined in two pairs covering two years, from August 2011 to August 2013, and the obtained results were validated with an inter-sensor consistency check and against the temporally closest available ground data.

First, the cross-consistency of the estimated displacement fields was checked by comparing the movements of the period 2011–2013 with the sum of those estimated for the time frames 2011–2012 and 2012–2013. The comparison was satisfactory, since the resulting displacement rates showed discrepancies on the order of a few centimeters per year.

The obtained displacement fields were then compared with available literature data, both at landslide scale and at point scale.

At landslide scale, the average displacement rates computed via ground-based SAR interferometry in the aforementioned kinematic regions identified in [18] were used as benchmark. As a result, the displacement fields estimated with satellite imagery were fully comparable with those provided by field surveys.

At point scale, literature data regarding 19 measurement points installed by the US Geological Survey were exploited. For that location, both GPS data and airborne remote sensing derived data were available. The comparison was satisfactory even in this case, since maximum deviations from literature data were on the order of fractions of a centimeter localized in the central part of the landslide.

The results discussed in this study demonstrated the reliability of space-borne radar imagery for large landslide-induced movement monitoring and its consistency with the results given by more expensive and time-demanding methodologies such as field surveys and airborne remote sensing. Future research will address the validation of the displacements estimated using this methodology over an extended time frame.

Author Contributions: Conceptualization, A.I., G.D.M., and D.A.; methodology, D.A., R.G., and D.D.M.; software, D.A. and G.D.M.; validation, D.A., D.D.M., and D.D.A.; resources, M.C., F.M. (Federico Minati), and F.M. (Fabio Malvarosa); writing—original draft preparation, D.A. and D.D.M.; writing—review and editing, G.D.M. and R.G.; supervision, A.I., M.C., F.M. (Federico Minati), and F.M. (Fabio Malvarosa).

Funding: This research received no external funding.

Acknowledgments: LiDAR data acquisition and processing were completed by the National Center for Airborne Laser Mapping (NCALM). NCALM funding was provided by the NSF's Division of Earth Sciences, Instrumentation and Facilities Program. EAR-1043051. The authors thank the company e-geos for providing free of charge the SAR datasets used in this study.

Conflicts of Interest: The authors declare no conflict of interest.

References

1. Coe, J.A. Regional moisture balance control of landslide motion: Implications for landslide forecasting in a changing climate. *Geology* **2012**, *40*, 323–326. [CrossRef]
2. Milillo, P.; Fielding, E.J.; Shulz, W.H.; Delbridge, B.; Burgmann, R. COSMO-SkyMed Spotlight Interferometry Over Rural Areas: The Slumgullion Landslide in Colorado, USA. *IEEE J. Sel. Top. Appl. Earth Obs. Remote Sens.* **2014**, *7*, 2919–2926. [CrossRef]
3. Lanari, R.; Mora, O.; Manunta, M.; Mallorqui, J.J.; Berardino, P.; Sansosti, E. A small-baseline approach for investigating deformations on full-resolution differential SAR interferograms. *IEEE Trans. Geosci. Remote Sens.* **2004**, *42*, 1377–1386. [CrossRef]
4. Di Martire, D.; Paci, M.; Confuorto, P.; Costabile, S.; Guastaferro, F.; Verta, A.; Calcaterra, D. A nation-wide system for landslide mapping and risk management in Italy: The second Not-ordinary Plan of Environmental Remote Sensing. *Int. J. Appl. Earth Obs. Geoinf.* **2017**, *63*, 143–157. [CrossRef]
5. Tapete, D.; Cigna, F. InSAR data for geohazard assessment in UNESCO World Heritage sites: State-of-the-art and perspectives in the Copernicus era. *Int. J. Appl. Earth Obs. Geoinf.* **2017**, *63*, 24–32. [CrossRef]
6. Di Martire, D.; Iglesias, R.; Monells, D.; Centolanza, G.; Sica, S.; Ramondini, M.; Pagano, L.; Mallorquí, J.J.; Calcaterra, D. Comparison between Differential SAR interferometry and ground measurements data in the displacement monitoring of the earth-dam of Conza della Campania (Italy). *Remote Sens. Environ.* **2014**, *148*, 58–69. [CrossRef]
7. Colesanti, C.; Ferretti, A.; Novali, F.; Prati, C.; Rocca, F. Sar monitoring of progressive and seasonal ground deformation using the permanent scatterers technique. *IEEE Trans. Geosci. Remote Sens.* **2003**, *41*, 1685–1701. [CrossRef]
8. Fumagalli, A.; Novali, F.; Prati, C.; Ferretti, A.; Rocca, F.; Rucci, A. A New Algorithm for Processing Interferometric Data-Stacks: SqueeSAR. *IEEE Trans. Geosci. Remote Sens.* **2011**, *49*, 3460–3470.
9. C Prati, C.; Ferretti, A.; Perissin, D. Recent advances on surface ground deformation measurement by means of repeated space-borne SAR observations. *J. Geodyn.* **2010**, *49*, 161–170. [CrossRef]
10. Hooper, A. A multi-temporal InSAR method incorporating both persistent scatterer and small baseline approaches. *Geophys. Res. Lett.* **2008**, *35*, 1–5. [CrossRef]
11. Singleton, A.; Li, Z.; Hoey, T.; Muller, J.-P. Evaluating sub-pixel offset techniques as an alternative to D-InSAR for monitoring episodic landslide movements in vegetated terrain. *Remote Sens. Environ.* **2014**, *147*, 133–144. [CrossRef]
12. Wasowski, J.; Bovenga, F. Investigating landslides and unstable slopes with satellite Multi Temporal Interferometry: Current issues and future perspectives. *Eng. Geol.* **2014**, *174*, 103–138. [CrossRef]
13. Manconi, A.; Casu, F.; Ardizzone, F.; Bonano, M.; Cardinali, M.; De Luca, C.; Gueguen, E.; Marchesini, I.; Parise, M.; Vennari, C.; et al. Brief Communication: Rapid mapping of landslide events: The 3 December 2013 Montescaglioso landslide, Italy. *Nat. Hazards Earth Syst. Sci.* **2014**, *14*, 1835–1841. [CrossRef]
14. Sun, L.; Muller, J.-P. Evaluation of the Use of Sub-Pixel Offset Tracking Techniques to Monitor Landslides in Densely Vegetated Steeply Sloped Areas. *Remote Sens.* **2016**, *8*, 695. [CrossRef]
15. Lüttig, C.; Neckel, N.; Humbert, A. A Combined Approach for Filtering Ice Surface Velocity Fields Derived from Remote Sensing Methods. *Remote Sens.* **2017**, *9*, 1062. [CrossRef]
16. Strozzi, T.; Luckman, A.; Murray, T.; Wegmuller, U.; Werner, C. Glacier motion estimation using SAR offset-tracking procedures. *IEEE Trans. Geosci. Remote Sens.* **2002**, *40*, 2384–2391. [CrossRef]
17. Riveros, N.; Euillades, L.; Euillades, P.; Moreiras, S.; Balbarani, S. Offset tracking procedure applied to high resolution SAR data on Viedma Glacier, Patagonian Andes, Argentina. *Adv. Geosci.* **2013**, *35*, 7–13. [CrossRef]
18. Huang, J.; Deng, K.; Fan, H.; Yan, S. An improved pixel-tracking method for monitoring mining subsidence. *Remote Sens. Lett.* **2016**, *7*, 731–740. [CrossRef]
19. Fan, H.; Gao, X.; Yang, J.; Deng, K.; Yü, Y. Monitoring Mining Subsidence Using A Combination of Phase-Stacking and Offset-Tracking Methods. *Remote Sens.* **2015**, *7*, 9166–9183. [CrossRef]
20. Parise, M.; Guzzi, R. Volume and Shape of the Active and Inactive Parts of the Slumgullion Landslide, Hinsdale County, Colorado. *Open-File Rep.*. 1992. Available online: https://pubs.er.usgs.gov/publication/ofr92216 (accessed on 11 February 2019).
21. Crandell, D.R.; Varnes, D.J. Slumgullion earthflow and earth slide near Lake City, Colorado. *Geol. Soc. Am. Bull.* **1960**, *71*, 1846.

22. Schulz, W.H.; Coe, J.A.; Ricci, P.P.; Smoczyk, G.M.; Shurtleff, B.L.; Panosky, J. Landslide kinematics and their potential controls from hourly to decadal timescales: Insights from integrating ground-based InSAR measurements with structural maps and long-term monitoring data. *Geomorphology* **2017**, *285*, 121–136. [CrossRef]

23. Guzzi, R.; Parise, M. Surface Features and Kinematics of the Slumgullion Landslide, near Lake City, Colorado. *US Geol. Surv.*. 1992; pp. 92–252. Available online: https://pubs.er.usgs.gov/publication/ofr92252 (accessed on 11 February 2019).

24. Coe, J.; Godt, J.; Ellis, W.; Savage, W.; Savage, J.; Powers, P.; Varnes, D.; Tachker, P. *Preliminary interpretation of seasonal movement of the Slumgullion landslide as determined from GPS observations, July 1998–July 1999*; Open-File Report; U.S. Geological Survey: Reston, VA, USA, 2000. Available online: http://pubs.usgs.gov/of/2000/ofr-00-0102/ (accessed on 11 February 2019).

25. Wang, C.; Mao, X.; Wang, Q. Landslide Displacement Monitoring by a Fully Polarimetric SAR Offset Tracking Method. *Remote Sens.* **2016**, *8*, 624. [CrossRef]

26. Wang, C.; Cai, J.; Li, Z.; Mao, X.; Feng, G.; Wang, Q. Kinematic Parameter Inversion of the Slumgullion Landslide Using the Time Series Offset Tracking Method with UAVSAR Data. *J. Geophys. Res. Solid Earth* **2018**, *123*, 8110–8124. [CrossRef]

27. Delbridge, B.G.; Fielding, E.; Hensley, S.; Schulz, W.H. Three-dimensional surface deformation derived from airborne interferometric UAVSAR: Application to the Slumgullion Landslide. *J. Geophys. Res. Solid Earth* **2016**, *121*, 3951–3977. [CrossRef]

28. Tepel, R.E. Landslides: Investigation and Mitigation. *Environ. Eng. Geosci.* **1998**, *VI*, 277–279. [CrossRef]

29. Smith, W. *Photogrammetric Determination of Movement on the Slumgullion Slide, Hinsdale County, Colorado 1985–1990*; Open-File Report; U.S. Geological Survey: Reston, VA, USA, 1993. [CrossRef]

30. Fleming, R.; Baum, R.L.; Giardino, M. *Map and Description of the Active Part of the Slumgullion Landslide, Hinsdale County, Colorado*; U.S. Geological Survey: Reston, VA, USA, 1999.

31. Delbridge, B.; Burgmann, R.; Fielding, E.; Hensley, S. Kinematics of the slumgullion landslide from UAVSAR derived interferograms. In Proceedings of the 2015 IEEE International Geoscience and Remote Sensing Symposium (IGARSS), Milan, Italy, 26–31 July 2015; pp. 3842–3845.

32. Franceschetti, G.; Lanari, R.; Fidler, J. *Synthetic Aperture Radar Processing*; CRC Press: Boca Raton, FL, USA, 2018.

33. Shapiro, L.; Stockman, G.C. *Computer Vision*; Prentice Hall: Upper Saddle River, NJ, USA, 2002.

34. Coe, J.A.; Godt, J.W.; Ellis, W.L.; Savage, W.Z.; Savage, J.E.; Powers, P.S.; Varnes, D.J.; Tachker, P. *Seasonal Movement of the Slumgullion Landslide as Determined from GPS Observations, July 1998–July 1999*; U.S. Geological Survey: Reston, VA, USA, 2000.

35. Hensley, S.; Zebker, H.; Jones, C.; Michel, T.; Muellerschoen, R.; Chapman, B. First deformation results using the NASA/JPL UAVSAR instrument. In Proceedings of the 2009 2nd Asian-Pacific Conference on Synthetic Aperture Radar, Xi'an, China, 36–30 October 2009; pp. 1051–1055.

36. Coe, J.A.; McKenna, J.P.; Godt, J.W.; Baum, R. Basal-topographic control of stationary ponds on a continuously moving landslide. *Earth Surf. Process. Landforms* **2009**, *34*, 264–279. [CrossRef]

37. Sun, L.; Muller, J.-P.; Chen, J. Time Series Analysis of Very Slow Landslides in the Three Gorges Region through Small Baseline SAR Offset Tracking. *Remote Sens.* **2017**, *9*, 1314. [CrossRef]

38. De Zan, F. Coherent Shift Estimation for Stacks of SAR Images. *IEEE Geosci. Remote Sens. Lett.* **2011**, *8*, 1095–1099. [CrossRef]

39. Yonezawa, C.; Takeuchi, S. Decorrelation of SAR data by urban damages caused by the 1995 Hyogoken-nanbu earthquake. *Int. J. Remote Sens.* **2001**, *22*, 1585–1600. [CrossRef]

40. Nitti, D.O.; Hanssen, R.F.; Refice, A.; Bovenga, F.; Nutricato, R. Impact of DEM-Assisted Coregistration on High-Resolution SAR Interferometry. *IEEE Trans. Geosci. Remote Sens.* **2011**, *49*, 1127–1143. [CrossRef]

41. Wempen, J.M.; McCarter, M.K. Comparison of L-band and X-band differential interferometric synthetic aperture radar for mine subsidence monitoring in central Utah. *Int. J. Min. Sci. Technol.* **2017**, *27*, 159–163. [CrossRef]

remote sensing

MDPI

Article

The Outlining of Agricultural Plots Based on Spatiotemporal Consensus Segmentation

Angel Garcia-Pedrero [1,*] , **Consuelo Gonzalo-Martín** [2] , **Mario Lillo-Saavedra** [3,4] and **Dionisio Rodríguez-Esparragón** [5]

1   Sustainable Forest Management Research Institute, Universidad de Valladolid & INIA, 42004 Soria, Spain
2   Computer School, Universidad Politécnica de Madrid, Campus de Montegancedo,
    28223 Pozuelo de Alarcón, Spain; consuelo.gonzalo@upm.es
3   Water Research Center for Agriculture and Mining, (CRHIAM), University of Concepción,
    4070386 Concepción, Chile; malillo@udec.cl
4   Faculty of Agricultural Engineering, University of Concepción, Campus Chillán, 3812120 Chillán, Chile
5   Instituto de Oceanografía y Cambio Global, IOCAG, Universidad de las Palmas de Gran Canaria,
    35017 Las Palmas de Gran Canaria, Spain; dionisio.rodriguez@ulpgc.es
*   Correspondence: angelmario.garcia@uva.es; Tel.: +34-975-129-400

Received: 17 October 2018; Accepted: 6 December 2018 ; Published: 8 December 2018

Abstract: The outlining of agricultural land is an important task for obtaining primary information used to create agricultural policies, estimate subsidies and agricultural insurance, and update agricultural geographical databases, among others. Most of the automatic and semi-automatic methods used for outlining agricultural plots using remotely sensed imagery are based on image segmentation. However, these approaches are usually sensitive to intra-plot variability and depend on the selection of the correct parameters, resulting in a poor performance due to the variability in the shape, size, and texture of the agricultural landscapes. In this work, a new methodology based on consensus image segmentation for outlining agricultural plots is presented. The proposed methodology combines segmentation at different scales—carried out using a superpixel (SP) method—and different dates from the same growing season to obtain a single segmentation of the agricultural plots. A visual and numerical comparison of the results provided by the proposed methodology with field-based data (ground truth) shows that the use of segmentation consensus is promising for outlining agricultural plots in a semi-supervised manner.

Keywords: agriculture parcel segmentation; superpixels; consensus

1. Introduction

The world's population has tripled over the last 100 years and is still growing dramatically while resources have remained the same, causing changes in the outlook of food supply. According to the Food and Agriculture Organization (FAO), global food production will need to grow by 70% in order to satisfy the food and feed demand of a population of 9 billion people by 2050. The agriculture sector faces a critical overall challenge: to ensure access to safe, healthy, and nutritious food while using natural resources more sustainably and making an effective contribution to climate change adaptation and mitigation [1].

This challenge implies a greater pressure than ever before on productive land. Accurate and up-to-date information about agricultural land, such as its status, acreage, ownership, and the type of crops, allows stakeholders to establish effective agricultural policies (e.g., for reducing greenhouse gas emissions, regulating water rights, and estimating subsidies and agriculture insurances) [2,3] and update agricultural geographical databases [4], among other important tasks. In order to have up-to-date information on agricultural land, it is essential that the outlines of the plots are correct and can be quickly updated.

For the last 40 years, the outlining of plots in agricultural lands has been addressed through different initiatives around the world. In the United States, the National Research Council has published two reports that examine the situation of the land parcel data in the U.S. and provide a series of recommendations that would foster a national data system for storing plot information [5,6]. On the other hand, the European Union has promoted the use of a common Land Parcel Identification System (LPIS, https://ec.europa.eu/jrc/en/scientific-tool/lpis-quality-assessment) among its members in order to maintain a record of the activities of farmers on their lands [7]. The success of all these initiatives depends enormously on a precise outlining of the parcels.

Traditionally, the outlining of the parcels has been carried out manually using photointerpretation or field campaigns and documented by surveying sketches and textual documentation, all of which is very expensive both in time and financial resources.

The application of Information Technology tools, such as remote sensing and geographical information systems, improves efficiency in the agricultural sector, enabling planning and decision making based on the spatially and temporally distributed data provided by these tools [8,9]. The new generation of optical remote sensors placed on aircraft, satellite platforms, and drones offers accessible and useful data of very high resolution for monitoring agricultural fields at a plot level [10,11]. The task of processing this data while maintaining accuracy and meeting the time requirements becomes a real challenge.

Several methods have been proposed in the remote sensing literature to try to solve the automatic or semi-automatic parcel outlining problem. Most of them are based on image segmentation, edge detection algorithms, classification models, or combinations of these techniques. An object-based approach for extracting human-made objects, particularly agricultural fields from high-resolution images, was proposed in [12]. This approach was able to extract regularly shaped objects by combining edge detection models with region-based segmentation. Da Costa et al. [13] developed an algorithm to outline vine plots automatically from very high resolution images by exploiting their textural properties. To differentiate between vine and non-vine pixels, they applied a thresholding method to the texture attributes of the image. In [14], a semi-automatic methodology for outlining field boundaries from satellite data was proposed. The authors first carried out segmentation using tonal and textural gradients, and the generated regions were then classified to obtain preliminary plot boundaries. Finally, they applied an active contour model [15] to refine the geometry of these boundaries. Turker et al. used perceptual grouping for automatically detecting sub-boundaries within existing agricultural fields from satellite imagery [16]. This approach combined field boundaries and image data to carry out a field-based analysis. A Canny edge detector was used to detect the edge pixels, and then lines were identified using a graph-based vectorization method.

From the analysis of the state of the art, it can be concluded that approaches based on segmentation methods have several drawbacks including that (1) they are sensitive to intra-plot variability, which can result in the production of more segmentation than desired, and (2) most of these methods depend heavily on the correct selection of parameters (e.g., the similarity measured used to group image pixels), which needs prior knowledge of the landscape or tuning by trial and error. Moreover, variability in the sizes and shapes of the plots means that certain configuration parameters do not allow plots with the different characteristics needed for a landscape to be outlined properly. Approaches based on edge detection tend to produce more of the desired edges, mainly due to the presence of spatial patterns and image noise. The oversegmentation problem presented by the two approaches is directly related to the high spatial resolution of the images used in the segmentation process. Nevertheless, in the case of outlining plots, a very high spatial resolution (VHSR) is critical for managing very differently shaped and sized parcels in the same scene. Methodologies based on the superpixel (SP) concept have been proposed to deal with VHSR. These methodologies aim to reduce the influence of noise and intra-class spectral variability, preserving most edges of the images and improving the computational speed of later steps, such as classification, clustering, and segmentation [17]. In fact, some approaches to outlining plots based on SPs are found in the literature. The authors in [18] combined a contour detection algorithm and simple linear iterative clustering (SLIC) to extract the cadastral boundaries from UAV orthoimages automatically. In [19], the problem of outlining plots was addressed as a machine learning problem. The authors first oversegmented a VHSR image, then labeled each pair of segments according to whether they belonged to an agricultural plot, and, finally, they trained a classifier using this information. The trained classifier was then used to segment the agricultural plots of other regions in the image automatically. The oversegmentation problem has been addressed in other areas, such as computer vision [20], video processing [21], and medicine [22], among others, by establishing a consensus between a set of different segments. This approach can also alleviate the parameter selection issue. Other works have addressed the problem of unsupervised parcel segmentation using time series. A procedure to identify different crops by combining information provided by an LPIS and a low spatial resolution image time series is presented in [23]. However, LPISs are not always available; in fact, as has already mentioned, a suitable definition and update of an LPIS involve at least the semi-automatic outlining of the parcels. Nevertheless, some works in the literature that have applied time series for improving classification in agriculture. Thus, in [24], an operational crop identification strategy based on the use of multispectral and multitemporal signatures was proposed for classification at the parcel level. It proposes a combination of synthetic-aperture radar (SAR) and optical data registered on particular dates with the objective of satisfying crop temporal constraints. The authors proved that the use of SAR time series reduces the crop classification delivery time when the optical image is replaced by several SAR images. The integration of spectral and temporal features was proposed in [25] for annual cropland mapping. Even though the results presented in this paper showed that the methodology proposed is independent of in situ data and that it is capable of differentiating the croplands effectively, this methodology requires spatial baseline land cover information provided by different sources. A different crop classification approach is proposed in [26]. Although an ensemble of multilayer perceptrons (MLPs) provides classified pixel-based and parcel-based maps from multitemporal satellite optical imagery, the labels of the training patterns were obtained through a ground survey. It should be noted that all these approaches are supervised, need additional information, and are not always available. To our knowledge, there is a lack of semi- or unsupervised methods that exploit temporal data for the purpose of outlining a high variety of parcels differing in size and shape. The use of temporal information to delineate agricultural plots appears promising since, during a growing season, plot boundaries in agricultural landscapes are relatively stable, while the phenological pattern of crops changes frequently [27]. Some successful examples of combining superpixels and temporal information for change detection can be found in the literature [28,29]. However, in this case, the parcel outlining problem could be seen as a special case of change detection, in which the pattern (parcel edges) tends to remain constant over time.

In this work, a new methodology based on consensus segmentation for outlining agricultural plots is presented. The proposed methodology combines segmentation on different scales—carried out using an SP method—with images registered on different dates to obtain a single segmentation of the agricultural plots.

This paper is organized as follows. Section 2 describes the imagery used and the ground truth built for the evaluation of the methodology proposed, which is explained in detail in Section 2.2. The results obtained are included and discussed in Section 3. The conclusions derived from this work are presented in Section 4.

2. Dataset and Methods

2.1. Dataset

The study area comprises approximately 160 km^2 (16,000 ha) of fragmented agricultural plots located in the Lolol Valley, O'Higgins Region, Chile (Figure 1). The region is characterized by a temperate climate, with an agricultural season between September and April (the spring-summer season). The rainfalls are concentrated in the winter months (June–August). The landscape is characterized by very diverse sizes, ranging from small-scale farms on small plots—smallholdings—(<5 ha on average) to large legally constituted entities with medium and large plots (>50 ha).

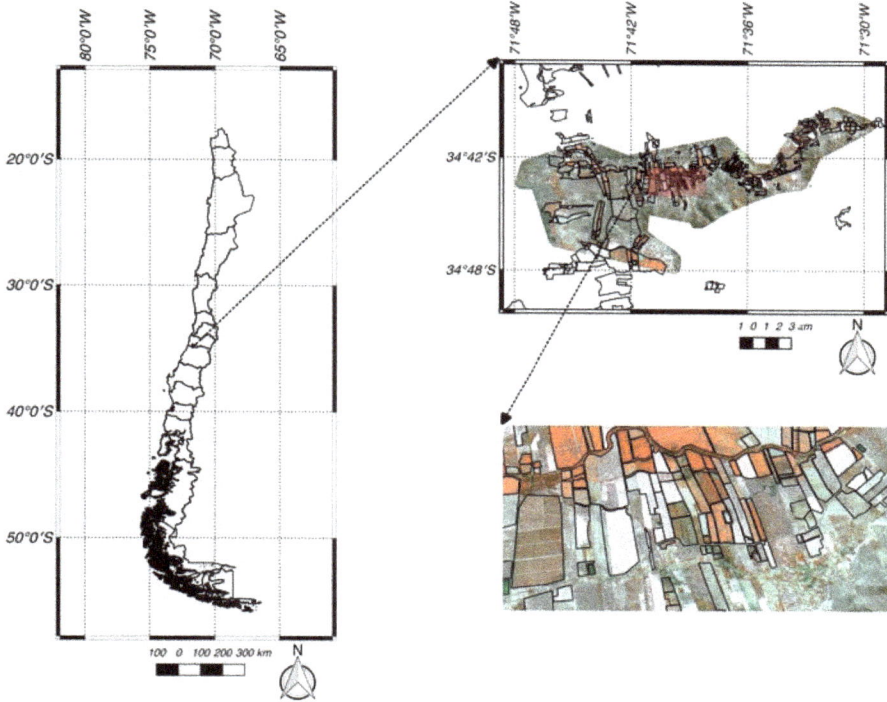

Figure 1. Study site, located in the Lolol Valley, O'Higgins Region, Chile.

Three multispectral Plèiades-1 satellite images are available for analyzing the study area. The corresponding dates are 4 December 2017, 30 January 2018, and 25 February 2018. Table 1 summarizes the spectral and spatial characteristics of this imagery.

Table 1. Spectral and spatial characteristics of the multispectral image taken by the Plèiades-1 satellite.

Band Name	Bandwidth (nm)	Spatial Resolution (m)
blue	430–550	
green	490–610	2
red	600–720	
near infrared	750–950	

The multispectral (MS) bands were geometrically corrected as well as co-registered. Moreover, the histograms of the images were adjusted to enhance the contrast. One scene containing mostly agricultural plots was selected and clipped from the three Plèiades images (as can be seen in Figure 2). The images under study correspond to a single agricultural season (2017–2018). As can be seen in Figure 2a–c, during an agricultural season, the main changes at the intra-plot scale correspond to the different phenological states of the crops. Furthermore, one ground truth map was obtained by manually outlining the agricultural plots in the clipped image for the date 30 January 2018. This map is used as ground truth to evaluate the result of the methodology proposed. Figure 2d sets out the polygons corresponding to the ground truth overlaid onto the color-composition image for the date 30 January 2018.

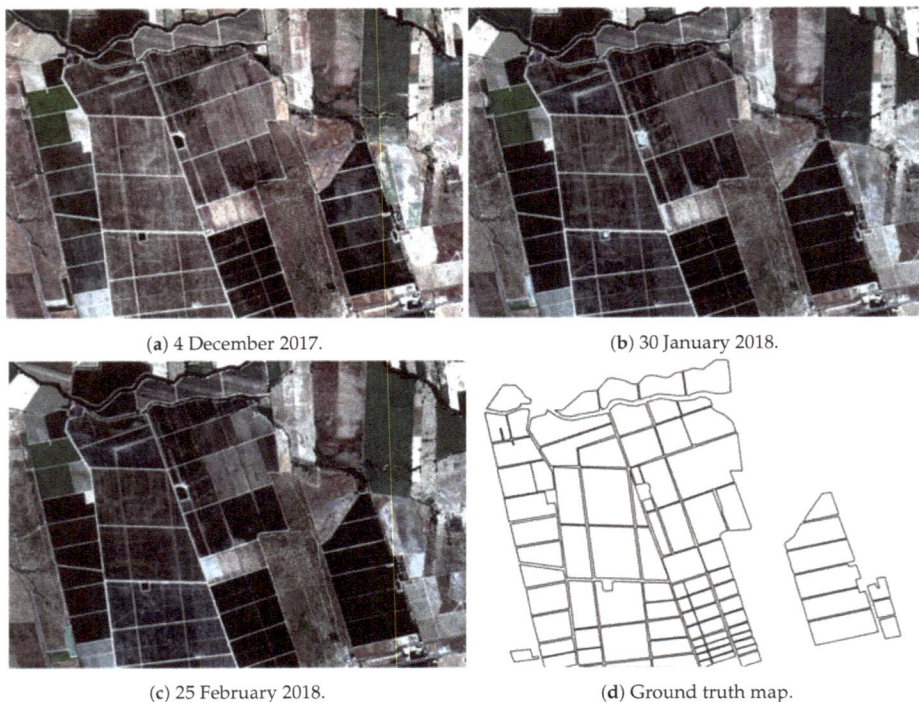

(a) 4 December 2017.   (b) 30 January 2018.

(c) 25 February 2018.   (d) Ground truth map.

Figure 2. Color compositions (3-2-1) of the Plèiades-1 imagery for three different dates and the ground truth of the study area.

2.2. Methodology

The methodology proposed in this paper integrates segmentation at different scales, carried out by an SP approach, and different dates to obtain a final single segmentation of the agricultural plots. An overview of the proposed methodology is shown in Figure 3.

The first step of this methodology is the image segmentation at different scales. Segmentation was carried out by means of the multispectral SLIC algorithm proposed in [30]. This method extends the original SLIC proposed in [31] by considering not only an RGB color space but a multispectral space. In this case, the clustering distance (Equation (1)) between two pixels p_i and p_j is composed of two components, one spatial (d_s) and the other spectral (d_c).

$$D_w = d_c + \frac{c}{g}d_s \tag{1}$$

$$d_s = \sqrt{(x_i - x_j)^2 + (y_i - y_j)^2} \tag{2}$$

$$d_c = \sqrt{\sum_{b=1}^{B}(p_i^b - p_j^b)^2} \tag{3}$$

where x and y denote the position of the pixels. B represents the total number of bands in the multispectral image. The constants g and c influence the size of the superpixel and its compactness, respectively. The higher the g value, the bigger the superpixel; on the other hand, the bigger the c value, the more compact the superpixel.

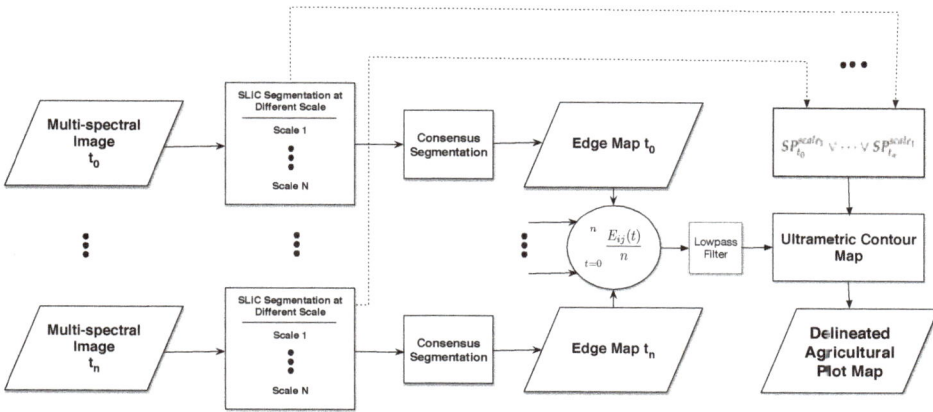

Figure 3. Overview of the methodology proposed.

Segments at different scales are generated by fixing the value of the compactness parameter but varying the size of the SPs. Similar to [30], the number of SPs was selected to follow a dyadic progression. In this work, the three original images were segmented into 10 different scales using the SLIC algorithm. The parameters used in the segmentation are a compactness factor (c), which is 0.04 times the maximum spectral value contained in each of the image, and the number of SPs, which ranges from 2^8 (scale$_1$) to 2^{17} (scale$_{10}$), on average.

The integration of the l segmentation for each date was carried out using a consensus process, which consists basically of a voting scheme that determines which pixels belong to the same region and which are part of the edges of objects in the image.

Thus, for each pair of adjacent pixels i and j, an E_{ij} value is obtained by Equations (4) and (5), which takes into account whether the pixels belong to the same region. If the two adjacent pixels (the ith and jth) do not belong to the same region, that means they are part of the edges between these regions. E_{ij} is defined as:

$$E_{ij} = \sum_{k=1}^{l} \Psi_{ij}^k \tag{4}$$

$$\Psi_{ij}^k = \begin{cases} S(r_a^k, r_b^k), & \text{if } i \in r_a^k, j \in r_b^k \text{ and } r_a^k \neq r_b^k; \\ 0, & \text{otherwise.} \end{cases} \tag{5}$$

where r_a^k and r_b^k represent the regions a and b, respectively, belonging to the kth segmentation, and $S(r_a^k, r_b^k)$ is an index of the similarity between the regions r_a^k and r_b^k. The larger the value of E_{ij}, the stronger the separation between these regions.

The similarity index, analogous to that proposed by [32], is defined as a combination between the similarity indices of color and texture:

$$S(r_a^k, r_b^k) = S_{color}(r_a^k, r_b^k) + S_{texture}(r_a^k, r_b^k) \tag{6}$$

where $S_{color}(r_a^k, r_b^k)$ measures the color similarity. For each region r_u^k, a color histogram is obtained using 25 bins for each spectral band. Then, from the color histograms, a feature vector $C_u = \{c_u^1, \ldots, c_u^{H_c}\}$ with a length of $H_c = 25 \times B$ is generated for region r_u^k (u can be a and b), where B is the number of bands of the multispectral images. The feature vector is normalized using the norm L^1 (also known as the Manhattan norm). Finally, the color similarity between the two regions r_a^k and r_b^k is calculated using the χ^2 statistic [33]:

$$S_{color}(r_a^k, r_b^k) = \sum_{h=1}^{H_c} \frac{(c_a^h - (c_a^h + c_b^h)/2)^2}{(c_a^h + c_b^h)/2} \tag{7}$$

where $S_{texture}(r_a^k, r_b^k)$ measures the texture similarity. Texture is calculated by means of steerable filters [34] using Gaussian derivatives in eight directions and $\sigma = 1$ as a basis. In this case, for each region r_u^k, a texture histogram is obtained using 10 bins for each spectral band and each filter direction, resulting in 80 features. Then, from the texture histograms, a feature vector $T_u = \{t_u^1, \ldots, t_u^{H_t}\}$ with a length $H_t = 80 \times B$ is generated for each region r_u^k, where B is the number of bands in the image. The similarity in texture between two regions r_a^k and r_b^k is calculated as follows:

$$S_{texture}(r_a^k, r_b^k) = \sum_{h=1}^{H_t} \frac{(t_a^h - (t_a^h + t_b^h)/2)^2}{(t_a^h + t_b^h)/2} \tag{8}$$

The result of the consensus process is an edge map for each date that is generated with the information of the voting scheme, which is normalized to a range of 0–1. These maps combine different information, such as (i) different scales of segmentation through the different sizes of SPs, (ii) the dissimilarity in both texture and color between neighboring regions, and (iii) the probability of a pixel belonging to an edge. The closer to 1 the value of a pixel, the greater the probability of it belonging to an edge in the multiple segmentation scales, and, therefore, the contrast between regions separated by such a pixel at different scales is greater.

The next step in the methodology is the integration of the edge maps for all dates into the labeled boundaries (I_{edge}). In this work, this integration was carried out by averaging the n edge maps.

$$I_{edge} = \sum_{t=1}^{n} \frac{E_{ij}(t)}{n} \tag{9}$$

Low-pass filtering should be applied to reduce the noise present in I_{edge}. In this work, a 3×3 median filter was used.

Since this averaged edge map generally includes open polygons, an ultrametric contour map (UCM) was calculated by means of the method proposed by [35]. A UCM is an edge map with the remarkable property, i.e., it produces a set of closed curves when any threshold is set [36]. The larger the threshold, the greater the contrast of the edges of the segments generated. The output of the UCM produces the outline of the final plots. In the end, a UCM is a soft representation of a segmentation that takes into account information from the edges of the image. The UCM has two inputs: the labeled boundaries (Equation (9)) and an image with boundary weights (IBW), which is defined by the Equation (10):

$$IBW_{edge} = SP_{t_0}^{scale_1} \vee \cdots \vee SP_{t_n}^{scale_1} \tag{10}$$

where $SP_{t_0}^{scale_1}$ and $SP_{t_n}^{scale_1}$ represent the SPs at scale 1 for the date t_0 and t_n, respectively; and $\vee$ is the logical operator OR. A detailed description of the UCM algorithm can be found in [35].

3. Results and Discussion

In Figure 4, the SP segmentation obtained for the image registered on 30 January 2018 and corresponding to scale$_4$ and scale$_8$ is overlaid onto the color-composition image. It is possible to see the good adherence of the SPs to image objects in the three displayed cases. However, it can also be observed that the smaller the SP size (Figure 4a), the better its adherence to the edges, as well as the homogeneity of the pixels that compose it. While, in the segmentation scales with larger SPs

(Figure 4a,b), the segments are less spectrally homogeneous, they still respect the borders between regions with a high contrast (the thickest edges).

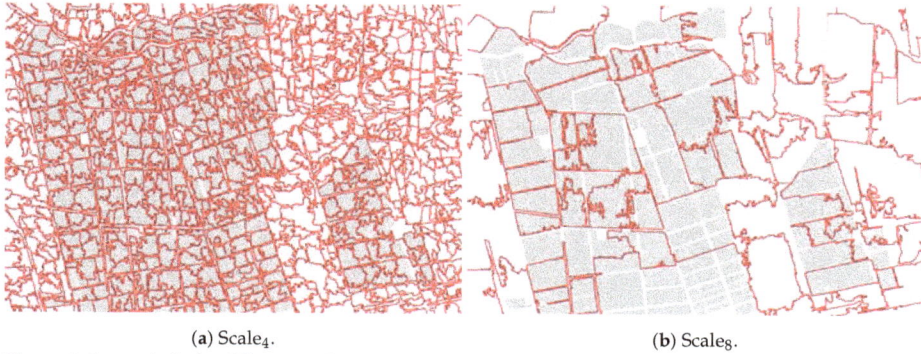

(a) Scale$_4$.   (b) Scale$_8$.

Figure 4. Superpixels for different scales are overlaid onto the ground truth of the study area with red color.

From the 10 segmentation scales for each date generated and consensus, the edge maps shown in Figure 5a–c were obtained.

As can be seen in Figure 5a–c, the edges of the plots tend to be more intense in all cases. In accordance with the methodology described, the next step is the integration of the three edge maps, one for each date, by calculating the average value. The average value is chosen as the basic form of integration because it allows for representing the information (borders) contained in all dates, thus reducing the effect of the appearance of borders in only one of the images, which is usually unusual for images of the same growing season (as mentioned in Section 1). In addition, the edges that appear as a result of phenological changes tend to decrease. The edge map obtained (Figure 5d) provides useful information for identifying the edges belonging to the plots; however, as mentioned above, it does not guarantee complete plots by applying a threshold, mainly due to edge-pixels not having the same value. Then, in order to obtain the final outlined agricultural plot map for the area considered, the UCM algorithm was applied to complete the edges.

Figure 6 shows the result of applying the UCM algorithm to the average of the three edge maps (Figure 5d). As can be observed, even though all edges have been closed, there is a lot of noise inside each plot. This noise is due to the fact that in the UCM, the totality of the probabilities of occurrence of the edges is represented. This is why it is necessary to determine a threshold value of the probability above which an edge is considered as such. Two indicators were used to determine this value objectively. The first was an error metric of the calculated edges with respect to a ground truth called boundary displacement error (BDE) [37,38]; the second metric considers the Shannon entropy, which represents the information content present in a border image: in this case, it is the number of edges present at the UCM. The BDE index measures the difference between two segmented images by averaging the displacement of boundary pixels. Specifically, the distance (d_E) between a boundary pixel (p_s) in the obtained boundary image (B_s) and the closest pixel (p_{gt}) in the ground-truth boundary image (B_{gt}) is used to define the error (disagreement) of each boundary pixel. The BDE index can be mathematically defined as in Equation (11).

$$BDE = \frac{1}{2}\left(\frac{1}{|B_s|}\sum_{ps_s \in B_s}\sum_{p_{gt} \in B_{gt}} min\{d_E(ps_s, p_{gt})\} + \frac{1}{|B_{gt}|}\sum_{p_{gt} \in B_{gt}}\sum_{p_s \in B_s} min\{d_E(p_{gt}, ps_s)\}\right) \quad (11)$$

where $|B|$ represents the number of boundary pixels in image B, and d_E is the Euclidean distance. The BDE index ranges within $[0, \infty)$, where the lower its value, the better. The BDE index is plotted against the edge entropy in Figure 5d for each threshold value. As shown in Figure 7, the smallest BDE metric error occurs for threshold values that are close to 1, while the worst values occur for values close to 0. Conversely, entropy behavior presents low values of information content for high threshold values and high information content for high threshold values. For the purpose of determining a specific threshold value, a compromise was established between the BDE value and the entropy value. As can be observed, a threshold value of 0.5 provides the balance between the error and the amount of edge information.

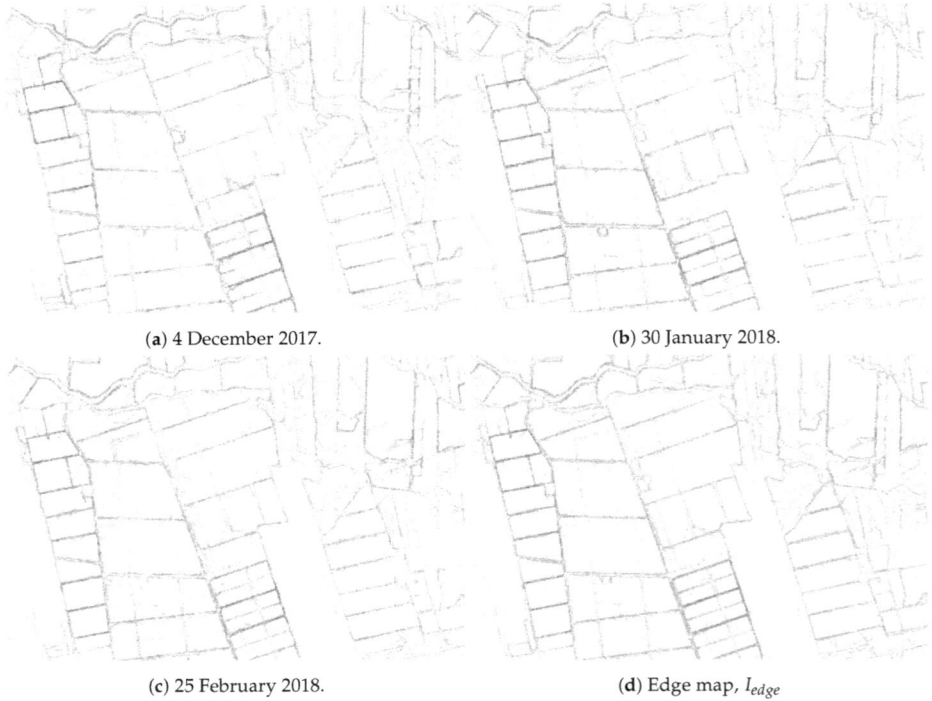

| (a) 4 December 2017. | (b) 30 January 2018. |

| (c) 25 February 2018. | (d) Edge map, I_{edge} |

Figure 5. Edge maps for three dates and the edge map, I_{edge}, obtained by averaging the three dates' edge maps.

Figure 6. Result of applying the ultrametric contour map (UCM) algorithm to the average of the three edge-maps.

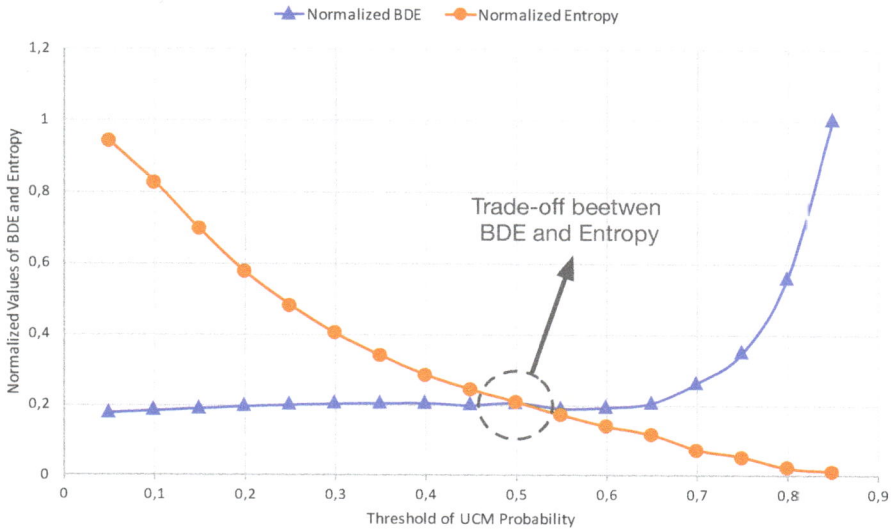

Figure 7. Representation of error versus the edge entropy for different threshold values.

Figure 8 shows the final outlined plot maps obtained for two different threshold values overlaid onto the ground truth map.

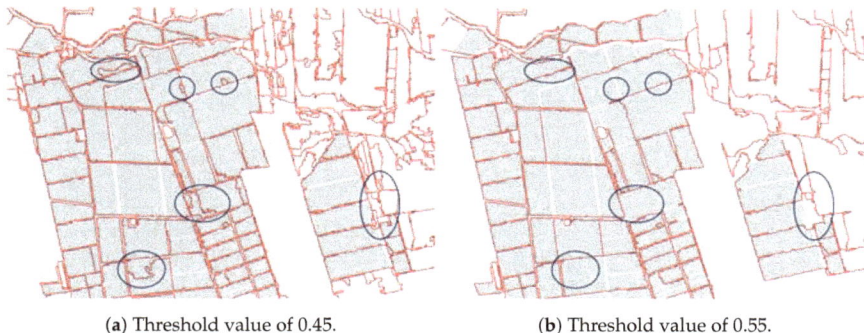

(a) Threshold value of 0.45. (b) Threshold value of 0.55.
Figure 8. Final outlined plot maps obtained for two different threshold values.

As can be seen, most of the edges agree between the two maps (Figure 8a,b), particularly in homogeneous areas with a high contrast between adjacent regions; however, discrepancies in plots with different tilled patterns can be also distinguished where the appearance of inner edges is present, as in plots within which anomalies are perceived due to poor agricultural practices, land heterogeneity, or crop diseases. As an example of the above, some areas where these changes occur are shown enclosed in circles. Consequently, only the edges stable over time—those that appear in the two images—remain in the final outlined plot map.

4. Conclusions

In this work, a new methodology based on consensus segmentation for outlining agricultural plots is presented. The methodology proposed combines segmentation at different scales—carried out using an SP method—and images registered at different dates to obtain a single segmentation of the agricultural plots. This methodology allows the outlining of agricultural plots of different sizes, shapes, colors, and textures. It is based on a consensus of segmentation (SP) at different scales and for different dates to determine the boundaries of the agricultural plots. The segmentation at different scales allows different-sized plots to be outlined. The use of SP for segmentation, which shows a good adherence to the edges for all scales, allows for differently shaped plots to be outlined. By highlighting the stable edges over time, the consensus of the segmentation reduces the intra-plot variability caused by the phenological stages present in a single growing season. The methodology also allows a threshold value to be determined in an objective way, which establishes a balance between the error and the amount of edge information. In particular, in this study, the threshold value determined for balance was 0.5. For lower threshold values, edges that are less stable over time appear on the outlined plot map, while for threshold values of up 0.5, only the more stable edges over time appear on the map.

Author Contributions: For this research article, A.G.-P., C.G.-M., and M.L.-S. conceived of and designed the experiments. A.G.-P. carried out the experiments. A.G.-P., C.G.-M., and M.L.-S. analyzed the data and results and wrote the paper, and D.R.-E. checked the paper style.

Funding: This work has been partially funded by the Agencia Estatal de Investigación (AEI) of Spain and the European Regional Development Fund (ERDF) through the ARTeMISat-2 project: Advanced Processing of Remote Sensing Data for Monitoring and Sustainable Management of Marine and Terrestrial Resources in Vulnerable Ecosystems (CTM2016-77733-R), and by the Water Research Center For Agriculture and Mining, CRHIAM (CONICYT–FONDAP–1513001).

Acknowledgments: A.G.-P. thanks the support of the SEÑALES Project (VA026P17) financed by the Junta de Castilla y León and the European Regional Development Fund (ERDF).

Conflicts of Interest: The authors declare no conflict of interest.

Abbreviations

The following abbreviations are used in this manuscript:

LPIS	Land Parcel Identification System
SLIC	Simple Linear Iterative Clustering
SP	Superpixels
UCM	Ultrametric Contour Map
VHSR	Very High Spatial Resolution

References

1. OECD/FAO. *OECD-FAO Agricultural Outlook 2017–2026*; Technical Report; OECD Publishing: Paris, France, 2017.
2. Mirón Pérez, J. Cadastre and the Reform of European Union's Common Agricultural Policy. Implementation of the SIGPAC, 2005. Available online: http://www.catastro.meh.es/documentos/publicaciones/ct/ct54/01-catastro_54_ing.pdf (accessed on 20 May 2017).
3. Van Der Molen, P. The use of the Cadastre Among the Members States: Property Rights, Land Registration and Cadastre in the European Union, 2002. Available online: http://www.catastro.meh.es/documentos/publicaciones/ct/ct45/02ingles.pdf (accessed on 20 May 2017).
4. Ciriza, R.; Sola, I.; Albizua, L.; Álvarez-Mozos, J.; González-Audícana, M. Automatic Detection of Uprooted Orchards Based on Orthophoto Texture Analysis. *Remote Sens.* **2017**, *9*, 492, doi:10.3390/rs9050492. [CrossRef]
5. National Research Council. *Need for a Multipurpose Cadastre*; National Academy Press: Washington, DC, USA, 1980.
6. National Research Council. *National Land Parcel Data: A Vision for the Future*; The National Academies Press: Washington, DC, USA, 2007.
7. Leo, O.; Lemoine, G. *Land Parcel Identification Systems in the Frame of Regulaton (EC) 1593/2000 Version 1.4*; Technical Report; Institute for Environment and Sustainability: Ispra, Italy, 2001.
8. Mulla, D.J. Twenty five years of remote sensing in precision agriculture: Key advances and remaining knowledge gaps. *Biosyst. Eng.* **2013**, *114*, 358–371, doi:10.1016/j.biosystemseng.2012.08.009. [CrossRef]

9. Shelestov, A.Y.; Kravchenko, A.N.; Skakun, S.V.; Voloshin, S.V.; Kussul, N.N. Geospatial information system for agricultural monitoring. *Cybern. Syst. Anal.* **2013**, *49*, 124–132. [CrossRef]

10. Waldner, F.; Canto, G.S.; Defourny, P. Automated annual cropland mapping using knowledge-based temporal features. *ISPRS J. Photogramm. Remote Sens.* **2015**, *110*, 1–13, doi:10.1016/j.isprsjprs.2015.09.013. [CrossRef]

11. Khot, L.R.; Espinoza, C.Z.; Jarolmasjed, S.; Sathuvalli, V.R.; Vandemark, G.J.; Miklas, P.N.; Carter, A.H.; Pumphrey, M.O.; Knowles, N.R.; Pavek, M.J. Low-altitude, high-resolution aerial imaging systems for row and field crop phenotyping: A review. *Eur. J. Agron.* **2015**, *70*, 112–123.

12. Mueller, M.; Segl, K.; Kaufmann, H. Edge- and region-based segmentation technique for the extraction of large, man-made objects in high-resolution satellite imagery. *Pattern Recognit.* **2004**, *37*, 1619–1628. [CrossRef]

13. Da Costa, J.P.; Michelet, F.; Germain, C.; Lavialle, O.; Grenier, G. Delineation of vine parcels by segmentation of high resolution remote sensed images. *Precis. Agric.* **2007**, *8*, 95–110. [CrossRef]

14. Tiwari, P.S.; Pande, H.; Kumar, M.; Dadhwal, V.K. Potential of IRS P-6 LISS IV for agriculture field boundary delineation. *J. Appl. Remote Sens.* **2009**, *3*, 033528. [CrossRef]

15. Kass, M.; Witkin, A.; Terzopoulos, D. Snakes: Active contour models. *Int. J. Comput. Vis.* **1988**, *1*, 321–331. [CrossRef]

16. Turker, M.; Kok, E.H. Field-based sub-boundary extraction from remote sensing imagery using perceptual grouping. *ISPRS J. Photogramm. Remote Sens.* **2013**, *79*, 106–121. [CrossRef]

17. Garcia-Pedrero, A.; Gonzalo-Martin, C.; Fonseca-Luengo, D.; Lillo-Saavedra, M. A GEOBIA Methodology for Fragmented Agricultural Landscapes. *Remote Sens.* **2015**, *7*, 767–787, doi:10.3390/rs70100767. [CrossRef]

18. Crommelinck, S.; Yang, M.Y.; Koeva, M.; Gerke, M.; Bennett, R.; Vosselman, G. Towards Automated Cadastral Boundary Delineation from UAV Data. *arXiv* **2017**, arXiv:1709.01813.

19. García-Pedrero, A.; Gonzalo-Martín, C.; Lillo-Saavedra, M. A machine learning approach for agricultural parcel delineation through agglomerative segmentation. *Int. J. Remote Sens.* **2017**, *38*, 1809–1819, doi:10.1080/01431161.2016.1278312. [CrossRef]

20. Rodolà, E.; Bulò, S.R.; Cremers, D. Robust Region Detection via Consensus Segmentation of Deformable Shapes. *Comput. Gr. Forum* **2014**, *33*, 97–106, doi:10.1111/cgf.12435. [CrossRef]

21. Faktor, A.; Irani, M. Video Segmentation by Non-Local Consensus Voting. In Proceedings of the British Machine Vision Conference, Nottingham, UK, 1–5 September 2014; pp. 1–12.

22. Carlier, T.; Haumont, C.; Bailly, C.; Bodet-Milin, C.; Ansquer, C.; Bodere, F. About few properties of PET segmentation using consensus approaches. *J. Nucl. Med.* **2017**, *58*, 611.

23. Inglada, J.; Dejoux, J.F.; Hagolle, O.; Dedieu, G. Multi-temporal remote sensing image segmentation of croplands constrained by a topographical database. In Proceedings of the Geoscience and Remote Sensing Symposium (IGARSS), Munich, Germany, 22–27 July 2012; pp. 6781–6784.

24. Blaes, X.; Vanhalle, L.; Defourny, P. Efficiency of crop identification based on optical and SAR image time series. *Remote Sens. Environ.* **2005**, *96*, 352–365, doi:10.1016/j.rse.2005.03.010. [CrossRef]

25. Matton, N.; Canto, G.S.; Waldner, F.; Valero, S.; Morin, D.; Inglada, J.; Arias, M.; Bontemps, S.; Koetz, B.; Defourny, P. An Automated Method for Annual Cropland Mapping along the Season for Various Globally-Distributed Agrosystems Using High Spatial and Temporal Resolution Time Series. *Remote Sens.* **2015**, *7*, 13208–13232, doi:10.3390/rs71013208. [CrossRef]

26. Kussul, N.; Lemoine, G.; Gallego, J.; Skakun, S.; Lavreniuk, M. Parcel Based Classification for Agricultural Mapping and Monitoring using Multi-temporal Satellite Image Sequences. In Proceedings of the Geoscience and Remote Sensing Symposium (IGARSS), Milan, Italy, 26–31 July 2015; pp. 165–168.

27. Liu, Y.; Li, M.; Mao, L.; Xu, F.; Huang, S. Review of remotely sensed imagery classification patterns based on object-oriented image analysis. *Chin. Geogr. Sci.* **2006**, *16*, 282–288. [CrossRef]

28. Huang, X.; Yang, W.; Xia, G.; Liao, M. Superpixel-based change detection in high resolution SAR images using region covariance features. In Proceedings of the 2015 8th International Workshop on the Analysis of Multitemporal Remote Sensing Images (Multi-Temp), Annecy, France, 22–24 July 2015; pp. 1–4.

29. Wu, Z.; Hu, Z.; Fan, Q. Superpixel-based unsupervised change detection using multi-dimensional change vector analysis and SVM-based classification. *ISPRS Ann. Photogramm. Remote Sens. Spat. Inf. Sci.* **2012**, *7*, 257–262. [CrossRef]

30. Gonzalo-Martín, C.; Lillo-Saavedra, M.; Menasalvas, E.; Fonseca-Luengo, D.; García-Pedrero, A.; Costumero, R. Local optimal scale in a hierarchical segmentation method for satellite images. *J. Intell. Inf. Syst.* **2016**, *46*, 517–529. doi:10.1007/s10844-015-0365-4. [CrossRef]
31. Achanta, R.; Shaji, A.; Smith, K.; Lucchi, A.; Fua, P.; Süsstrunk, S. SLIC superpixels compared to state-of-the-art superpixel methods. *IEEE Trans. Pattern Anal. Mach. Intell.* **2012**, *34*, 2274–2282. [CrossRef] [PubMed]
32. Uijlings, J.R.; van de Sande, K.E.; Gevers, T.; Smeulders, A.W. Selective search for object recognition. *Int. J. Comput. Vis.* **2013**, *104*, 154–171. [CrossRef]
33. Rubner, Y.; Tomasi, C.; Guibas, L.J. The earth mover's distance as a metric for image retrieval. *Int. J. Comput. Vis.* **2000**, *40*, 99–121. [CrossRef]
34. Freeman, W.T.; Adelson, E.H. The design and use of steerable filters. *IEEE Trans. Pattern Anal. Mach. Intell.* **1991**, *13*, 891–906. [CrossRef]
35. Arbelaez, P.; Maire, M.; Fowlkes, C.; Malik, J. From contours to regions: An empirical evaluation. In Proceedings of the IEEE Conference on Computer Vision and Pattern Recognition CVPR 2009, Miami Beach, FL, USA, 20–25 June 2009; pp. 2294–2301.
36. Dollár, P.; Zitnick, C.L. Structured forests for fast edge detection. In Proceedings of the IEEE International Conference on Computer Vision, Sydney, Australia, 1–8 December 2013; pp. 1841–1848.
37. Freixenet, J.; Muñoz, X.; Raba, D.; Marti, J.; Cufí, X. Yet another survey on image segmentation: Region and boundary information integration. In *European Conference on Computer Vision*; Springer: Berlin/Heidelberg, Germany, 2002; pp. 408–422.
38. Yang, A.Y.; Wright, J.; Ma, Y.; Sastry, S.S. Unsupervised segmentation of natural images via lossy data compression. *Comput. Vis. Image Underst.* **2008**, *110*, 212–225. [CrossRef]

remote sensing

MDPI

Article

Building Detection from VHR Remote Sensing Imagery Based on the Morphological Building Index

Yongfa You [1,2] , Siyuan Wang [1,*], Yuanxu Ma [1], Guangsheng Chen [3], Bin Wang [4], Ming Shen [1,2] and Weihua Liu [1,2]

[1] Key Laboratory of Digital Earth Science, Institute of Remote Sensing and Digital Earth, Chinese Academy of Sciences, Beijing 100094, China; youyf@radi.ac.cn (Y.Y.); mayx@radi.ac.cn (Y.M.); shenming@radi.ac.cn (M.S.); liuwh@radi.ac.cn (W.L.)

[2] College of Resources and Environment, University of Chinese Academy of Sciences, Beijing 100049, China

[3] Ecosystem Dynamics and Global Ecology Laboratory, School of Forestry and Wildlife Sciences, Auburn University, Auburn, AL 36837, USA; auburncgs@126.com

[4] School of Civil Engineering, Beijing Jiaotong University, Beijing 100044, China; bwang@bjtu.edu.cn

* Correspondence: w_siyuan@126.com; Tel.: +86-10-8217-8170

Received: 26 June 2018; Accepted: 9 August 2018; Published: 15 August 2018

Abstract: Automatic detection of buildings from very high resolution (VHR) satellite images is a current research hotspot in remote sensing and computer vision. However, many irrelevant objects with similar spectral characteristics to buildings will cause a large amount of interference to the detection of buildings, thus making the accurate detection of buildings still a challenging task, especially for images captured in complex environments. Therefore, it is crucial to develop a method that can effectively eliminate these interferences and accurately detect buildings from complex image scenes. To this end, a new building detection method based on the morphological building index (MBI) is proposed in this study. First, the local feature points are detected from the VHR remote sensing imagery and they are optimized by the saliency index proposed in this study. Second, a voting matrix is calculated based on these optimized local feature points to extract built-up areas. Finally, buildings are detected from the extracted built-up areas using the MBI algorithm. Experiments confirm that our proposed method can effectively and accurately detect buildings in VHR remote sensing images captured in complex environments.

Keywords: building detection; built-up areas extraction; local feature points; saliency index; morphological building index

1. Introduction

Buildings are the places where human beings live, work, and recreate [1]. The distribution of buildings is useful in many applications such as disaster assessment, urban planning, and environmental monitoring [2,3], and the precise location of buildings can also help municipalities in their efforts to better assist and protect their citizens [4]. Therefore, it is very important to accurately detect buildings. With the development of sensor technology, High spatial Resolution/Very High spatial Resolution (VHR) remote sensing images with multispectral channels can be acquired. In the context of this paper, images with a spatial resolution lower than one meter in the panchromatic channel are referred to as VHR imagery, and images with a spatial resolution greater than one meter and lower than ten meters in the panchromatic channel are referred to as High Resolution imagery [5]. Since these High Resolution/VHR remote sensing images contain a large amount of spectral, structure, and texture information, they provide more potential for accurate building detection. However, manual processing of these images to extract buildings requires continuous hard work and attention from humans, and it is impractical when applied to regional or global scales. Therefore, it is necessary to develop methods that can automatically or semi-automatically detect buildings from High Resolution/VHR remote sensing images. In the past decades, a large number of studies in this area have been conducted. Depending on whether or not the auxiliary information is used, we can divide the methods developed into two categories. The first category uses monocular remote sensing images to detect buildings, and the second category combines remote sensing images with auxiliary data such as height information to detect buildings. Several review articles can be found in [6–10]. Among them, Unsalan and Boyer [7] extended the work in [6] by comparing and analyzing the performance of different methods proposed until late 2003. Baltsavias [8] provided a review of different knowledge-based object extraction methods. Haala and Kada [9] discussed previous works on building reconstruction from a method and data perspective. More recently, Cheng and Han [10] systematically analyzed the existing methods devoted to object detection from optical remote sensing images. Since this study is dedicated to detecting buildings from a single VHR remote sensing imagery, our discussion of previous studies will focus on this area.

The development of low-orbit earth imaging technology has made available VHR remote sensing images with multispectral bands. In order to make full use of this spectral and spatial information, a large number of studies have used classification methods to detect buildings. For example, Lee et al. [11] combined supervised classification, iterative self-organizing data analysis technique algorithm (ISODATA) and Hough transformation to automatically detect buildings from IKONOS images. In their study, the classification process was designed to obtain the approximation locations and shapes of candidate building objects, and ISODATA segmentation followed by Hough transformation were performed to accurately extract building boundaries. Later, Inglada [12] used a large number of geometric features to characterize the man-made objects in high resolution remote sensing images and then combined them with support vector machine classification to extract buildings. In a different study, Senaras et al. [13] proposed a decision fusion method based on a two-layer hierarchical ensemble learning architecture to detect buildings. This method first extracted fundamental features such as color, texture, and shape features from the input image to train individual base-layer classifiers, and then fused the outputs of multiple base-layer classifiers by a meta-layer classifier to detect buildings. More recently, a new method based on a modified patch-based Convolutional Neural Network (CNN) architecture has been proposed for automatic building extraction [14]. This method did not require any pre-processing operations and it replaced the fully connected layers of the CNN model with the global average pooling. In summary, although these classification methods are effective for building extraction, it should be noted that these methods require a large volume of training samples, which is quite laborious and time-consuming.

Graph theory, as an important branch of mathematics, has also been used for building detection. For example, Unsalan and Boyer [7] developed a system to extract buildings and streets from satellite images using graph theory. In their work, four linear structuring elements were used to construct the binary balloons and then these balloons were represented in a graph framework to detect buildings and streets. However, due to the assumptions involved in the detection process, this method is only applicable to the type of buildings in North America. Later, Sirmacek and Unsalan [15] combined scale invariant feature transform (SIFT) with graph theory to extract buildings, where the vertices of the graph were represented by the SIFT key points. They validated this method on 28 IKONOS images and obtained promising results with a building detection accuracy of 88.4%. However, it should be noted that this method can only detect buildings that correspond to preset templates and are spatially isolated. In a different work, Ok et al. [16] developed a novel approach for automatic building detection based on fuzzy logic and the GrabCut algorithm. In their work, the directional spatial relationship between buildings and their shadows was first modeled to generate fuzzy landscapes, and then the buildings were detected based on the fuzzy landscapes and shadow evidence using the GrabCut partitioning algorithm. Nevertheless, the performance of this method is limited by the accuracy of shadow extraction. Later, Ok [17] extended their previous work by introducing a new shadow detection method and a two-level graph partitioning framework to detect buildings more accurately. However, buildings whose shadows are not visible cannot be detected by this method.

On the other hand, some studies have also used active contour models to detect buildings. For example, Peng and Liu [18] proposed a new building detection method using a modified snake model combined with radiometric features and contextual information. Nevertheless, this method cannot effectively extract buildings in complex image scenes. In a different work, Ahmadi et al. [19] proposed a new active contour model based on level set formulation to extract building boundaries. An experiment conducted in an aerial image showed that this model can achieve a completeness ratio of 80%. However, it should be noted that this model fails to extract buildings with similar radiometric values to the background. More recently, Liasis and Stavrou [20] used the HSV color components of the input image to modify the traditional active contour segmentation model to detect buildings. However, some non-building objects such as roads and bridges are also incorrectly labeled as buildings by this method when applied to high-density urban environments.

In recent years, a number of feature indices that can predict the presence of buildings have also been proposed. For example, Pesaresi et al. [21] developed a novel texture-derived built-up presence index (PanTex) for automatic building detection based on fuzzy composition of anisotropic textural co-occurrence measures. The construction of the PanTex was based on the fact that there was a high local contrast between the buildings and their surrounding shadows. Therefore, they used the contrast textural measures derived from the gray-level co-occurrence matrix to calculate the PanTex. Later, Lhomme et al. [22] proposed a semi-automatic building detection method using a new feature index called "Discrimination by Ratio of Variance" (DRV). The DRV was defined based on the gray-level variations of the building's body and its periphery. More recently, Huang and Zhang [23] proposed the morphological building index (MBI) to automatically detect buildings from GeoEye-1 images. The fundamental principle of the MBI was to represent the intrinsic spectral-structural properties of buildings (e.g., brightness, contrast, and size) using a set of morphological operations (e.g., top-hat by reconstruction, directionality, and granulometry). Furthermore, some improved methods for the original MBI, aiming at reducing the commission and omission errors in urban areas, have also been proposed [24,25]. The original MBI and its improved methods are effective for the detection of buildings in urban areas, but they fail to detect buildings in non-urban areas (e.g., mountainous, agricultural, and rural areas) where many irrelevant objects such as farmland, bright barren land, and impervious roads will cause large numbers of interferences to the detection of buildings. To solve this problem, a postprocessing framework for the MBI algorithm was proposed in [26] to extend the detection of buildings to non-urban areas by additionally considering the geometrical, spectral, and contextual information of the input image. However, it should be noted that this method is limited by the performance of these additional information extractions.

In this study, a new building detection method based on the MBI algorithm is proposed to detect buildings from VHR remote sensing images captured in complex environments. The proposed method can effectively solve the problem that many irrelevant objects with similar spectral characteristics to buildings will cause large numbers of interferences to the detection of buildings. Specifically, the proposed method first extracts built-up areas from the VHR remote sensing imagery, and then detects buildings from the extracted built-up areas. For the extraction of built-up areas (first step), the spatial voting method [27] based on the local feature points is used in this study. The term "local feature point" is defined as a small point of interest that is distinct from the background [28]. Among the literature, various local feature point detectors have been used to extract built-up areas, such as the Gabor-based detector [27], the SIFT-based detector [15], the Harris-based detectors [29,30], and the FAST-based detector [31]. However, it should be mentioned that these local feature point detectors have a common problem when used for built-up areas extraction. Since they are mainly designed to detect local feature points over areas with complex textures or salient edges, they not only detect local feature points in built-up areas, but also detect local feature points in non-built-up areas. However, these local feature points in non-built-up areas (referred to as false local feature points in this study) will weaken the extraction accuracy of built-up areas, so it is necessary to design a method that can effectively eliminate these false local feature points. To this end, a saliency index is proposed in this study, which is constructed based on the density and the distribution evenness of the local feature points in a local circle window. In addition, we adopt the idea of voting based on superpixels in [32] to improve the original spatial voting method [27]. Through these processes, we can extract the built-up areas more accurately. On the other hand, for the detection of buildings (second step), since the original MBI algorithm is susceptible to large numbers of interferences from irrelevant objects (e.g., bright barren land, farmland, and impervious roads) in non-built-up areas, it has poor performance when detecting buildings in non-urban areas, such as mountainous, agricultural, and rural areas. To solve this problem, we propose applying the MBI algorithm in the extracted built-up areas (first step) to detect buildings, which can directly eliminate large numbers of interferences caused by irrelevant objects in non-built-up areas. In addition, to further eliminate some errors in built-up areas, we also build a rule based on the shadow, spectral, and geometric information for the postprocessing of the initial building detection results. Through these processes, our proposed method can effectively detect buildings in images captured in complex environments.

The remainder of this paper is arranged as follows: Section 2 provides a detailed description of the proposed method; Section 3 analyzes and compares the experimental results; Section 4 presents the discussion; and Section 5 provides the conclusion.

2. Proposed Method

The proposed method is mainly composed of three key steps. First, the local feature points are detected using the Gabor wavelet transform results of the input VHR remote sensing imagery, and then these local feature points are optimized using a proposed saliency index. Next, a spatial voting matrix is computed based on these optimized local feature points to extract built-up areas. Finally, buildings are detected from the built-up areas using the MBI algorithm, and then the initial building detection results are further optimized by the built rule. The flow chart of the proposed method is shown in Figure 1.

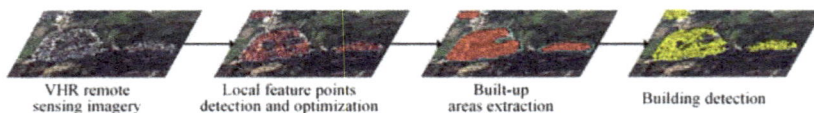

| VHR remote sensing imagery | Local feature points detection and optimization | Built-up areas extraction | Building detection |

Figure 1. The flow chart of the proposed method.

2.1. Local Feature Points Detection and Optimization

2.1.1. Local Feature Points Detection

Built-up areas are mainly composed of man-made objects such as buildings and roads. Compared with natural objects, these man-made objects usually produce a large number of local feature points. Since the density of local feature points in built-up areas is higher than that of non-built-up areas, many studies have used the density map of local feature points to identify built-up areas [27,32,33]. In addition, some studies have shown that 2D Gabor wavelets [34] are able to detect salient cues such as local feature points from images. Therefore, we use the Gabor wavelets to extract local feature points.

In order to obtain a complete representation of the image, the input VHR remote sensing image is first decomposed by Gabor wavelets at multi-scales along multi-directions, and then the magnitudes of the decomposition of all scales in each direction are summed up to obtain the Gabor energy map. Given that (x, y) represents the coordinate of a pixel in the image, u represents the scale, v represents the direction, and U and V represent the number of scales and directions, respectively. the Gabor energy map can be defined as

$$GE_v(x,y) = \sum_{u=1}^{U} D_{u,v}(x,y) \tag{1}$$

where $D_{u,v}(x,y)$ denotes the magnitude of the decomposition at scale u and direction v, and $GE_v(x,y)$ denotes the Gabor energy map at direction v. After the Gabor energy maps in all directions are obtained, the local feature point detection method proposed by Sirmacek and Unsalan [27] is used to detect local feature points from the Gabor energy maps. More specifically, this method first searches for the local maxima within the eight-connected neighborhoods of all pixels in the Gabor energy map, and these local maxima are taken as candidates for local feature points. Then, these candidate points are optimized according to their magnitude of $GE_v(x,y)$, and only the candidate whose magnitude is greater than a threshold, which is automatically obtained by performing Otsu's method on $GE_v(x,y)$ [35], will be retained as the local feature point of direction v. Finally, these procedures are applied to the Gabor energy maps in all V directions to obtain a local feature point set, noted as Ω_d. Figure 2a shows an example of using this method to detect local feature points. As shown in Figure 2a, this method not only detects a large number of local feature points in the built-up areas, but also detects many false local feature points in the non-built-up areas. Since these false local feature points will weaken the extraction accuracy of built-up areas, it is necessary to develop a method that can effectively eliminate them.

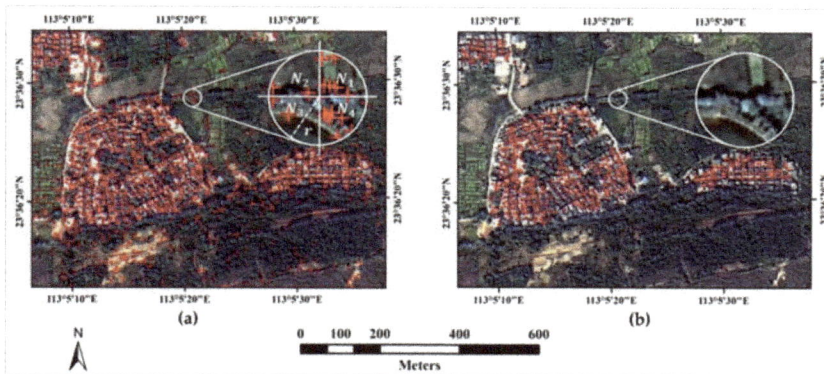

Figure 2. Local feature points detection and optimization. (**a**) Originally detected local feature points; (**b**) Optimized results by the proposed saliency index.

2.1.2. Local Feature Points Optimization

In order to eliminate these false local feature points located in non-built-up areas and obtain a reliable local feature point set, a saliency index is proposed in this study. The proposal of the saliency index is inspired by the texture saliency index proposed in [36]. These two indices are similar, but their implementation and purpose are completely different. The construction of our proposed saliency index is based on the fact that the local feature points are more densely and evenly distributed in built-up areas than in non-built-up areas. Therefore, we use the density and the distribution evenness of the local feature points in a local circle window to calculate the saliency index. To more clearly describe the derivation process of the saliency index, we use the enlarged circle in Figure 2a as an example for illustration. Given that N_i represents the number of the local feature points in the ith quadrant of the local circle window, the spatial distribution evenness parameter P_e can be defined as

$$P_e = min(N_1, N_2, N_3, N_4)/mean(N_1, N_2, N_3, N_4) \tag{2}$$

where $min(\cdot)$ represents the minimum operation, and $mean(\cdot)$ represents the averaging operation. If there is no local feature point in any of the four quadrants, then $min(N_1, N_2, N_3, N_4)$ is equal to 0, which in turn causes P_e to be equal to 0. In addition, given that r represents the radius of the local circle window, N_p represents the number of the local feature points in the local circle window, and N_w represents the number of pixels in the local circle window, the point density parameter P_d can be defined as

$$P_d = N_p / N_w \tag{3}$$

where $N_p = N_1 + N_2 + N_3 + N_4$, and $N_w = \pi * r^2$. This equation indicates that the more local feature points in the local circle window, the larger the value of P_d. The suggested value of r is twice the average size of buildings in the image. For example, the average size of buildings in Figure 2a is about 13 pixels, so the recommended value of r is 26 pixels. The sensitivity of built-up areas extraction to the parameter r setting will be discussed in Section 4.1.

For each local feature point in Ω_d, its saliency index SI can be calculated by the product of P_e and P_d

$$SI = P_d \times P_e \tag{4}$$

where the combination of P_d and P_e ensures that only those local feature points that are densely and evenly distributed in the local circle window will have large SI values. Since the SI value of the local feature points in built-up areas is higher than that in non-built-up areas, we use the Otsu' method [35] to automatically calculate the threshold to segment these SI values to optimize the initial local feature point set Ω_d, and those local feature points whose SI values are less than the threshold will be eliminated. Figure 2b shows the result of the optimization of the local feature points using our proposed saliency index. As shown in Figure 2b, those false local feature points located in non-built-up areas are effectively eliminated by our method. More specifically, from the enlarged circle in Figure 2a, we can see that the local feature points located in non-built-up areas are unevenly distributed in the local circle window. The number of the detected local feature points in the first and fourth quadrants of the local circle window is six, and the number of the detected local feature points in the second and third quadrants is two, which causes the calculated P_e value to be relatively low, resulting in a relatively low SI value. Therefore, we can effectively eliminate these local feature points with low SI values through threshold processing, as shown in the enlarged circle of Figure 2b. As we obtain the optimized local feature point set, the next step is to use these local feature points to extract built-up areas.

2.2. Built-Up Areas Extraction

Based on the assumption that the probability of existence of built-up areas around the local feature points is high, Sirmacek and Unsalan [27] proposed a spatial voting approach to calculate the voting matrix to measure the probability that each pixel belongs to a built-up area. However, the calculation of the voting matrix is time-consuming. To solve this problem, we adopt the idea of voting based on superpixels in [32] to improve the original voting method, which is achieved by replacing the primary computational unit from pixels to a homogeneous object. In addition, since the cardinality of the optimized local feature point set Ω_d is so large that using these local feature points to calculate the voting matrix is still time-consuming, we also introduce a local feature point sparse representation method to reduce the cardinality of Ω_d to further speed up the calculation process. The main steps for extracting built-up areas using the improved method are as follows.

(1) Superpixel segmentation: The simple linear iterative clustering (SLIC) method is used here to partition the input VHR image into superpixels [37]. Given the number parameter q and the compactness parameter c, the input image will be partitioned into q homogeneous objects. In order to automatically handle different images, we set $c = 20$ and use the width w and height h of the input image to calculate the parameter q with the expression $q = \sqrt{10 \times w \times h}$.

(2) Local feature point sparse representation: This method first searches for the connected components Φ in Ω_d, and then uses the centroid of the connected component to represent

all the local feature points it contains. In this way, the optimized local feature point set Ω_d can be represented by a sparse local feature point set, noted as $\Omega_s = \{(X_i, Y_i)|i = 1, 2, \ldots, Q\}$, where (X_i, Y_i) denotes the centroid coordinate of the *i*th connected component Φ_i, and Q denotes the number of connected components. The centroid coordinate of Φ_i is defined as

$$X_i = \frac{1}{W_i} \sum_{p=1}^{W_i} x_p \ Y_i = \frac{1}{W_i} \sum_{p=1}^{W_i} y_p \tag{5}$$

where (x_p, y_p) represents the coordinate of the *p*th local feature point in Φ_i, and W_i represents the number of the local feature points belonging to Φ_i.

(3)   Calculation of the voting matrix: In order to improve the computational efficiency and extraction accuracy, our improved spatial voting method uses the *q* homogeneous objects $\{h_j|j = 1, 2, \ldots, q\}$ as the basic calculation units and combines them with the sparse local feature point set $\Omega_s = \{(X_i, Y_i)\}$ to calculate the voting matrix, which is defined as

$$VM(j) = \sum_{i=1}^{Q} \frac{1}{2\pi\sigma_i^2} exp\left(-\frac{(X^j - X_i)^2 + (Y^j - Y_i)^2}{2\sigma_i^2}\right) \tag{6}$$

where $VM(j)$ represents the voting value of the homogeneous object h_j, σ_i represents the tolerance parameter of the *i*th connected component Φ_i, which is calculated by the expression $\sigma_i = 20 \times W_i$, (X^j, Y^j) represents the centroid coordinate of h_j, and (X_i, Y_i) represents the centroid coordinate of Φ_i. Figure 3a shows the voting matrix calculated by the improved spatial voting method. As shown in Figure 3a, the calculated voting matrix can clearly indicate the location of the built-up areas. The high voting value (marked in red) in the voting matrix corresponds to the built-up area, while the low voting value (marked in blue) corresponds to the non-built-up area.

(4)   Built-up areas extraction: Since the voting value of the built-up area is higher than that of the non-built-up area, we use the Otsu' method [35] to segment the voting matrix to extract built-up areas. Figure 3b shows the built-up areas (marked with red-colored area) extracted using the voting matrix shown in Figure 3a. As shown in Figure 3b, the extracted built-up areas match very well with the reference data (marked with cyan-colored polygons), which demonstrates the effectiveness of our improved spatial voting method.

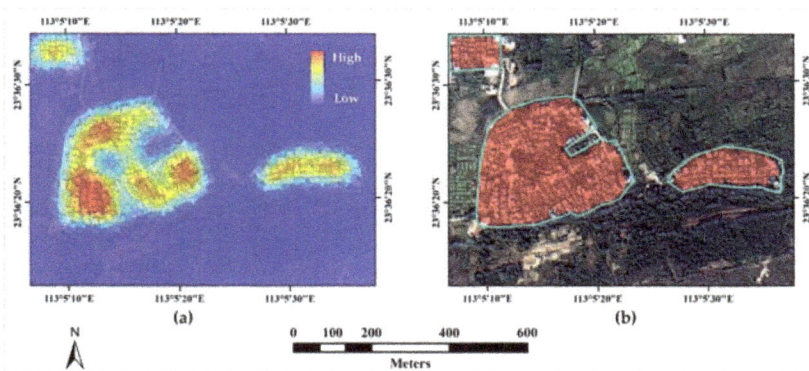

Figure 3. Improved spatial voting results. (**a**) The calculated voting matrix; (**b**) The extracted built-up areas (marked with red-colored area) and their corresponding reference data (marked with cyan-colored polygons).

2.3. Building Detection via the MBI Algorithm

The original MBI algorithm [23] is specifically designed for the detection of buildings in urban areas where the density of buildings is high. It fails to detect buildings in rural, agricultural, and mountainous regions. In addition, many irrelevant objects (e.g., open areas, bright barren land, and impervious roads) that have similar spectral characteristics to buildings will generate large numbers of interferences when detecting buildings, and these interferences are difficult to eliminate by conventional methods. In order to solve these problems, we first use the aforementioned method described in Section 2.2 to extract built-up areas from the input image, and then apply the MBI algorithm to detect buildings from the extracted built-up areas.

The calculation of the MBI is briefly described as follows. First, the brightness image, defined by the maximum of all the visible bands, is used as the basic input for building detection. Next, the white top-hat (WTH) transformation of the brightness image is performed in a reconstruction manner to highlight the high local contrast characteristics of buildings, and then the differential morphological profiles (*DMP*) are constructed based on the multi-scale and multi-directional WTH transformation to represent the complex spatial patterns of buildings in different scales and directions. Since buildings generally exhibit high local contrast and isotropic characteristics, they have larger DMP values than most other objects such as roads in most directions and scales. Therefore, the multi-scale and multi-directional DMP are averaged to calculate the MBI, which is defined as

$$\text{MBI} = \frac{\sum_{d,s} DMP(d,s)}{N_d \times N_s} \tag{7}$$

where d and s represent the direction and scale of the WTH transformation, and N_d and N_s represent the total number of directions and scales, respectively. Since a large MBI value means that there is a high possibility of the presence of a building structure, we use a preset threshold t to binarize the MBI feature image to obtain buildings, where the locations with MBI values greater than t will be extracted as buildings. The selection of the value of t is based on [23], and a larger t value means that more building candidates will be removed.

Detecting buildings from the extracted built-up areas can effectively eliminate most of the interference coming from other irrelevant objects. However, in the initial building detection results, there are still some small errors caused by open areas, vegetation, roads, and small noises. Therefore, to further eliminate these errors, we build a rule based on the shadow, spectral, and geometric information. Given that β represents the connected component in the binarization map, the rule is defined as

IF IBR(β) $\cup$ *dilate*($S(\beta)$) = $\varnothing$ **OR** NDVI(β) > t_{NDVI} **OR** LWR(β) > t_{LWR} **OR** Count(β) < t_{Count},
THEN β should be removed,

where IBR($\cdot$) denotes the initial building results, $S(\cdot)$ denotes the shadow feature map produced by the shadow index proposed in [38], dilate($S(\cdot)$) represents the morphological dilation of S by a linear structural element in the opposite direction to the solar illumination angle, NDVI($\cdot$), LWR($\cdot$), and Count($\cdot$) denote the normalized difference vegetation index, the length–width ratio, and the number of pixels of the connected component, respectively, and t_{NDVI}, t_{LWR}, and t_{Count} represent their corresponding thresholds, respectively. The selection of the values of these thresholds will be discussed in Section 3.2. Since buildings in high-resolution remote sensing images usually cast shadows around them, the rule uses the spatial relationship between buildings and shadows to eliminate false alarms caused by the connected components that are not adjacent to shadows, such as open areas and parking lots. If there is no overlap between IBR(β) and *dilate*($S(\beta)$), then β will be removed. Furthermore, the rule uses the normalized difference vegetation index to eliminate false alarms caused by the bright vegetation, and it also uses the length–width ratio and the area to eliminate false alarms caused by the elongated and narrow roads and small noises. In this way, our proposed method can not only eliminate the interferences caused by irrelevant objects in non-built-up areas but also remove the false alarms caused by open areas, vegetation, roads, and small noises in built-up areas.

Figure 4a shows the original VHR remote sensing imagery, and Figure 4b shows the building map (marked with yellow-colored areas) detected by our proposed method. As shown in Figure 4b, our method can effectively detect buildings in the image, and it can also eliminate a large amount of interference caused by irrelevant objects that are easily confused with buildings. More specifically, from the enlarged square area shown in Figure 4b, we can see that the detected buildings match very well with the true distribution of the buildings displayed in the enlarged square area in Figure 4a.

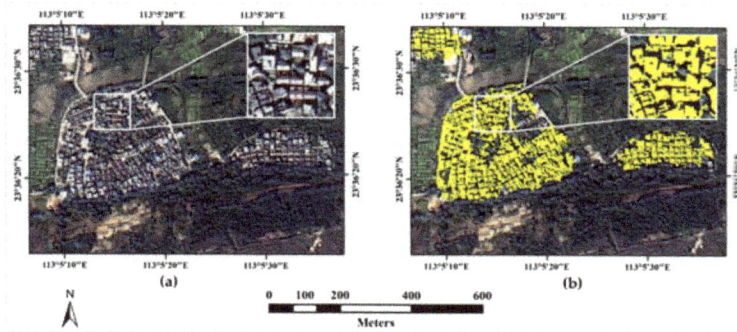

Figure 4. Building detection results. (**a**) Original very high resolution (VHR) remote sensing imagery; (**b**) Building map (marked with yellow-colored areas) detected by our proposed method.

3. Experiments

3.1. Data Set Description

The GaoFen2 satellite is a Chinese high-resolution optical satellite equipped with two panchromatic/multispectral (PAN/MSS) cameras. It was launched on 19 August 2014. The main characteristics of the GaoFen2 satellite are shown in Table 1. In order to evaluate the accuracy of our proposed method, we used five representative image patches selected from three pansharpened GaoFen2 satellite images to perform experiments. The three pansharpened GaoFen2 satellite images are obtained by merging the high spatial resolution panchromatic images with the high spectral resolution multispectral images using the NNDiffuse algorithm [39], and detailed information about them is given in Table 2. As shown in Table 2, the image patches R1 and R4 are selected from the GaoFen2 satellite image with the Scene ID of 3609415; the image patch R2 is selected from the GaoFen2 satellite image with the Scene ID of 3131139, and the image patches R3 and R5 are selected from the GaoFen2 satellite image with the Scene ID of 2097076. All five image patches R1–R5 include four spectral bands (red, green, blue, and near-infrared) with a spatial resolution of 1 m and a size of 1000×1000 pixels, and they cover different complex image scenes such as mountainous, agricultural, and rural areas. These image patches are shown in Figure 5. As can be seen from Figure 5, these image patches include a variety of land-cover types such as buildings, impervious roads, bright barren land, mudflats, farmland, vegetation, and water. Among them, impervious roads, bright barren land, and mudflats have similar spectral characteristics to buildings, so they usually cause large numbers of interferences to the detection of buildings. Therefore, using these images for experiments can fully verify the performance of our proposed method. Table 3 shows the number of samples selected for accuracy assessment and the major error sources in each image patch. All the samples were manually labeled by visual interpretation.

Table 1. The main characteristics of the GaoFen2 satellite.

Sensor Specification	Panchromatic	Multispectral
Spatial Resolution	1 m	4 m
Swath Width	45 km	45 km
Repetition Cycle	5 days	5 days
Spectral Range	450~900 nm	Blue: 450~520 nm Green: 520~590 nm Red: 630~690 nm Near-infrared: 770~890 nm

Table 2. The GaoFen2 satellite images used for the detection of buildings in this study.

Scene ID	Acquisition Date and Time (UTC)	Image Locations	Image Patches
3609415	30 April 2017, 11:27:58	Tai'an City, China	R1, R4
3131139	18 December 2016, 11:31:01	Zhongshan City, China	R2
2097076	14 February 2016, 12:11:49	Xichang City, China	R3, R5

Table 3. Number of samples and major error sources for R1–R5.

Image	Number of Samples		Major Error Sources
	Building	Background [1]	
R1	11218	11814	bright barren land, impervious roads
R2	12989	14816	impervious roads, open areas, farmland
R3	10334	12115	bright barren land, impervious roads, farmland
R4	11028	12395	bright barren land, impervious roads, farmland
R5	7120	7986	bright barren land, farmland, mudflats

[1] Background refers to the area that is not a building.

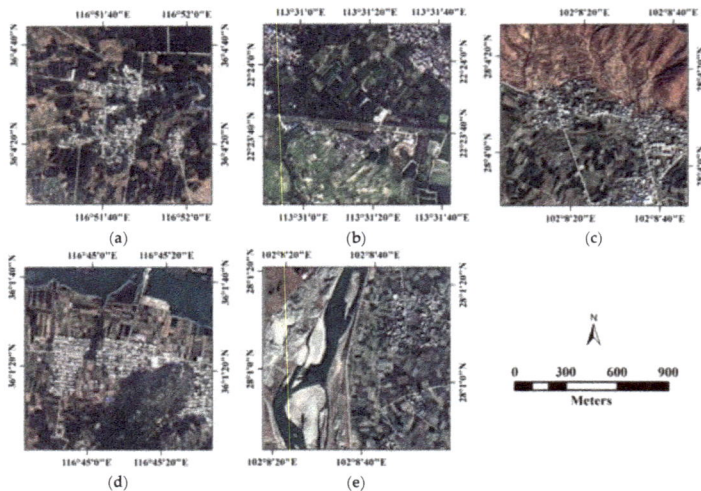

Figure 5. Five selected test image patches. (**a**) R1; (**b**) R2; (**c**) R3; (**d**) R4; (**e**) R5.

3.2. Accuracy Assessment Metrics and Parameter Settings

3.2.1. Accuracy Assessment Metrics

In order to quantitatively evaluate the building detection results, four widely accepted evaluation measures [40] were used in this paper, which are commission error (CE), omission error (OE), overall accuracy (OA), and Kappa coefficient. The CE represents pixels that belong to the background but are labeled as buildings, and the OE represents pixels that belong to a building but are labeled as the background. The OA and the Kappa coefficient are comprehensive indicators for assessing the classification of buildings and backgrounds, which are calculated by the confusion matrix. In addition, since the accurate extraction of built-up areas is a prerequisite for the good performance of our method, we also quantitatively assessed the built-up areas extraction accuracy using the following three metrics: Precision (P), Recall (R), and F-measure (F). Among them, P and R metrics correspond to the correctness and completeness of the built-up areas extraction results, respectively, and F is a comprehensive measure of P and R. The three metrics are defined as

$$P = \frac{TP}{TP + FP} \tag{8}$$

$$R = \frac{TP}{TP + FN} \tag{9}$$

$$F = 2 \times \frac{P * R}{P + R} \tag{10}$$

where TP denotes the number of pixels labeled as built-up area by both reference data and our method, FP denotes the number of pixels incorrectly labeled as built-up area by our method while they are truly labeled as non-built-up area by reference data, and FN denotes the number of pixels incorrectly labeled as non-built-up area by our method while they are truly labeled as built-up area by reference data. Meanwhile, a higher F value indicates better performance.

3.2.2. Parameter Settings

Our proposed method mainly involves the following parameters: the parameters U, V, and r for the local feature points detection and optimization, the parameters N_d, N_s, and t for the calculation of the MBI, and the parameters t_{NDVI}, t_{Count}, and t_{LWR} for the postprocessing of the building detection results. A detailed description of the selection of these parameters is as follows.

(1) Local feature points detection and optimization parameters: A large number of experiments show that when the scale number U and the direction number V of the Gabor wavelets are set to 5 and 4, respectively, most of the local feature points in the image can be detected. Therefore, in this study, the values of U and V are fixed as 5 and 4, respectively. Meanwhile, for the radius r of the local circle window, the suggested value is twice the average size of buildings in the image, so it should be tuned according to different test images.

(2) MBI parameters: As analyzed in [23], the four-directional MBI is sufficient to estimate the presence of buildings, and the accuracy of building extraction does not improve significantly as N_d increases. Therefore, in this study, the value of N_d is fixed as 4. For the scale parameter N_s, the suggested value of it is calculated by the expression $N_s = ((L_{max} - L_{min})/5) + 1$, where L_{max} and L_{min} represent the maximum and minimum sizes of buildings in the image, respectively. Therefore, it needs to be changed according to the test image. For the threshold t, its recommended range is [1,6], and a large t value will result in a large omission error and a small commission error. This parameter should also be adjusted for different test images.

(3) Postprocessing parameters: For the threshold t_{NDVI}, according to the author's experience, its appropriate range is between 0.1 and 0.3, and we can adjust it within this range according to the test images to obtain the best performance. In this study, the parameter t_{NDVI} is fixed as 0.2. For the thresholds t_{Count} and t_{LWR}, since they are relevant to the geometric characteristics of the building, we should also adjust them according to different test images. The appropriate value of t_{Count} should be less than the area of the smallest building in the image to avoid erroneous removal of the building. In this study, the value of t_{Count} is fixed as 20. In addition, after many trials, we determined that the appropriate value of t_{LWR} should be greater than 3. In this study,

the values of t_{LWR} for R1–R5 are 5, 4, 3.5, 4, and 3.5, respectively. Since the postprocessing of the building detection results is to further eliminate some small errors, it does not play a pivotal role in our method. Therefore, these postprocessing parameters (t_{NDVI}, t_{Count}, and t_{LWR}) are also not critical to our method.

To sum up, the critical parameters of our method are U, V, r, N_d, N_s, and t. Among them, the values of U, V, and N_d are fixed to 5, 4, and 4, respectively, and these values can be used for most images. The parameters that need to be tuned according to different test images are r, N_s, and t, and their values for the five test images R1–R5 are given in Table 4. The sensitivity of these three parameters will be discussed in Section 4.1.

Table 4. Critical parameter settings of our method.

Image	r	N_s	t
R1	26	7	4
R2	26	7	5
R3	24	7	5
R4	26	7	4
R5	24	7	5

3.3. Results and Analysis

3.3.1. Built-Up Areas Extraction Results and Analysis

In order to extract built-up areas, images R1–R5 are first decomposed by Gabor wavelets at multi-scales along multi-directions, and then these decomposition results are further processed to obtain local feature points. Next, the proposed saliency index is used to optimize these local feature points. Finally, a voting matrix is calculated based on these optimized local feature points to extract built-up areas. The originally detected local feature points of R1–R5 are shown in the third column of Figure 6, and the results of optimizing these local feature points using the proposed saliency index are shown in the first column of Figure 6. To more clearly display the extraction results of the local feature points, the original VHR remote sensing images are converted into grayscale images for display. From the third column of Figure 6, we can see that a large number of local feature points (marked with red-colored points) are detected over areas with complex textures or salient edges. Among them, many are located in non-built-up areas, which will impair the extraction accuracy of the built-up areas. In contrast, in their optimized results (shown in the first column of Figure 6), there are only a few local feature points in non-built-up areas, which indicates that our proposed saliency index can effectively eliminate the false local feature points located in non-built-up areas and obtain a reliable local feature point set.

The results of the built-up areas extracted using the optimized local feature points are shown in the second column of Figure 6. In addition, to further verify the effectiveness of our proposed saliency index, we also used the originally detected local feature points that were not optimized by the saliency index to extract built-up areas for comparison. The built-up areas extracted using the originally detected local feature points are shown in the fourth column of Figure 6. In the fourth column of Figure 6 we can see that the boundaries of the extracted built-up areas (marked with red-colored area) are very inaccurate, and they incorrectly identify many non-built-up areas as built-up areas. By contrast, as shown in the second column of Figure 6, the built-up areas extracted using the optimized local feature points show very good performance in all test images, which match very well with the reference data (marked with cyan-colored polygons).

The accuracy evaluation results of the built-up areas extracted under these two conditions are provided in Table 5. As shown in Table 5, for the results of the built-up areas extracted using the saliency index, their average precision value is 0.841, their average recall value is 0.936, and their average F-measure value is 0.885, which suggest that our method can effectively extract built-up areas from the image with high correctness and completeness. On the other hand, from Table 5 we can find that the accuracy of the results extracted using the saliency index is better than the accuracy of the results extracted without using the saliency index. Specifically, for each image, the precision value and the F-measure value of the result extracted using the saliency index are higher than that of the result extracted without using the saliency index. For example, when the saliency index is used, the precision value and the F-measure value of R1 are 0.925 and 0.913, respectively, and when the saliency index is not used, the precision value and the F-measure value of R1 are 0.605 and 0.751, respectively. Furthermore, in terms of the average value, the average precision value and the average F-measure value of the results extracted using the saliency index increase by 0.33 and 0.221, respectively, as compared with the results extracted without using the saliency index. The significant improvements of precision and F-measure metrics indicate that our proposed saliency index is very effective in improving the performance of built-up areas extraction. However, it should be noted that the recall value of the results extracted using the saliency index is slightly lower than that of the results extracted without using the saliency index, which is mainly because some small scattered buildings cannot be extracted as built-up areas. According to the above qualitative and quantitative analysis of the built-up areas extraction results, we can prove that our method can accurately extract built-up areas from images by means of the saliency index.

Table 5. The accuracy evaluation results of built-up areas extraction.

Image	With the Saliency Index [1]			Without the Saliency Index [2]		
	Precision	Recall	F-Measure	Precision	Recall	F-Measure
R1	0.925	0.901	0.913	0.605	0.989	0.751
R2	0.812	0.930	0.867	0.435	0.998	0.606
R3	0.849	0.978	0.909	0.720	0.996	0.836
R4	0.857	0.940	0.896	0.458	0.998	0.628
R5	0.762	0.931	0.838	0.335	0.993	0.501
Average	0.841	0.936	0.885	0.511	0.995	0.664

[1] The results of the built-up areas extracted using the saliency index. [2] The results of the built-up areas extracted without using the saliency index.

3.3.2. Building Detection Results and Analysis

After obtaining the built-up areas of the test images R1–R5, we first use the MBI algorithm to detect buildings from the extracted built-up areas, and then we use the built rule to further eliminate some small errors in the initial building detection results. The final building results of R1–R5 detected by our method are shown in the third column of Figure 7, and their corresponding reference map obtained by visual interpretation (marked with yellow-colored areas) are shown in the first column of Figure 7. In addition, to further verify the effectiveness of our method, we also compared the building detection results of our method with the original MBI algorithm [23]. To ensure the fairness of the comparison, the parameter settings of the original MBI algorithm are consistent with our method. The building results detected by the original MBI algorithm are shown in the second column of Figure 7. As shown in Figure 7, the original MBI algorithm (the second column) performed poorly for all test images as compared with the reference map (the first column). The detected buildings include a large number of irrelevant objects such as farmland, barren land, and impervious roads. In contrast, our method (the third column) can effectively eliminate the interferences from these irrelevant objects and achieve satisfactory results. Taking the test image R5 for illustration, there are several land cover types in the image, including mudflats, bright barren land, impervious roads, buildings, farmland, and river. Among them, bright objects such as mudflats and bright barren land have similar spectral characteristics to buildings and are brighter than their surroundings, which satisfies the brightness hypothesis of the MBI algorithm. Therefore, these bright objects are incorrectly extracted as buildings by the original MBI algorithm. However, our method can eliminate these irrelevant objects and achieve satisfactory performance.

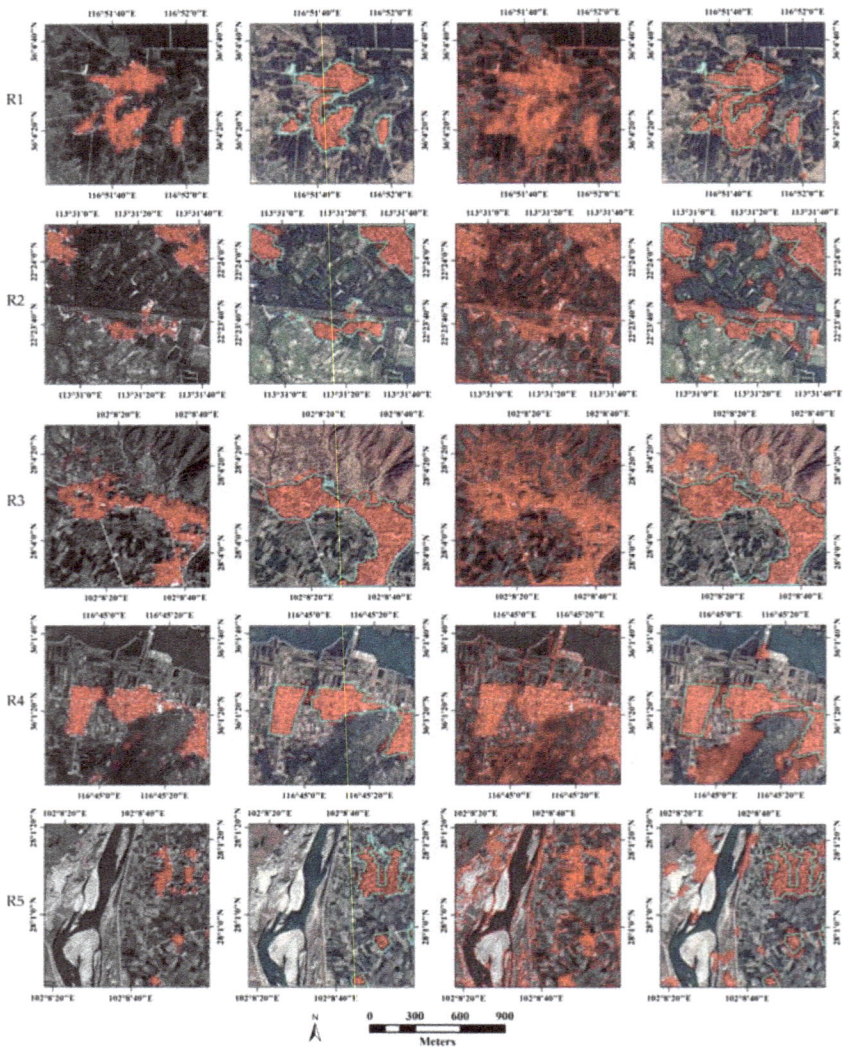

Figure 6. Results of local feature points and built-up areas. (**first column**) Local feature points (marked with red-colored points) optimized using the proposed saliency index; (**second column**) The reference data (marked with cyan-colored polygons) and built-up areas (marked with red-colored area) extracted using the optimized local feature points; (**third column**) Originally detected local feature points (marked with red-colored points); (**fourth column**) The reference data (marked with cyan-colored polygons) and built-up areas (marked with red-colored area) extracted using the originally detected local feature points.

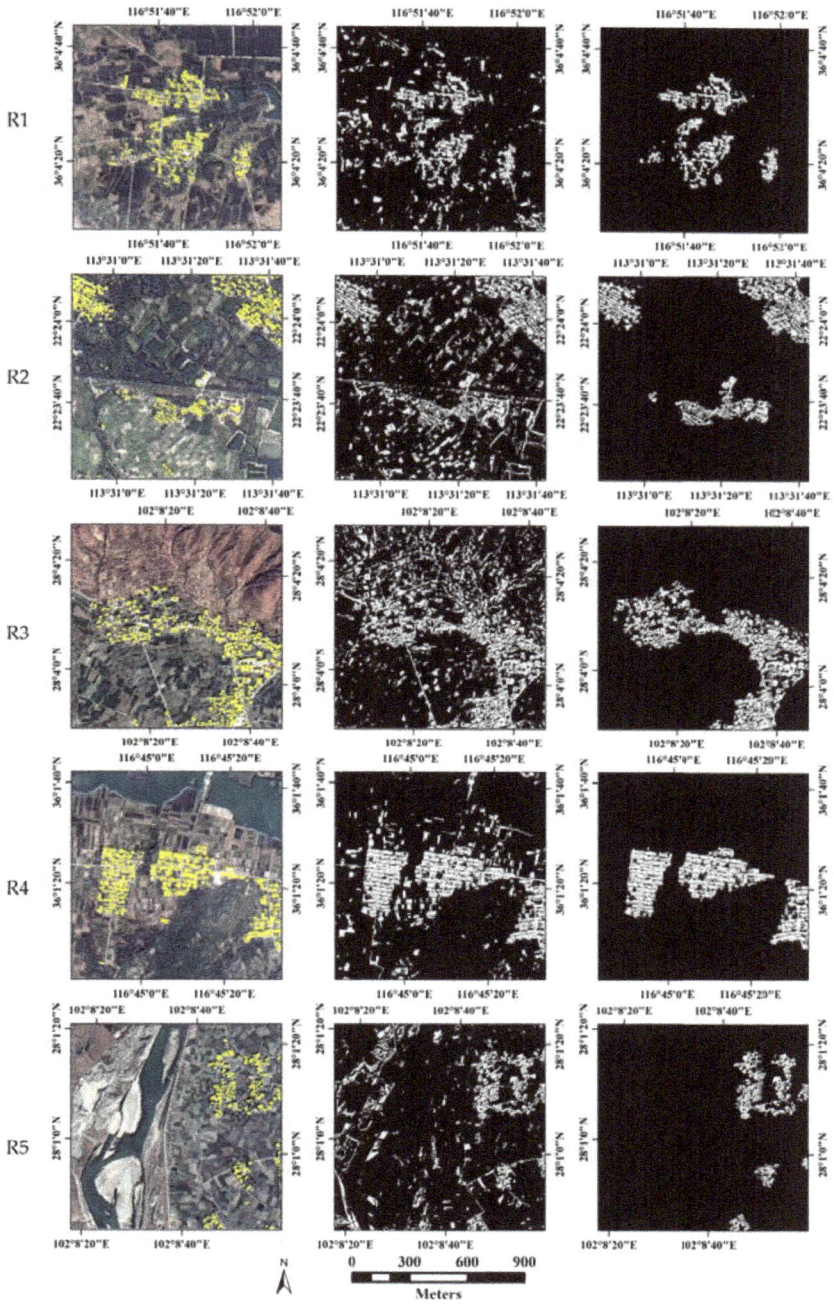

Figure 7. Building detection results. (**first column**) Reference maps of building distribution (marked with yellow-colored areas); (**second column**) Building maps extracted with the original morphological building index (MBI) algorithm; (**third column**) Building maps extracted with our method.

The accuracy evaluation results of building detection are given in Table 6. As depicted in Table 6, the OA of all the test images of our proposed method is greater than 93%, the Kappa coefficient is greater than 0.85, the maximum CE is 11.58%, and the maximum OE is 9.01%. This suggests that our proposed method can effectively distinguish between buildings and backgrounds, and it can also detect buildings with high accuracy and completeness. In addition, from Table 6 we can find that the average CE of the original MBI algorithm is as high as 32%, while our proposed method reduces the average CE to 5.9%, which demonstrates that our proposed method can effectively eliminate large numbers of interferences caused by irrelevant objects. Meanwhile, compared with the original MBI algorithm, the average OA and the average Kappa coefficient of our proposed method increase by 17.55% and 0.342, respectively. The significant improvements of CE, OA, and Kappa coefficient prove that our proposed method remarkably outperforms the original MBI algorithm and can more effectively detect buildings in complex image scenes such as mountainous, agricultural, and rural areas. On the other hand, it should be noted that the OE of our proposed method is slightly larger than the original MBI algorithm, which is mainly because some small scattered buildings have not been recognized during the extraction of built-up areas (first step), so these buildings are not detected in the second step.

Table 6. Accuracy assessment of building detection for the original MBI algorithm and our method.

Image	Accuracy Assessment							
	The Original MBI Algorithm				Our Proposed Method			
	CE (%)	OE (%)	OA (%)	Kappa	CE (%)	OE (%)	OA (%)	Kappa
R1	28.82	7.50	78.11	0.565	4.33	8.60	93.79	0.876
R2	32.02	3.96	77.02	0.549	4.58	6.57	94.84	0.896
R3	37.86	8.30	70.46	0.426	6.29	9.01	93.05	0.860
R4	32.92	1.92	76.43	0.539	11.58	2.36	92.87	0.858
R5	28.64	3.48	80.10	0.608	2.74	7.36	95.29	0.905
Average	32.05	5.03	76.42	0.537	5.90	6.78	93.97	0.879

4. Discussion

4.1. Parameter Sensitivity Analysis

Our method has three critical parameters (r, N_s, and t) that need to be adjusted according to different test images. In this section, we analyzed the effect of the different values of these three parameters on the accuracy of the results.

4.1.1. Sensitivity Analysis of the Parameter r

As shown in Figure 2, the proposed saliency index is very effective for the optimization of the local feature point set. It can effectively eliminate a large number of false local feature points located in non-built-up areas, thereby making the extracted built-up areas more accurate. The saliency index contains a tunable parameter, the radius r, which is used to control the size of the local circle window. Figure 8 shows the sensitivity of built-up areas extraction to the parameter r setting. As depicted in Figure 8, when the value of r is about twice the average size of buildings, the corresponding precision–recall curves are very close to each other, which suggests that the extraction of built-up areas is not dramatically sensitive to the value of r. In addition, the precision–recall curves also indicate that a good performance can be achieved when the value of r is about twice the average size of buildings. Taking the test image R1 for illustration, the average size of buildings in R1 is about 13 pixels. From the precision–recall curves of R1 shown in Figure 8a, we can see that the curve of "$r = 26$" is close to the curve of "$r = 22$" and the curve of "$r = 30$", and a good performance can be achieved when the value of r is close to 26.

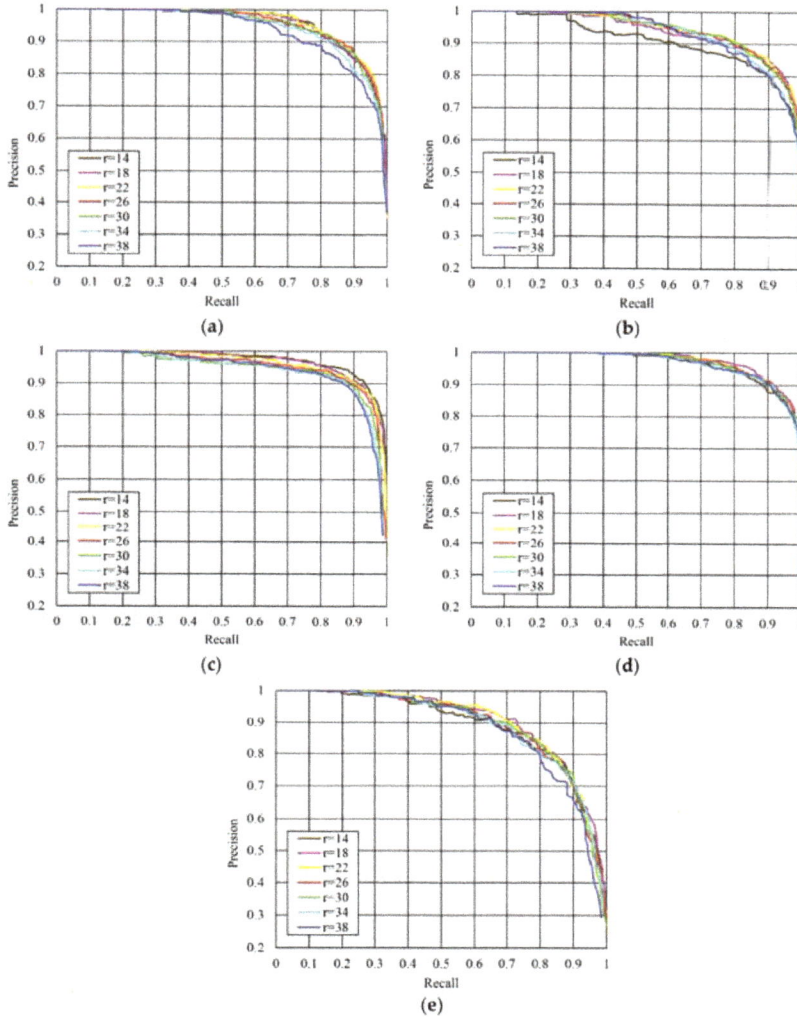

Figure 8. Built-up areas predication performance under different *r*. (**a**) The precision–recall curves of R1, where the average size of buildings is 13 pixels; (**b**) The precision–recall curves of R2, where the average size of buildings is 13 pixels; (**c**) The precision–recall curves of R3, where the average size of buildings is 12 pixels; (**d**) The precision–recall curves of R4, where the average size of buildings is 13 pixels; (**e**) The precision–recall curves of R5, where the average size of buildings is 12 pixels.

4.1.2. Sensitivity Analysis of the Binary Threshold *t*

The threshold *t* is used to segment the calculated MBI feature image to obtain buildings. Figure 9 shows the effect of different *t* values on building detection accuracy. As shown in Figure 9, as the threshold *t* increases, the CE of R1–R5 gradually decreases and the OE gradually increases, which suggests that a larger value of *t* will remove more uncertain candidate buildings. At the same time, from Figure 9 we can see that, when the value of *t* is between 1 and 6, the OA and the Kappa coefficient for all test images are very high, and the difference between them is small, but as the value of t continues to increase, the OA and the Kappa coefficient begin to decrease, which indicates that the appropriate range of t is 1 to 6, and a value of t greater than 6 will impair the accuracy of building detection.

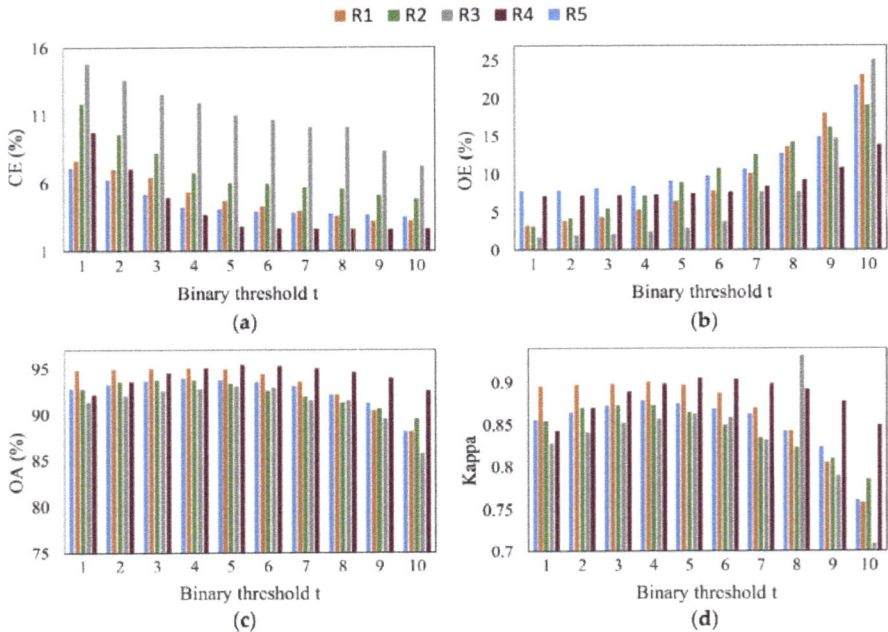

Figure 9. Building detection accuracy under different values of t. (**a**) commission error (CE); (**b**) omission error (OE); (**c**) overall accuracy (OA); (**d**) Kappa coefficient.

4.1.3. Sensitivity Analysis of the Parameter N_s

The parameter N_s is the number of scales of the WTH transformation, which is determined by the size of buildings in the image. The building sizes of R1–R5 range from 4 pixels to 36 pixels, and its corresponding N_s value is 7. Figure 10 shows the effect of different N_s values on building detection accuracy. As shown in Figure 10, as the value of N_s increases, the CE of R1–R5 increases slowly, and the OE decreases slowly, which indicates that a larger value of N_s will extract more buildings. On the other hand, as can be seen from Figure 10, when the value of N_s increases from 4, the OA and the Kappa coefficient of R1–R5 also increase slowly, but when the value of N_s increases to more than 7, the OA and the Kappa begin to decrease slightly, which suggests that the optimal performance can be achieved when the value of N_s is close to the actual size of buildings in the image.

4.2. Merits, Limitations, and Future Work

In this study, we proposed a new building detection method based on the MBI algorithm to detect buildings in complex image scenes. Experiments performed in several representative images demonstrate that the proposed method can effectively detect buildings in VHR remote sensing images and significantly improve on the original MBI algorithm [23]. It achieved good performance with an average OA greater than 93%. In addition, our method has two main advantages. On the one hand, for the detection of buildings, our method can effectively eliminate a large number of false alarms caused by irrelevant objects, which can greatly improve the accuracy of building detection. More specifically, our method first extracts built-up areas from the image and then detects buildings from the extracted built-up areas, which can directly remove a large number of false alarms located in non-built-up areas. Moreover, our method does not rely heavily on the accurate extraction of auxiliary information such as shadows to eliminate false alarms, which is different from most methods. On the other hand, for the extraction of built-up areas, our proposed saliency index solves a common problem in the local feature point extraction, which can greatly improve the extraction accuracy of built-up areas. In addition to these advantages, our method has some limitations that are worth noting.

First, some small scattered buildings cannot be extracted by our method. This is mainly because these small scattered buildings usually produce only a few local feature points, which are easily erroneously eliminated by the saliency index. Therefore, these small scattered buildings cannot be recognized during the extraction of built-up areas (first step), causing them to be missed in the second step. This is also a limitation of our proposed saliency index. In our future work, we will consider incorporating additional information such as short straight lines into the built-up areas extraction process as a supplement to the local feature points to extract these small scattered buildings.

Second, buildings with dark roofs cannot be extracted by our method. Since the MBI algorithm assumes that the building is a bright structure with high local contrast, those buildings with dark roofs will be treated as backgrounds and correspond to low MBI values, which will be removed when binarizing the MBI feature image. In our future work, we will consider combining the spatial features of the building, such as edges, to further judge those areas with relatively low MBI values to avoid erroneous removal.

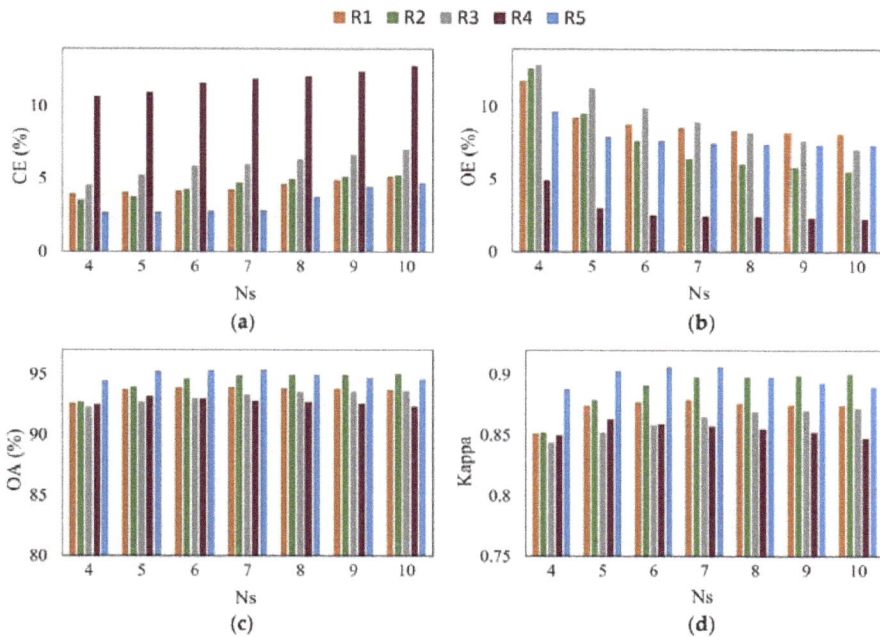

Figure 10. Building detection accuracy under different values of N_s. (**a**) CE; (**b**) OE; (**c**) OA; (**d**) Kappa coefficient.

5. Conclusions

In this paper, we have proposed a new building detection method based on the MBI algorithm to detect buildings from VHR remote sensing images captured in complex environments. This method improves the original MBI algorithm and can effectively detect buildings in non-urban areas. Three key steps are included in our proposed method: local feature points detection and optimization, built-up areas extraction, and building detection. First, the Gabor wavelet transform results of the VHR remote sensing imagery are used to extract local feature points, and then these local feature points are optimized by the proposed saliency index to eliminate the false local feature points located in non-built-up areas. Second, a spatial voting matrix is calculated based on these optimized local feature points to extract built-up areas. Finally, buildings are detected from the extracted built-up areas using the MBI algorithm. Experiments on several representative image patches of GaoFen2 satellite validate the effectiveness of our proposed method for building detection. At the same time, the comparative experiments of built-up areas extraction also proved the effectiveness of our proposed saliency index. In the future, we plan to add additional information to the extraction of built-up areas and the binarization of the MBI feature image to overcome the limitations of our method.

Author Contributions: Y.Y. and S.W. conceived and designed the algorithm and the experiments; Y.M., B.W., G.C., M.S., and W.L. provided help and suggestions in the experiment; G.C. provided help in the paper revision; Y.Y. conducted the experiments and wrote the paper.

Funding: This research was funded by the Strategic Priority Research Program of the Chinese Academy of Sciences: CAS Earth Big Data Science Project (No. XDA19030501), the National Natural Science Foundation of China (Nos. 91547107, 41271426, and 41428103), the Major Program of High Resolution Earth Observation System (No. 30-Y20A37-9003-15/17), and the Science and Technology Research Project of Xinjiang Military.

Acknowledgments: The authors would like to show their gratitude to the editors and the anonymous reviewers for their insightful and valuable comments.

Conflicts of Interest: The authors declare no conflict of interest.

Abbreviations

The abbreviations used in this paper are as follows:

MBI	morphological building index
ISODATA	iterative self-organizing data analysis technique algorithm
CNN	convolutional neural network
SIFT	scale invariant feature transform
DRV	discrimination by ratio of variance
SLIC	simple linear iterative clustering
WTH	white top-hat
DMP	differential morphological profiles
IBR	initial building results
NDVI	normalized difference vegetation index
LWR	length–width ratio
CE	commission error
OE	omission error
OA	overall accuracy

References

1.   Ridd, M.K.; Hipple, J.D.; Photogrammetry, A.S.f.; Sensing, R. *Remote Sensing of Human Settlements*; American Society for Photogrammetry and Remote Sensing: Las Vegas, NV, USA, 2006.

2.   Pesaresi, M.; Guo, H.; Blaes, X.; Ehrlich, D.; Ferri, S.; Gueguen, L.; Halkia, M.; Kauffmann, M.; Kemper, T.; Lu, L. A Global Human Settlement Layer From Optical HR/VHR RS Data: Concept and First Results. *IEEE J. Sel. Top. Appl. Earth Obs. Remote Sens.* **2013**, *6*, 2102–2131. [CrossRef]

3.   Florczyk, A.J.; Ferri, S.; Syrris, V.; Kemper, T.; Halkia, M.; Soille, P.; Pesaresi, M. A New European Settlement Map From Optical Remotely Sensed Data. *IEEE J. Sel. Top. Appl. Earth Obs. Remote Sens.* **2016**, *9*, 1978–1992. [CrossRef]

4.   Konstantinidis, D.; Stathaki, T.; Argyriou, V.; Grammalidis, N. Building Detection Using Enhanced HOG–LBP Features and Region Refinement Processes. *IEEE J. Sel. Top. Appl. Earth Obs. Remote Sens.* **2016**, *10*, 1–18. [CrossRef]

5. Solano-Correa, Y.T.; Bovolo, F.; Bruzzone, L. An Approach for Unsupervised Change Detection in Multitemporal VHR Images Acquired by Different Multispectral Sensors. *Remote Sens.* **2018**, *10*, 533. [CrossRef]
6. Mayer, H. Automatic object extraction from aerial imagery—A survey focusing on buildings. *Comput. Vis. Image Underst.* **1999**, *74*, 138–149. [CrossRef]
7. Unsalan, C.; Boyer, K.L. A system to detect houses and residential street networks in multispectral satellite images. *Comput. Vis. Image Underst.* **2005**, *98*, 423–461. [CrossRef]
8. Baltsavias, E.P. Object extraction and revision by image analysis using existing geodata and knowledge: Current status and steps towards operational systems. *ISPRS-J. Photogramm. Remote Sens.* **2004**, *58*, 129–151. [CrossRef]
9. Haala, N.; Kada, M. An update on automatic 3D building reconstruction. *ISPRS-J. Photogramm. Remote Sens.* **2010**, *65*, 570–580. [CrossRef]
10. Cheng, G.; Han, J.W. A survey on object detection in optical remote sensing images. *ISPRS-J. Photogramm. Remote Sens.* **2016**, *117*, 11–28. [CrossRef]
11. Lee, D.S.; Shan, J.; Bethel, J.S. Class-guided building extraction from Ikonos imagery. *Photogramm. Eng. Remote Sens.* **2003**, *69*, 143–150. [CrossRef]
12. Inglada, J. Automatic recognition of man-made objects in high resolution optical remote sensing images by SVM classification of geometric image features. *ISPRS-J. Photogramm. Remote Sens.* **2007**, *62*, 236–248. [CrossRef]
13. Senaras, C.; Ozay, M.; Vural, F.T.Y. Building Detection With Decision Fusion. *IEEE J. Sel. Top. Appl. Earth Observ. Remote Sens.* **2013**, *6*, 1295–1304. [CrossRef]
14. Alshehhi, R.; Marpu, P.R.; Woon, W.L.; Dalla Mura, M. Simultaneous extraction of roads and buildings in remote sensing imagery with convolutional neural networks. *ISPRS-J. Photogramm. Remote Sens.* **2017**, *130*, 139–149. [CrossRef]
15. Sirmacek, B.; Unsalan, C. Urban-Area and Building Detection Using SIFT Keypoints and Graph Theory. *IEEE Trans. Geosci. Remote Sens.* **2009**, *47*, 1156–1167. [CrossRef]
16. Ok, A.O.; Senaras, C.; Yuksel, B. Automated Detection of Arbitrarily Shaped Buildings in Complex Environments From Monocular VHR Optical Satellite Imagery. *IEEE Trans. Geosci. Remote Sens.* **2013**, *51*, 1701–1717. [CrossRef]
17. Ok, A.O. Automated detection of buildings from single VHR multispectral images using shadow information and graph cuts. *ISPRS-J. Photogramm. Remote Sens.* **2013**, *86*, 21–40. [CrossRef]
18. Peng, J.; Liu, Y.C. Model and context-driven building extraction in dense urban aerial images. *Int. J. Remote Sens.* **2005**, *26*, 1289–1307. [CrossRef]
19. Ahmadi, S.; Zoej, M.J.V.; Ebadi, H.; Moghaddam, H.A.; Mohammadzadeh, A. Automatic urban building boundary extraction from high resolution aerial images using an innovative model of active contours. *Int. J. Appl. Earth Obs. Geoinf.* **2010**, *12*, 150–157. [CrossRef]
20. Liasis, G.; Stavrou, S. Building extraction in satellite images using active contours and colour features. *Int. J. Remote Sens.* **2016**, *37*, 1127–1153. [CrossRef]
21. Pesaresi, M.; Gerhardinger, A.; Kayitakire, F. A Robust Built-Up Area Presence Index by Anisotropic Rotation-Invariant Textural Measure. *IEEE J. Sel. Top. Appl. Earth Obs. Remote Sens.* **2008**, *1*, 180–192. [CrossRef]
22. Lhomme, S.; He, D.C.; Weber, C.; Morin, D. A new approach to building identification from very-high-spatial-resolution images. *Int. J. Remote Sens.* **2009**, *30*, 1341–1354. [CrossRef]
23. Huang, X.; Zhang, L.P. A Multidirectional and Multiscale Morphological Index for Automatic Building Extraction from Multispectral GeoEye-1 Imagery. *Photogramm. Eng. Remote Sens.* **2011**, *77*, 721–732. [CrossRef]
24. Zhang, Q.; Huang, X.; Zhang, G.X. A Morphological Building Detection Framework for High-Resolution Optical Imagery Over Urban Areas. *IEEE Geosci. Remote Sens. Lett.* **2016**, *13*, 1388–1392. [CrossRef]
25. Huang, X.; Zhang, L. Morphological Building/Shadow Index for Building Extraction From High-Resolution Imagery Over Urban Areas. *IEEE J. Sel. Top. Appl. Earth Obs. Remote Sens.* **2012**, *5*, 161–172. [CrossRef]
26. Huang, X.; Yuan, W.L.; Li, J.Y.; Zhang, L.P. A New Building Extraction Postprocessing Framework for High-Spatial-Resolution Remote-Sensing Imagery. *IEEE J. Sel. Top. Appl. Earth Obs. Remote Sens.* **2017**, *10*, 654–668. [CrossRef]

27. Sirmacek, B.; Unsalan, C. Urban Area Detection Using Local Feature Points and Spatial Voting. *IEEE Geosci. Remote Sens. Lett.* **2010**, *7*, 146–150. [CrossRef]

28. Xu, W.; Huang, X.; Zhang, W. A multi-scale visual salient feature points extraction method based on Gabor wavelets. In Proceedings of the IEEE International Conference on Robotics and Biomimetics, Guilin, China, 19–23 December 2009; pp. 1205–1208.

29. Kovacs, A.; Sziranyi, T. Improved Harris Feature Point Set for Orientation-Sensitive Urban-Area Detection in Aerial Images. *IEEE Geosci. Remote Sens. Lett.* **2013**, *10*, 796–800. [CrossRef]

30. Hu, Z.; Li, Q.; Zhang, Q.; Wu, G. Representation of Block-Based Image Features in a Multi-Scale Framework for Built-Up Area Detection. *Remote Sens.* **2016**, *8*, 155. [CrossRef]

31. Shi, H.; Chen, L.; Bi, F.K.; Chen, H.; Yu, Y. Accurate Urban Area Detection in Remote Sensing Images. *IEEE Geosci. Remote Sens. Lett.* **2015**, *12*, 1948–1952. [CrossRef]

32. Li, Y.S.; Tan, Y.H.; Deng, J.J.; Wen, Q.; Tian, J.W. Cauchy Graph Embedding Optimization for Built-Up Areas Detection From High-Resolution Remote Sensing Images. *IEEE J. Sel. Top. Appl. Earth Obs. Remote Sens.* **2015**, *8*, 2078–2096. [CrossRef]

33. Tao, C.; Tan, Y.H.; Zou, Z.R.; Tian, J.W. Unsupervised Detection of Built-Up Areas From Multiple High-Resolution Remote Sensing Images. *IEEE Geosci. Remote Sens. Lett.* **2013**, *10*, 1300–1304. [CrossRef]

34. Lee, T.S. Image representation using 2D gabor wavelets. *IEEE Trans. Pattern Anal. Mach. Intell.* **1996**, *18*, 959–971.

35. Otsu, N. A Threshold Selection Method from Gray-Level Histograms. *IEEE Trans. Syst. Man Cybern.* **1979**, *9*, 62–66. [CrossRef]

36. Hu, X.Y.; Shen, J.J.; Shan, J.; Pan, L. Local Edge Distributions for Detection of Salient Structure Textures and Objects. *IEEE Geosci. Remote Sens. Lett.* **2013**, *10*, 466–470. [CrossRef]

37. Achanta, R.; Shaji, A.; Smith, K.; Lucchi, A.; Fua, P.; Susstrunk, S. SLIC Superpixels Compared to State-of-the-Art Superpixel Methods. *IEEE Trans. Pattern Anal. Mach. Intell.* **2012**, *34*, 2274–2281. [CrossRef] [PubMed]

38. Teke, M.; Başeski, E.; Ok, A.Ö.; Yüksel, B.; Şenaras, Ç. Multi-spectral False Color Shadow Detection. In *Photogrammetric Image Analysis*; Springer: Berlin/Heidelberg, Germany, 2011; pp. 109–119.

39. Sun, W.; Messinger, D. Nearest-neighbor diffusion-based pan-sharpening algorithm for spectral images. *Opt. Eng.* **2013**, *53*, 013107. [CrossRef]

40. Congalton, R.G.; Green, K. *Assessing the Accuracy of Remotely Sensed Data: Principles and Practices*; CRC Press: Boca Raton, FL, USA, 2008.

remote sensing

MDPI

Article

Method Based on Edge Constraint and Fast Marching for Road Centerline Extraction from Very High-Resolution Remote Sensing Images

Lipeng Gao [1,2], Wenzhong Shi [1,2,3,*], Zelang Miao [4,5,*] and Zhiyong Lv [6]

[1] School of Remote Sensing and Information Engineering, Wuhan University, Wuhan 430079, China; gaolipengcumt@gmail.com
[2] Collaborative Innovation Center of Geospatial Technology, Wuhan University, Wuhan 430079, China
[3] Department of Land Surveying and Geo-Informatics, Hong Kong Polytechnic University, Kowloon, Hong Kong, China
[4] School of Geosciences & Info-Physics, Central South University, Changsha 410083, China
[5] Key Laboratory of Metallogenic Prediction of Nonferrous Metals and Geological Environment Monitoring, Central South University, Changsha 410083, China
[6] School of Computer Science and Engineering, Xi'An University of Technology, Xi'an 710048, China; Lvzhiyong_fly@hotmail.com
* Correspondence: john.wz.shi@polyu.edu.hk (W.S.); zelang.miao@csu.edu.cn (Z.M.); Tel.: +86-158-0251-7206 (Z.M.)

Received: 11 April 2018; Accepted: 31 May 2018; Published: 7 June 2018

Abstract: In recent decades, road extraction from very high-resolution (VHR) remote sensing images has become popular and has attracted extensive research efforts. However, the very high spatial resolution, complex urban structure, and contextual background effect of road images complicate the process of road extraction. For example, shadows, vehicles, or other objects may occlude a road located in a developed urban area. To address the problem of occlusion, this study proposes a semiautomatic approach for road extraction from VHR remote sensing images. First, guided image filtering is employed to reduce the negative effects of nonroad pixels while preserving edge smoothness. Then, an edge-constraint-based weighted fusion model is adopted to trace and refine the road centerline. An edge-constraint fast marching method, which sequentially links discrete seed points, is presented to maintain road-point connectivity. Six experiments with eight VHR remote sensing images (spatial resolution of 0.3 m/pixel to 2 m/pixel) are conducted to evaluate the efficiency and robustness of the proposed approach. Compared with state-of-the-art methods, the proposed approach presents superior extraction quality, time consumption, and seed-point requirements.

Keywords: road extraction; very high-resolution image; fast marching method; semiautomatic; edge constraint

1. Introduction

Accurate and up-to-date road network information is extremely critical for various urban applications, such as navigation and infrastructure maintenance [1–3]. The advent of modern remote sensing has enabled the extraction of information from very high-resolution (VHR) and highly detailed optical images of roads to update urban road networks [4,5]. High spatial resolution enriches feature details but complicates object extraction [6–10]. Although considerable effort has been devoted to road-feature extraction from VHR images, a completely practical road-feature extraction technology remains unrealistic.

A considerable number of articles have been published on road-feature extraction from remote sensing images. Generally, state-of-the-art methods for road-feature extraction from VHR images fall into two categories: Automatic and semiautomatic methods. Automatic approaches require no prior information and can be executed by a series of image-processing algorithms, such as mathematical morphology [11,12], active snake model [13], dynamic programming [14], neural networks [15–17], probabilistic graphical models [18], filtering-based methods [19], and object-oriented methods [20]. In general, however, the unsatisfactory performance of the automatic method in road-feature extraction from images presenting complex natural road scenarios (e.g., image noise and tree and shadow occlusion) restricts its practical applications [21]. The limitation of automatic methods has encouraged the proliferation of studies on semiautomatic methods. In contrast to automatic methods, semiautomatic methods require user input or other prior information to achieve robust and stable results.

Two technical ideas are present for semi-automatic road extraction; the first involves treating the extraction as a problem of image segmentation (divided into road and non-road) and then obtaining the final result by post-processing [22–26]. This method is easily affected by vegetation occlusion or large shadows, which leads to low recognition rates. In addition, due to the introduction of a post-processing algorithm, other features are easily misjudged as roads.

The other idea involves treating the extraction as a network optimization problem. The road network is obtained by the connection of road seed points, and the final result is acquired with the use of graph theory or dynamic programming techniques [27–32]. The local features of the road (such as extensibility, edge characteristics, and topological structure of the road network, etc.) are fully considered in this method, and a reliable initial road seed point is obtained through human–computer interaction. Therefore, the accuracy of road extraction results is relatively high. However, the extraction effect of shaded and occluded roads is poor because of the different methods of connecting road seed points. In addition, the number of seed points needed for U- or S-shaped roads is more than that for linear roads, thereby requiring considerable manual work.

According to this analysis, the method based on image segmentation is more efficient but less accurate than that based on road seed points, which is less efficient but more precise. Inspired by Reference [33], we propose to treat road seed point connection as a shortest-path problem to improve the efficiency of road extraction on the basis of seed points. The fast marching method was recently developed for connecting road seed points [33]; it is a particular case of level set methods, which were developed by Osher and Sethian [34] as an efficient computational numerical algorithm for tracking and modeling the motion of a physical wave interface (front). This method has been applied to different research fields, including computer graphics, medical imaging, computational fluid dynamics, image processing, and computation of trajectories [33,35,36]. In Reference [33], this method showed high stability and general advantages and suitability for processing low-/medium-resolution remote sensing images. However, it is difficult to extract unbiased road centerline information from VHR remote sensing images by using the fast marching method alone.

In VHR remote sensing images, "noise" is produced by the improvement of resolution, which leads to inconspicuous useful edge information. Complex image backgrounds also produce a large number of finely divided edges, which are difficult to process and thus result in the difficulty of road edge extraction. Extracting straight roads and planar roads is challenging due to the existence of the same objects with different spectra and different objects with the same spectrum, which make the extraction of roads effectively by using the road spectral feature alone a difficult task. Thus, this study presents a semiautomatic edge-constraint fast marching (ECFM) method to extract road centerlines from VHR images. Edge information, road spectral feature, and the road centerline probability map are utilized and an edge-constraint-based weighted fusion model is introduced to assist the fast marching method. The proposed method enables accurate and unbiased road centerline extraction and shows high generalization capability in processing complex road scenarios, such as S-shaped, U-shaped, and shaded roads. The contributions of the method are as follows.

(a) Edge information of remote sensing imagery has been studied extensively and widely used in the extraction and tracking of linear objects, such as roads and rivers, in medium-/low-resolution remote sensing imagery. The present study indicates that the synergy of edge information, road centerline probability map, and road spectral feature can overcome the shortcomings of the bias of the road centerline extracted by the fast marching method, which uses spectral feature only. Moreover, our method is robust to road extraction in shaded areas.

(b)   Another contribution of this study is that the proposed method needs only a few road seed points when extracting an S-shaped or U-shaped road. This characteristic leads to the efficiency of the widespread practical application of road centerline extraction from remote sensing images.

The remainder of this paper is organized as follows. In Section 2, related state-of-the-art methods are reviewed. Section 3 provides an introduction to the proposed method. Experimental results are given in Section 4 and discussed in Section 5. The conclusion is presented in Section 6.

2. Related Work

Many approaches have been proposed in the last decades for extracting road segmentation from aerial and satellite images. Low-level features can be extracted and heuristic rules (such as connectivity and shape) can be defined in numerous ways to classify structures similar to roads. A geometric stochastic road model based on road width, length, curvature, and pixel intensity was applied in Reference [37]. Hinz and Baumgartner [38] used road models and their contexts, including their knowledge of radiation, geometry, and topology. The disadvantage of these rule-based heuristic models is that obtaining the optimal set of rules and parameters is difficult because of the wide variety of roads. Therefore, these methods can work only in areas where the features used (such as image edges) occur mainly on roads (e.g., rural areas).

Most approaches consider road extraction a binary segmentation problem. The path trajectory point [22] and the angle-based texture feature [23] of a particular pixel can be defined to quantify road probability on the basis of shape. Das et al. [24] adopted the spectral and local linear features of multispectral road images. By combining the probabilistic support vector machine (PSVM) method, dominant singular value method, local gradient function, and vertical central axis transformation method to classify the region, the authors detected the road edge, linked the broken road, and eliminated the non-road area. The advantages of this method were verified by experiments on many road images. In Reference [25], the image was initially segmented by fused multiscale collaborative representation and graph cuts, and the initial contour of the road was then obtained by filtering the road shape. Finally, the road centerline was obtained through tensor voting. In Reference [26], the image was first divided into road and non-road through SVM soft classification; then, the probability of each pixel belonging to the road was obtained simultaneously; the final road was acquired through the graph cut method. However, these methods work well in multispectral images only and can detect only the main roads in urban areas. Thus, extracting roads from areas with dense buildings or other areas which are similar to road grayscale is challenging.

Another semiautomatic road extraction method regards road extraction as the connection and tracking problem of road seed points. Hu et al. [27] proposed a segmented parabolic model to delineate road centerline networks. The method first uses seed points to generate parabolic segments and then applies least-squares template matching to calculate parameters for precise parabola extraction. Miao et al. [28] proposed a kernel density estimation method combined with the geodesic method to decrease the number of seed points required for road extraction. Zhou et al. [29] used particle filtering to track road segments between seed points. However, particle filtering is limited by its incapability to effectively deal with road branches. To extend the generalization capability of particle filtering to complex scenarios, Movaghati et al. [30] integrated particle filtering with extended Kalman filtering. Lv et al. [31] proposed a multifeature sparsity-based model that can utilize multifeature complementation to extract roads from high-resolution imagery. Dal Poz et al. [32] proposed a semiautomatic method to extract urban/suburban roads from stereoscopic satellite images. This method uses seed points to construct the road model in the object space. Optimal road segments between seed points are then generated through dynamic programming. Road extraction based on seed points can achieve high precision, but the efficiency is low. The main reason is that the input of road seed points needs human intervention. A large number of required seed points will affect the efficiency of road extraction.

3. Methodology

As shown in Figure 1, the proposed approach consists of three main steps. These steps include: (1) Road feature enhancement: The VHR image can reveal ground objects in great detail and depict the color, shape, size, and structure of objects. However, its spectra may contain considerable noise, which may reduce the reliability of the road extraction result. Thus, the image is first filtered through guided filtering to enhance road features; (2) Road probability estimation: Three road features are extracted, and an edge-constraint-based weighted fusion model is introduced for multifeature fusion and road probability estimation; (3) Seed-point connection: The fast marching method is used to link road seed points on the basis of the potential road map. To test the accuracy and efficiency of the proposed approach, the performance of the proposed method on four VHR images is compared with that of other road extraction approaches.

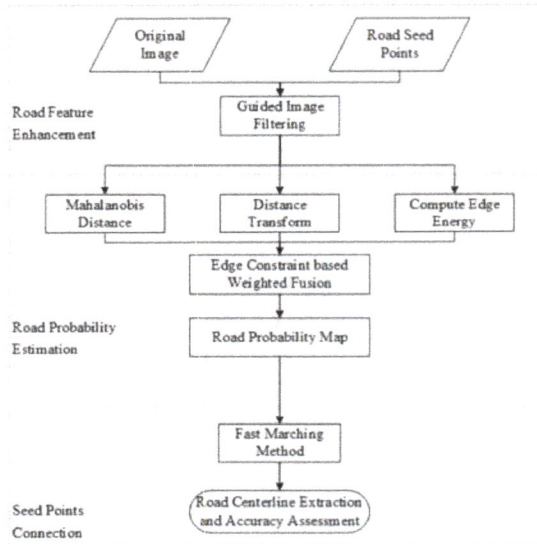

Figure 1. Flowchart of the proposed ECFM method.We use a real example to illustrate the detailed flow of the presented method. The example is shown in Figure 2. A detailed description of the method is provided in the following subsections.

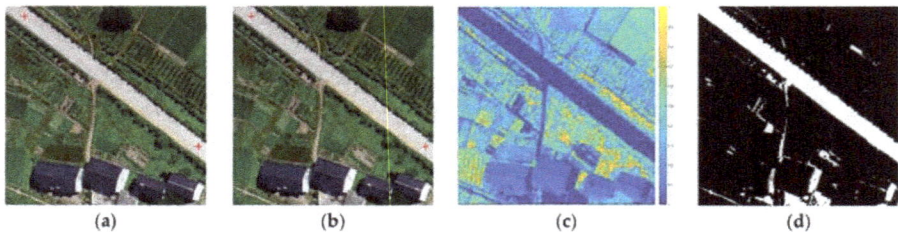

(a)   (b)   (c)   (d)

Figure 2. *Cont.*

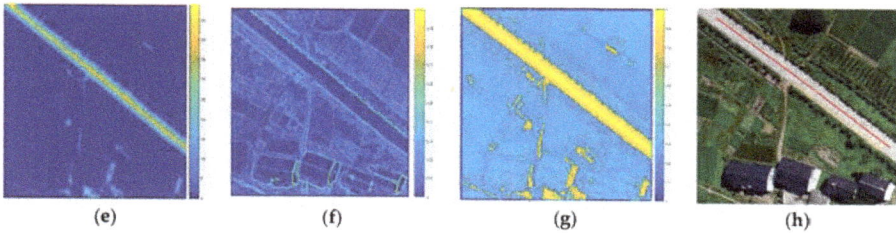

Figure 2. (a) Test image. Seed points are represented by red crosses; (b) Image preprocessed through guided filtering; (c) Mahalanobis distance map; (d) Thresholding result, in which 1 and 0 represent road and nonroad classes, respectively; (e) Distance transform result; (f) Edge-energy information; (g) Road probability map obtained through multifeature fusion; (h) Minimal path extracted from the road probability map through the fast marching method.

3.1. Road Feature Enhancement

The principle of this step is the compression of nonroad pixel signals in advance. In VHR images, roads are assumed to be locally homogeneous and elongated areas. However, some roads in VHR images are contaminated by numerous nonroad pixels, such as cars and traffic lines. Thus, image filtering is necessary to reduce the negative effect of nonroad pixels. Guided filtering performs edge-preserving smoothing on an image while guided by a second image, the so-called guidance image [39]. Similar to other filtering operations, guided image filtering is a neighborhood operation. However, it accounts for the statistics of the neighboring pixels of a central pixel in the guidance image when calculating the output value.

The commonly used linear translation-variant filtering can be formulated as follows:

$$q_i = \sum_j W_{ij}(I)p_j, \tag{1}$$

where i and j are pixel indices, and I, p, and q denote the input, guidance, and output images, respectively. The filter kernel W_{ij} is the weighted average function of I and p, which is defined as:

$$W_{ij}(I) = \frac{1}{|\omega|^2} \sum_{k:(i,j)\in\omega_k} \left(1 + \frac{(I_i - \mu_k)(I_j - \mu_k)}{\sigma_k^2 + \varepsilon}\right), \tag{2}$$

where ω_k is an overlapping window centered at pixel k; $|\omega|$ is the number of pixels in ω_k; μ_k and σ_k^2 denote the mean and variance of I in ω_k, respectively; and ε is a regularization parameter that controls the smoothness degree.

The key assumption of guided filtering is a local linear model between I and q [39]. This model is defined as:

$$q_i = a_k I_i + b_k, \forall i \in \omega_k, \tag{3}$$

where (a_k, b_k) are some linear coefficients assumed to be constant in ω_k. The two parameters are computed with a linear ridge regression model:

$$E(a_k, b_k) = \sum_{i\in\omega_k} \left((a_k I_i + b_k - p_i)^2 + \varepsilon a_k^2\right), \tag{4}$$

where

$$a_k = \frac{\frac{1}{|\omega|}\sum_{i\in\omega_k} I_i p_i - \mu_k \bar{p}_k}{\sigma_k^2 + \varepsilon}, \tag{5}$$

$$b_k = \bar{p}_k - a_k \mu_k. \tag{6}$$

Here, $\bar{p}_k$ is the mean of p in ω_k. After computing (a_k, b_k) for all windows ω_k in the image, the output of guided filtering is expressed as:

$$q_i = \frac{1}{|\omega|} \sum_{k|i \in \omega_k} (a_k I_i + b_k). \tag{7}$$

The guided image filtering result is shown in Figure 2b.

3.2. Road Probability Estimation

This step aims to exploit multiple features of roads to overcome the shortcomings of the traditional fast marching method, which only considers spectral information. Thus, to estimate road probability, road spectral information, centerline probability, and edge-energy features are combined through a weighted fusion approach.

3.2.1. Mahalanobis Distance

The initial road seed point generated by users is taken as the central pixel, and its neighboring pixels (i.e., the 5×5 window used in this study) are taken as training samples. The Mahalanobis distance [40] is subsequently applied to compute the road probability of pixel x, as follows:

$$D_M(x) = \sqrt{(I(x) - m)^T C^{-1}(I(x) - m)} \tag{8}$$

where $D_M(x)$ is the value of Mahalanobis distance at pixel x; $I(x)$ is the vector datum of the spectral value of pixel x; and m and C indicate the mean values and the covariance matrix of the training samples, respectively. After computing the Mahalanobis distance values of all pixels, simple thresholding is used to divide the image into the foreground (i.e., road) and background (i.e., nonroad) regions. The thresholding is defined as:

$$Label(x) = \begin{cases} 1, & \text{if } D_M(x) \leq T \\ 0, & \text{otherwise} \end{cases} \tag{9}$$

where $Label(x)$ is the class label of the pixel x, and T is the area ratio of the road area to the entire image region. In this study, a road area ratio of 0.2 is obtained through trial-and-error, and 1 and 0 represent the road and nonroad classes, respectively. Figure 2c,d show the Mahalanobis distance matrix and the corresponding thresholding result, respectively.

Then, the road spectral feature can be computed by applying a Gaussian filter, as follows:

$$S_{i,j} = \frac{1}{2\pi\sigma^2} e^{-\frac{(i-k-1)^2-(j-k-1)^2}{2\sigma^2}} \tag{10}$$

where $S_{i,j}$ is the spectral feature value at the pixel location of (i,j), σ is the standard deviation, and k is the slide window size.

The obtained road class is processed through distance transformation [41] to produce a distance map $D_{i,j}$ that can be taken as the road centerline probability map. The result is shown in Figure 2e. Although the Mahalanobis distance method misclassifies some nonroad pixels as road pixels, this error negligibly affects the generation of the road centerline probability map because the connection of seed points in this study relies on the fast marching method, which is robust to noise.

3.2.2. Edge Energy

The edge information of remote sensing images has been extensively studied and widely used in the extraction and tracking of linear objects, such as roads and rivers, in medium/low-resolution remote sensing images. Thus, edge information can be used as a constraint for accurate road centerline extraction.

Image edge energy can be computed through an edge-filtering operation. The edge-filter operator filters the image on the basis of spectral variance and local similarity by considering the 3×3 neighborhood of a pixel.

$$\begin{bmatrix} v_{i-1,j-1} & v_{i-1,j} & v_{i-1,j+1} \\ v_{i,j-1} & v_{i,j} & v_{i,j+1} \\ v_{i+1,j-1} & v_{i+1,j} & v_{i+1,j+1} \end{bmatrix}$$

where (i, j) is the spatial coordinate of each pixel in the image, and $v_{i,j}$ is the spectral value of the pixel.

The Laplace operator is one of the most commonly used operators in edge extraction. To enhance the ability of the Laplacian operator to detect changes in the grayscale on the diagonal [42], a redesigned template that assigns different weights to the vertical, horizontal, and diagonal is defined as follows:

$$2\begin{bmatrix} 0 & -1 & 0 \\ -1 & 4 & -1 \\ 0 & -1 & 0 \end{bmatrix} + \begin{bmatrix} -1 & 0 & -1 \\ 0 & 4 & 0 \\ -1 & 0 & -1 \end{bmatrix} = \begin{bmatrix} -1 & -2 & -1 \\ -2 & 12 & -2 \\ -1 & -2 & -1 \end{bmatrix} \tag{11}$$

The image is convolved by the above 3×3 neighborhood to obtain the edge detection result $E_{i,j}$, as shown in the following equation:

$$E_{i,j} = \frac{1}{12}\left(\begin{array}{l} SA\left(\vec{v}_{i-1,j-1}, \vec{v}_{i,j}\right) + 2SA\left(\vec{v}_{i,j-1}, \vec{v}_{i,j}\right) + SA\left(\vec{v}_{i+1,j-1}, \vec{v}_{i,j}\right) \\ +2SA\left(\vec{v}_{i-1,j}, \vec{v}_{i,j}\right) + 2SA\left(\vec{v}_{i+1,j}, \vec{v}_{i,j}\right) \\ +SA\left(\vec{v}_{i-1,j+1}, \vec{v}_{i,j}\right) + 2SA\left(\vec{v}_{i,j+1}, \vec{v}_{i,j}\right) + SA\left(\vec{v}_{i+1,j+1}, \vec{v}_{i,j}\right) \end{array} \right) \tag{12}$$

and

$$SA\left(\vec{v}, \vec{w}\right) = \cos^{-1}\left(\frac{\vec{v} \cdot \vec{w}}{\|\vec{v}\|\|\vec{w}\|} \right) \tag{13}$$

where SA stands for spectral angle and is a measure of similarity between two pixels, and $\vec{v}, \vec{w}$ represents the spectral values of two pixels.

The edge filter operator has the following characteristics: (1) Small spectral variation in the homogeneous region. This characteristic leads to low edge operator values in the homogeneous region; (2) Sharp changes in the spectral range of the adjacent boundary area. This characteristic leads to high edge operator value in the boundary area. These two characteristics can be used to obtain the edge energy of an image, as shown in Figure 2f.

3.2.3. Road Probability Estimation

The information fusion of road features aims to estimate road candidates, to discard as many false positives as possible, and to improve the consistency of the extracted roads. Most existing fusion methods are feature fusion-based methods that combine multiple features derived from road areas.

Thus, an edge-constraint-based weighted fusion model, which consists of three items, was proposed to integrate road features detected through the approaches presented in Sections 3.2.1 and 3.2.2:

$$\hat{P} = \frac{1}{Z}(\alpha f_S + \beta f_D + \lambda(f_k - 1)f_E / Curv_E) \tag{14}$$

where $\hat{P}$ is the road probability map; Z is a normalization constant; $f_S, f_D,$ and f_E denote the road spectral feature map, centerline probability, and edge energy information, respectively; f_k is a metric calculated through the KDE method [43] to evaluate the distance of any given pixel from the boundary with a range of [0, 1]; α, β, and λ are the weights of the three terms in the model; and $Curv_E$ is the curvature measure of current pixels and depends on the relative direction of neighboring vectors [44]. It is defined as

$$Curv_k = \left| \frac{\vec{\mu}_k}{\left|\vec{\mu}_k\right|} - \frac{\vec{\mu}_{k+1}}{\left|\vec{\mu}_{k+1}\right|} \right| \tag{15}$$

where $\vec{\mu}_k = (i_k - i_{k-1}, j_k - j_{k-1})$, $\vec{\mu}_{k+1} = (i_{k+1} - i_k, j_{k+1} - j_k)$, and i and j are the row and column numbers of the current pixel, respectively.

This model is based on the assumption that the pixel with high spectral intensity, low edge intensity, and small curvature has a dominant role in extraction. The constraints used in the computational model can maximize extraction reliability and accuracy. Figure 2g shows an example of road probability estimation.

3.3. Seed-Point Connection

For a given image I and two road seed points p_1 and p_2, the road potential map P is obtained by the edge-constraint-based weighted fusion model:

$$P(x) = 1/\widehat{P}(x) \tag{16}$$

The road has a small value on the potential energy map and thus has a large traveling speed term $1/P$. Let $S = \{s_1, s_2 \ldots s_n\}$ be the set of paths between p_1 and p_2, and let l be the length parameter. The energy term is formulated as follows:

$$E(s) = P(s(l))dl \tag{17}$$

The shortest path S_i between p_1 and p_2 is denoted as C_{p_1,p_2}. Thus, the energy term $E(s)$ has a global minimum value. For any given pixel x in image I, the value in the minimal energy map of p_1 is defined as:

$$U(x) = \min\{P(s(l))dl\}, x \in I, s = C_{p_1,x} \tag{18}$$

where $U(x)$ is an Eikonal equation:

$$\begin{cases} \nabla U(x) = P(x), x \in I \\ U(p_1) = 0 \end{cases} \tag{19}$$

The minimal path C_{p_1,p_2} can be obtained by solving the following difference equation:

$$\begin{cases} \frac{dC_{p_1,p_2}}{dl}(l) = -\nabla U(C_{p_1,p_2}(l)) \\ C_{p_1,p_2}(0) = p_2 \end{cases} \tag{20}$$

Here, the fast marching method [34] is used to connect the seed points. The fast marching method is a particular case of level set methods and is a numerical solution of the Eikonal equation.

During fast marching, the pixel with the shortest arrival time is used as the point of the current front, and the minimum arrival time of its four neighborhood points is updated in accordance with the minimum arrival time of the point. Once the loop terminates, the final minimum arrival time of each point in the image is obtained. Then, the road centerline that connects two seed points will be generated. Figure 2h presents an example of seed-point connection through the fast marching method.

4. Experimental Study

An experimental study was performed with eight VHR remote sensing images to validate the effectiveness and adaptability of the proposed method in road extraction. A discussion of the experimental study is presented in this section, which is divided into three subsections. The first subsection provides a description of the study. The second subsection presents a discussion of the four experimental set-ups. The detailed parameter settings applied in the experimental set-ups are also given in this subsection. Finally, the results of the four experiments are provided in the last subsection.

4.1. Datasets

To assess the effectiveness and adaptability of the presented method, experiments were conducted with eight VHR remote sensing images. The images are described below.

The first image is shown in the first row of Figure 3. It is an aerial image with a spatial resolution of 0.3 m/pixel and a spatial size of 400 pixels × 400 pixels. It was downloaded from Computer Vision Lab [45].

(a)   (b)   (c)

Figure 3. Comparison of the results of road centerline extraction. (a) Red represents the result obtained with edge constraint; (b) Yellow represents the result obtained without edge constraint; (c) Superimposition of the two results. Seed points are shown as blue crosses.

The second image has a spatial resolution of 0.6 m/pixel and a spatial size of 512 pixels × 512 pixels. It was collected by the QuickBird satellite and was downloaded from VPLab [46]. The image is shown in the second row of Figure 3.

The third and fourth remote sensing images have spatial sizes of 400 pixels × 400 pixels and are shown in Figure 4. The images were downloaded from Computer Vision Lab [45]. They have a spatial resolution of 0.6 m/pixel and show an area that is mainly covered by vegetation, roads, and buildings.

The fifth image is shown in Figure 5 and has a spatial size of 3500 pixels × 3500 pixels and a spatial resolution of 1 m/pixel. It was collected by the IKONOS satellite and shows an area of Hobart, Australia. This image includes different types of noises, such as vehicle occlusion, sharp roadway curves, and building shadows.

Figure 4. Two cases of U-shaped road extraction. (**a**) Case 1; (**b**) Case 2.

Figure 5. Road extraction result provided by the proposed ECFM method for an IKONOS image.

The sixth image, which is shown in Figure 7, was collected by the QuickBird satellite. The image shows an area in Hong Kong. It has a spatial resolution of 0.6 m PAN band and a size of 1200 pixels × 1600 pixels. It includes various road conditions, such as road material changes, vehicle occlusion, and overhanging trees.

The seventh image has a spatial size of 3000 pixels × 3000 pixels and a spatial resolution of 2 m/pixel, as shown in Figure 8. This image was collected by the WorldView-2 satellite and shows an area of Shenzhen, China, covering a variety of roads with different materials. The image also includes several types of noise, such as zebra crossings, traffic-marking lines, and toll stations.

The eighth image, as shown in Figure 10, is a grayscale image with a spatial size of 725 pixels × 1018 pixels and a spatial resolution of 1 m/pixel. This image was collected by the IKONOS satellite and shows an area of Hobart, Australia, depicting several road conditions, such as overhanging trees, vehicle occlusion, and roads with large curvatures.

Road extraction from these datasets is challenging because of their very high spatial resolution of 1 m or higher. In addition, as seen from each image, roads, buildings, vehicles, and shade may be conflated with one another. Hence, uncertainties may be encountered during road centerline extraction from these datasets.

4.2. Experimental Setup and Parameter Setting

The accuracy and efficiency of the proposed ECFM road extraction method was investigated through the following six experimental setups with the eight VHR remote sensing images shown above.

The first experiment was designed to test the effect of the edge constraint in the proposed approach. Two VHR remote sensing images were used in the experiment, as shown in Figure 3. Two road seed points were marked by the user, and the road centerline was extracted through our proposed method with edge constraint and through a method without edge constraint. The parameters of the proposed method were T = 0.2, α = 0.9, β = 0.7, and λ = 0.5.

The second experiment aimed to assess the performance of the proposed approach in extracting the centerlines of U-shaped roads. Two VHR remote sensing images showing U-shaped roads were adopted in the experiment, as depicted in Figure 4. The images have a resolution of 0.6 m. To ensure fair comparison, we compared the proposed ECFM method with (1) Hu et al.'s method [27] and (2) Miao et al.'s method [28] because these two methods rely on user-selected seed points. We used the endpoints at both ends of the U-shaped road as the seed points for road extraction. If the two seed points failed to provide the correct road extraction results, we added some intermediate points to ensure the integrity of the road extraction results. The optimal parameters of each experiment were identified through the trial-and-error method. The parameters of these approaches were as follows: (1) In Hu's method, the window size of the step-edge template was set at h = 5; (2) In Miao's method, the threshold parameter was set at T = 0.002; (3) In the proposed method, the parameters were set as T = 0.2, α = 0.9, β = 0.9, and λ = 0.4.

The third and fourth experiments were designed to investigate the accuracy and efficiency of the proposed ECFM method. This experiment employed satellite images with high spatial resolution and had two objectives. First, similar to the first experiment, it aimed to test the efficiency of the proposed method. Second, it aimed to verify the robustness of our proposed method for the centerline extraction of shadowed roads. We compared the proposed ECFM method with (1) Hu et al.'s method [27] and (2) Miao et al.'s method [28]. The parameter details of each approach are as follows: (1) In Hu's method, the window size of the step-edge template was set at h = 5; (2) In Miao's method, the threshold parameter was set at T = 0.002; (3) In the proposed method, the parameters were varied in accordance with the shading condition of the road. When the road was not shaded, the parameters were set as T = 0.2, α = 0.9, β = 0.9, and λ = 0.4. By contrast, when the road was shaded, the parameters were set as T = 0.2, α = 0.5, β = 0.5, and λ = 0.05.

The experiments were designed as follows:

(1) For all methods, as few seed points are selected as possible to improve the efficiency of road extraction while ensuring integrity.
(2) For an occluded road area, road seed points that are not occluded by shadows or automobiles are selected as much as possible to ensure the accuracy of road extraction.

The fifth and sixth experiments aimed to test the road extraction efficiency and accuracy of different methods under the same seed points. The fifth experiment used a Worldview-2 color image, and the sixth experiment used an IKONOS grayscale image. This design had two purposes. The first was to verify the efficiency and accuracy of different methods under the condition of using the same seed points, and the other was to verify the robustness of the methods proposed in this work on images with different color modes (color images and grayscale images). Seed points for these two groups of experiments were obtained by artificial marking. To ensure fairness, road extraction should be conducted according to the collection sequence of artificial seed points when different methods are adopted. (1) Hu et al.'s method [27] and (2) Miao et al.'s method [28] were used here for comparison. The parameters used in these experiments were the same as those applied in the third and fourth experiments.

4.3. Results and Quantitative Evaluation

Four accuracy measures [27,47] were used to evaluate the performance of the presented method. These measures included: (1) Completeness = TP/(TP + FN); (2) Correctness = TP/(TP + FP); (3) Quality = TP/(TP + FP + FN), where TP, FN, and FP represent true positive, false negative, and false positive, respectively; (4) Seed-point number. The ground truth was produced through the hand-drawing method, and the buffer width was set to four pixels.

4.3.1. Test of the Edge Constraint

Two remote sensing images were selected to test the edge constraint effect on road centerline extraction. The results are presented in Figure 3. The method using edge constraint provided better results than those provided by the method without edge constraint. The results obtained through the method without edge constraint easily deviated from the true road centerline, whereas those obtained through the proposed method with edge constraint could preserve the road centerline. The proposed method using edge constraint is more accurate than other methods because of the two following advantages: First, edge-energy computation and distance transformation can provide the ridgeline of the road segment, as shown in Figure 2g. Second, the fast marching method can trace the road centerline along the ridgeline. The visual comparison of the results, as presented in Figure 3, illustrates the advantages of the proposed method in road centerline extraction.

4.3.2. Experiment on Centerline Extraction from U-Shaped Roads

The results of the three methods are compared in Figure 4. This figure shows that all the three methods extracted the expected road centerlines. Compared with that of Hu's method, the performance of Miao's method and the proposed ECFM method improved with the number of road seed points. The proposed ECFM method, however, provided better results for both images than Hu's and Miao's methods. Table 1 shows the quantitative evaluation results of the three methods. Among the three tested methods, the presented method achieved the highest quality values for the two cases. These values coincided with the extraction results presented in Figure 4. Although Hu's method accurately extracted centerlines, it consumed more road seed points than the other two methods because it requires intermediate road seed points when extracting centerlines from S- or U-shaped road segments. By contrast, the proposed method extracts centerlines from S- or U-shaped roads with only two road seed points.

Table 1. Comparison of Three Semiautomatic Road Centerline Extraction Methods.

	Hu et al.'s Method	Miao et al.'s Method	Proposed ECFM Method
Case 1			
Completeness (%)	89.76	83.26	90.95
Correctness (%)	93.54	84.04	94.93
Quality (%)	85.34	79.78	85.91
Number of seed points	8	2	2
Case 2			
Completeness (%)	96.68	97.81	99.82
Correctness (%)	97.47	98.67	99.91
Quality (%)	94.32	96.54	99.73
Number of seed points	9	2	2

4.3.3. Experiment on An IKONOS Image

Figure 5 shows that the proposed ECFM method extracted most of the road segments and provided satisfactory results. A visual comparison between the extraction results is shown in Figure 6a–d. This figure shows that the proposed method performed better than the other methods. Table 2 shows the quantitative results of the three methods. The results shown in Table 2 indicate that the three methods successfully extracted a relatively complete road centerline with relatively high extraction quality. Nevertheless, the efficiency of the proposed ECFM method is superior to that of Hu's and Miao's methods. For example, the proposed method used the fewest seed points among all three tested methods. Given that the solution of Hu's method for parabola parameters is heavily dependent on the radiometric features of dual edges, this method will encounter problems when extracting features from images with unclear edges. Specifically, Hu's method will not provide the desired result if the road boundary is unclear. Miao's method exploits the geodesic method to connect road seed points. Its performance, however, is affected by road occlusions. The presented method achieved the highest quality values among all tested methods, indicating that it achieves the best balance between road extraction quality and seed-point consumption. Although Hu's method can extract relatively complete centerlines, its quality values are lower than those of the presented method because the result obtained through Hu's method is biased to the ground truth, whereas that obtained through the presented method is considerably closer to the ground truth.

Figure 6. *Cont.*

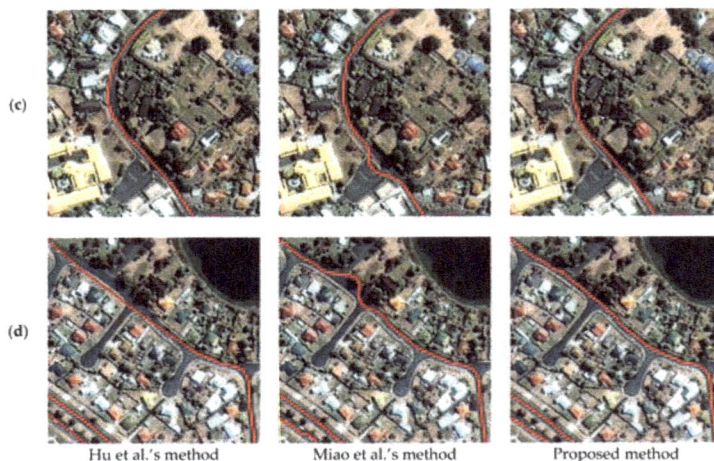

Figure 6. Comparison of the results provided by different road extraction methods for an IKONOS image. (**a**) Case 1; (**b**) Case 2; (**c**) Case 3; (**d**) Case 4.

Table 2. Quantitative Evaluation of Different Centerline Extraction Methods.

	Hu et al.'s Method	Miao et al.'s Method	Proposed ECFM Method
Experiment on IKONOS image			
Completeness (%)	96.24	97.94	98.39
Correctness (%)	96.99	97.07	97.83
Quality (%)	93.45	95.13	96.30
Number of seed points	442	279	264
Experiment on QuickBird image			
Completeness (%)	94.91	90.80	95.58
Correctness (%)	95.16	93.57	97.82
Quality (%)	90.54	85.47	93.60
Number of seed points	8	8	5

4.3.4. Experiment on A QuickBird Image

Figure 7 shows that Miao's method cannot efficiently manage abrupt changes, such as road junctions and sudden material changes or conflations, in images. This limitation is attributed to the method's requirement for an intermediate step to measure initial road centerline probability, which is computed on the basis of seed-point information, from the binary road image. Miao's method could not extract the expected road centerline if road segments between seed points were occluded by shadows or by a vehicle. By contrast, the proposed method utilizes edge energy and curvature to reduce the effect of shadows and vehicles on the road. The performance of Hu's method was comparable with that of the proposed method. However, the road seed-point consumption of the proposed method was superior to that of Hu's method. Table 2 shows the quantitative evaluation results of three methods. Although the proposed method used fewer seed points than the other two methods, it obtained higher completeness, correctness, and quality values. These values coincided with the extraction results presented in Figure 7. The experimental results illustrate that the proposed method is robust to noise and has considerable potential applications in road extraction from VHR remote sensing images.

| Hu et al.'s method | Miao et al.'s method | Proposed method |

Figure 7. Comparison of the results provided by different road extraction methods for a QuickBird image.

4.3.5. Experiment on A WorldView-2 Image

Figure 8 shows that the proposed ECFM method can be used to reliably and accurately extract roads in a wide range of high-resolution remote sensing images. Figure 9 shows the local comparison of roads extracted by different methods. Overall, all three methods can achieve satisfactory results. The comparison in Figure 9a shows that ECFM and Hu's methods have good anti-noise performance when encountering toll stations, and compared with Miao's method, the road centerline extracted is closer to the center. This difference is due to the fact that Miao's method considers only the spectral features of roads while our and Hu's methods not only consider the spectral features but also combine the edge features. Figure 9b shows that in road sections where road materials change greatly, all three methods can extract the road centerline accurately. Nevertheless, comparison indicates that the road centerline extracted by the ECFM method is smooth, and the technique can maintain high accuracy in sections with large road curvatures. Figure 9c shows the differences among three methods of extracting roads near road intersections. According to the figure, the road centerlines extracted by ECFM and Hu's methods are relatively smooth. The road centerline extracted by Miao's method is easily influenced by vehicles on the road, so the extraction results in the vehicle-intensive area are not smooth enough. Figure 9d shows the results of different methods in the case of shadow occlusion. Comparison shows that Hu's extraction result is relatively smooth because the technique adopts the piecewise parabolic model, which can obtain a relatively smooth curve. However, according to the figure, the road centerline acquired by this method can easily shift. Miao's method is influenced by shadows and cars, which lead to the unsmooth extraction results. The ECFM method proposed in this paper has achieved a relatively balanced performance, and it is better than the compared techniques in terms of road smoothness and accuracy. The statistical results in Table 3 are also consistent with those in Figure 9. Table 3 shows that the ECFM method performs well in terms of completeness, correctness, and quality under the condition of using the same number and location of road seed points.

Table 3. Quantitative Evaluation of Different Centerline Extraction Methods.

	Hu et al.'s Method	Miao et al.'s Method	Proposed ECFM Method
Experiment on WorldView-2 image			
Completeness (%)	95.63	94.01	97.56
Correctness (%)	95.25	92.03	96.84
Quality (%)	91.28	86.94	94.55
Number of seed points	249	249	249
Experiment on IKONOS grayscale image			
Completeness (%)	93.16	91.28	92.53
Correctness (%)	86.01	88.62	90.23
Quality (%)	80.90	81.71	84.20
Number of seed points	67	67	67

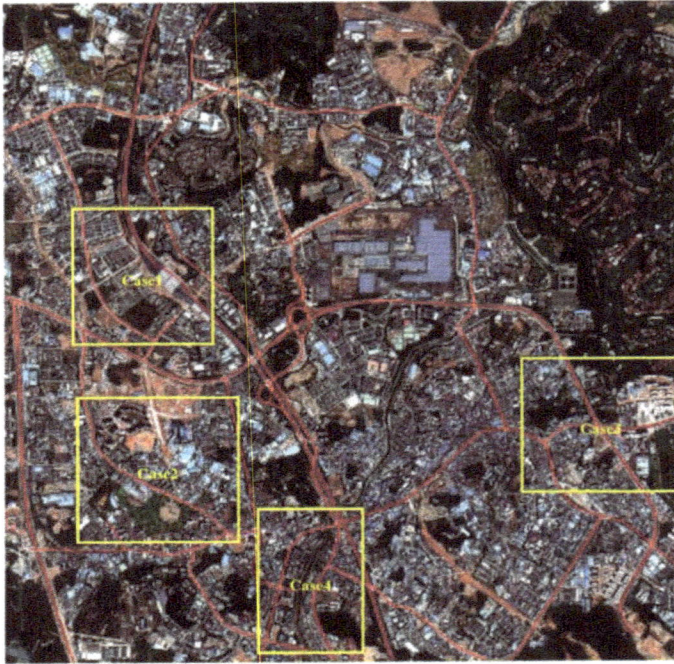

Figure 8. Road extraction result provided by the proposed ECFM method for a WorldView-2 image.

Figure 9. *Cont.*

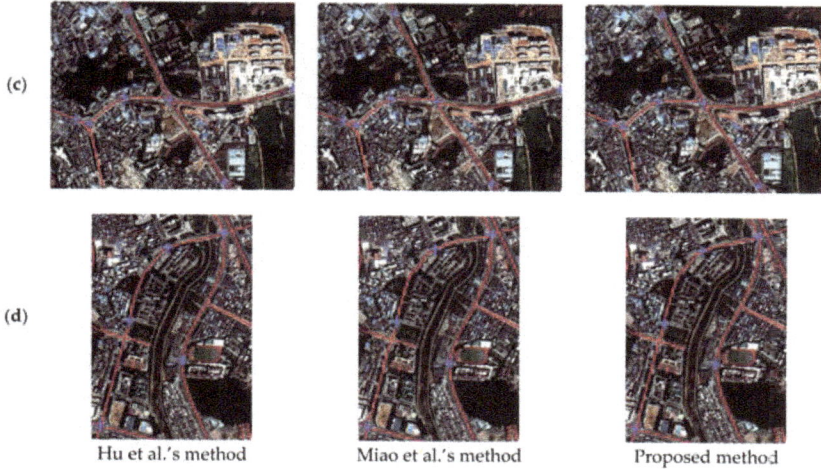

Figure 9. Comparison of the results provided by different road extraction methods for a WorldView-2 image. Road seed points are marked with blue crosses. (**a**) Case 1; (**b**) Case 2; (**c**) Case 3; (**d**) Case 4.

4.3.6. Experiment on An IKONOS Grayscale Image

Figure 10 shows the results of three different methods for extracting the road centerline from an IKONOS grayscale remote sensing image. As can be seen from the figure, all roads can be extracted completely by the three methods. The road centerline extracted by Hu's method is the smoothest, but the limitation of the piecewise parabolic model it uses causes the extracted results in areas with large changes in road curvature to tend to deviate from the road center. Miao's method and the ECFM method can avoid this problem. Compared with Miao's technique (which considers only the spectral features of roads), the ECFM method (which fuses the edge features and spectral features, thereby potentially overcoming the influence of spectral changes placed by shadows on road extraction results to a certain extent) shows better performance on shadow and vegetation occlusion. As can be seen from the statistical results in Table 3, the extraction completeness of all three methods is high when the same number and location of road seed points are used. However, our method achieves the best performance in terms of extracting correctness indicators. Similarly, our method demonstrates the best quality.

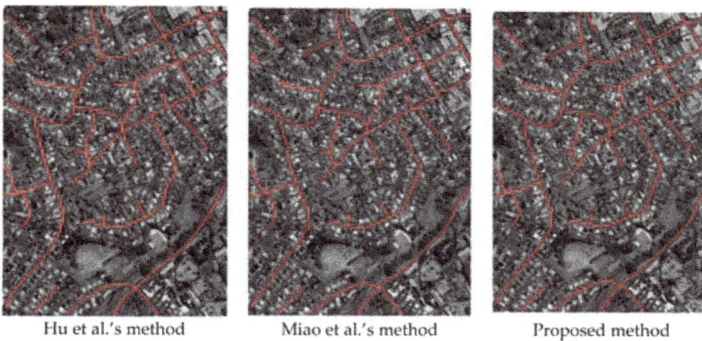

Figure 10. Comparison of the results provided by different road extraction methods for an IKONOS grayscale image.

5. Discussion

In this section, we present our analysis and discussion of the parameter sensitivity and computational costs of Experiments 3 and 4. Then, from Experiments 5 and 6, we discuss the influence of the number and location of seed points on road extraction results. The details are provided in the following subsections.

5.1. Parameter Sensitivity Analysis

We analyzed the effects of parameters α, β, and λ used in the edge-constraint-based weighted fusion model. These parameters have various effects on road extraction performance. The QuickBird satellite image shown in Figure 7 was tested, and the three parameters were set from 0 to 1 with an interval of 0.075. As shown in Figure 11a, road extraction quality was less than 5% when α was small. However, when α exceeded 0.15, performance suddenly increased and was maintained at approximately 90%. This result indicated that spectral information plays a dominant role in the fusion model. β was proportional to recognition quality, as shown in Figure 11b. Thus, increasing the weight of the road centerline probability feature improves extraction accuracy. Figure 11c shows that the effect of the edge constraint is not proportional to extraction quality. Recognition rate will decrease if λ is excessively small or large. The proposed method yielded a good extraction result when λ was approximately 0.4.

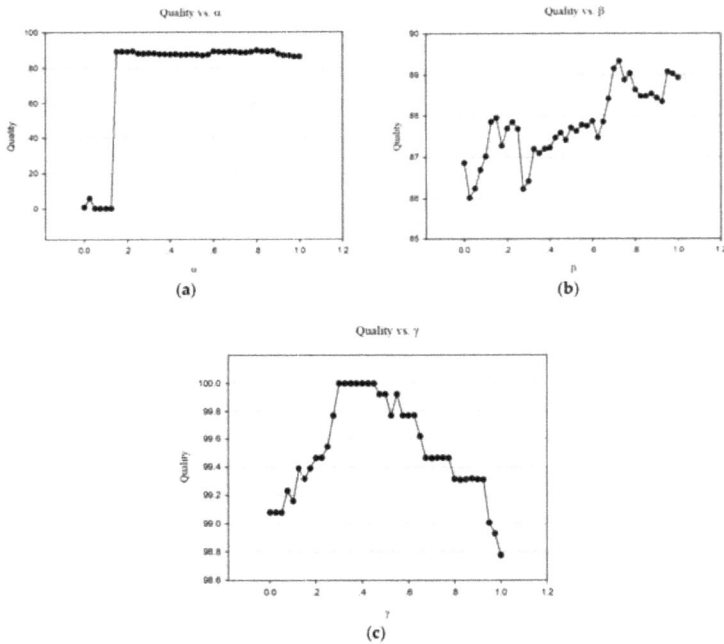

Figure 11. Quality of the results provided by the proposed method for the QuickBird image under different α, β, and λ values. (**a**) Quality vs. α; (**b**) Quality vs. β; (**c**) Quality vs. λ.

5.2. Computational Cost Analysis

In this section, we present a discussion of the computational cost of the proposed approach. All the experiments were performed on a personal computer with a 3.1-GHz Pentium dual-core CPU and 16-GB memory. Each experiment was repeated five times, and the average running time and seed-point number of the proposed approach with IKONOS and QuickBird satellite images are presented in Table 4. The proposed method ensured correct road extraction and consumed less computational time than the other two methods. According to this analysis, without considering the number and location of seed points, the road extraction efficiency of the method proposed in this work is the highest, thereby introducing an effective way for the extensive practical application of extracting road centerline from remote sensing images. In general, the presented method is moderately efficient.

Table 4. Computation Cost of Different Centerline Extraction Methods.

	Hu et al.'s Method	Miao et al.'s Method	Proposed ECFM Method
Experiment on IKONOS image (size: 3500 pixels × 3500 pixels)			
Time (s)	1643 s	1307 s	1273 s
Number of seed points	442	279	264
Experiment on QuickBird image (size: 1200 pixels × 1600 pixels)			
Time(s)	40 s	43 s	29 s
Number of seed points	8	8	5

5.3. Number and Location of Seed Points Analysis

In the third and fourth experiments, we adopted the strategy of obtaining the highest extraction quality by multiple extractions regardless of the number and location of seed points to compare different methods. As can be seen from Table 4, on the premise of ensuring the extraction quality, the number and time of seed points required by different methods remarkably vary. On the premise of obtaining the highest quality, our method requires the fewest number of seed points and time. However, the location and number of seed points have a considerable influence on different methods, and whether they are key factors affecting the experimental results needs to be analyzed.

Therefore, in the fifth and sixth experiments, to verify the influence of the number and location of seed points on the road extraction results of different methods, we used the same number and location of seed points to conduct comparative experiments. Seed points were obtained by artificial marking before the start of comparative experiment. The experimental results and statistical results show that the ECFM method produces good results in both groups of experiments. The statistical results in Table 5 indicate that when the same number of seed points is used, Miao's method consumes the shortest time, followed by the ECFM method, and Hu's method consumes the longest time. This finding is due to the fact that Miao's method uses the simplest features, while Hu's method uses the piecewise parabolic model and least-squares template matching, thereby prolonging the optimization of the road curve. Meanwhile, our method uses three features (spectral feature, edge feature, and road centerline probability), and the time required is also increased compared with Miao's method.

Table 5. Computation Cost of Different Centerline Extraction Methods

	Hu et al.'s Method	Miao et al.'s Method	Proposed ECFM Method
Experiment on WorldView-2 image (size: 3000 pixels × 3000 pixels)			
Time (s)	759 s	621 s	720 s
Number of seed points	249	249	249
Experiment on IKONOS grayscale image (size: 725 pixels × 1018 pixels)			
Time(s)	108 s	79 s	97 s
Number of seed points	67	67	67

A comparison of the data in Tables 4 and 5 shows that when a similar number of seed points is applied, the time required by our method to extract roads from different remote sensing images is remarkably different. This result has a substantial relationship with the resolution of the image used, the size of the area, and the density of the road network. A high resolution of the remote sensing image, large area, and high road network density result in a long extraction time.

6. Conclusions

This study presents a semiautomatic approach that uses road seed points to extract road centerlines from VHR remote sensing images. An edge-constraint-based weighted fusion model was introduced to overcome the influence of road occlusion and noise on road extraction. Finally, an edge-constraint fast marching method was proposed to improve the accuracy and quality of the road extraction results.

Six experiments were conducted on eight VHR remote sensing images that are related to different road conditions, including vehicle occlusion, sharp roadway curves, and building shadows. The advantages of the proposed method are as follows: (1) favorable road extraction accuracy and efficiency and (2) robustness to extracting road centerlines from VHR remote sensing images. Overall, the presented method is a superior and practical solution to road extraction from VHR optical remote sensing images.

In future work, the performance of the proposed method on additional types of remote sensing images, such as unmanned aerial vehicle images with very high spatial resolution, will be extensively investigated. The application of the proposed method to roads constructed from different materials and the automatic selection of road seed points are interesting future research directions.

Author Contributions: L.G. was primarily responsible for the original idea and experimental design. W.S. provided important suggestions for improving the paper's quality. Z.M. provided ideas to improve the quality of the paper. Z.L. contributed to the experimental analysis and revised the paper. The contribution of Z.L. was equal to the first author (L.G.).

Acknowledgments: The authors would like to thank the editor-in-chief, the anonymous associate editor, and the reviewers for their insightful comments and suggestions. This work was supported by the National Natural Science Foundation of China (41331175 and 61701396), the Scientific Research Foundation for Distinguished Scholars, Central South University (502045001), the Natural Science Foundation of Shaan Xi Province (2017JQ4006), and Engineering Research Center of Geospatial Information and Digital Technology, NASG.

Conflicts of Interest: The authors declare no conflict of interest.

References

1. Shi, W.; Zhu, C.; Wang, Y. Road feature extraction from remotely sensed image: Review and Prospects. *Acta Geod. Cartogr. Sin.* **2001**, *30*, 257–262. [CrossRef]
2. Shi, W.; Miao, Z.; Debayle, J. An integrated method for urban main-road centerline extraction from optical remotely sensed imagery. *IEEE Trans. Geosci. Remote Sens.* **2014**, *52*, 3359–3372. [CrossRef]
3. Mena, J.B. State of the art on automatic road extraction for GIS update: A novel classification. *Pattern Recognit. Lett.* **2003**, *24*, 3037–3058. [CrossRef]
4. Wang, W.; Yang, N.; Zhang, Y.; Wang, F.; Cao, T.; Eklund, P. A review of road extraction from remote sensing images. *J. Traffic Transp. Eng. (Engl. Ed.)* **2016**, *3*, 271–282. [CrossRef]
5. Miao, Z.; Shi, W.; Zhang, H.; Wang, X. Road centerline extraction from high-resolution imagery based on shape features and multivariate adaptive regression splines. *IEEE Geosci. Remote Sens. Lett.* **2013**, *10*, 583–587. [CrossRef]
6. Huang, X.; Zhang, L.; Li, P. Classification and extraction of spatial features in urban areas using high-resolution multispectral imagery. *IEEE Geosci. Remote Sens. Lett.* **2007**, *4*, 260–264. [CrossRef]
7. Li, M.; Stein, A.; Bijker, W.; Zhan, Q. Region-based urban road extraction from VHR satellite images using Binary Partition Tree. *Int. J. Appl. Earth Obs. Geoinf.* **2016**, *44*, 217–225. [CrossRef]
8. Li, Z.; Shi, W.; Wang, Q.; Miao, Z. Extracting man-made objects from high spatial resolution remote sensing images via fast level set evolutions. *IEEE Trans. Geosci. Remote Sens.* **2015**, *53*, 883–899. [CrossRef]
9. Miao, Z.; Shi, W.; Gamba, P.; Li, Z. An object-based method for road network extraction in VHR satellite images. *IEEE J. Sel. Top. Appl. Earth Obs. Remote Sens.* **2015**, 1–10. [CrossRef]
10. Huang, X.; Zhang, L. Road centreline extraction from high-resolution imagery based on multiscale structural features and support vector machines. *Int. J. Remote Sens.* **2009**, *30*, 1977–1987. [CrossRef]

11. Courtrai, L.; Lefèvre, S. Morphological path filtering at the region scale for efficient and robust road network extraction from satellite imagery. *Pattern Recognit. Lett.* **2016**, *83*, 195–204. [CrossRef]

12. Zang, Y.; Wang, C.; Cao, L.; Yu, Y.; Li, J. Road network extraction via aperiodic directional structure measurement. *IEEE Trans. Geosci. Remote Sens.* **2016**, *54*, 3322–3335. [CrossRef]

13. Butenuth, M.; Heipke, C. Network snakes: Graph-based object delineation with active contour models. *Mach. Vis. Appl.* **2012**, *23*, 91–109. [CrossRef]

14. Gruen, A.; Li, H. Road extraction from aerial and satellite images by dynamic programming. *ISPRS-J. Photogramm. Remote Sens.* **1995**, *50*, 11–20. [CrossRef]

15. Mokhtarzade, M.; Zoej, M.J.V. Road detection from high-resolution satellite images using artificial neural networks. *Int. J. Appl. Earth Obs. Geoinf.* **2007**, *9*, 32–40. [CrossRef]

16. Panboonyuen, T.; Jitkajornwanich, K.; Lawawirojwong, S.; Srestasathiern, P.; Vateekul, P. Road segmentation of remotely-sensed images using deep convolutional neural networks with landscape metrics and conditional random fields. *Remote Sens.* **2017**, *9*, 680. [CrossRef]

17. Mirnalinee, T.T.; Das, S.; Varghese, K. An integrated multistage framework for automatic road extraction from high resolution satellite imagery. *J. Indian Soc. Remote Sens.* **2011**, *39*, 1–25. [CrossRef]

18. Wegner, J.D.; Montoya-Zegarra, J.A.; Schindler, K. A higher-order CRF model for road network extraction. In Proceedings of the 2013 IEEE Conference on Computer Vision and Pattern Recognition (CVPR), Portland, OR, USA, 23–28 June 2013; IEEE CSP: Washington, DC, USA, 2013; pp. 1698–1705.

19. Zang, Y.; Wang, C.; Yu, Y.; Luo, L.; Yang, K. Joint enhancing filtering for road network extraction. *IEEE Trans. Geosci. Remote Sens.* **2017**, *55*, 1511–1524. [CrossRef]

20. Maboudi, M.; Amini, J.; Hahn, M.; Saati, M. Road network extraction from VHR satellite images using context aware object feature integration and tensor voting. *Remote Sens.* **2016**, *8*, 637. [CrossRef]

21. Miao, Z.; Shi, W.; Samat, A.; Lisini, G.; Gamba, P. Information fusion for urban road extraction from VHR optical satellite images. *IEEE J. Sel. Top. Appl. Earth Obs. Remote Sens.* **2016**, 1–14. [CrossRef]

22. Hu, J.; Razdan, A.; Femiani, J.C.; Cui, M.; Wonka, P. Road network extraction and intersection detection from aerial images by tracking road footprints. *IEEE Trans. Geosci. Remote Sens.* **2007**, *45*, 4144–4157. [CrossRef]

23. Lin, X.G.; Zhang, J.X.; Liu, Z.J.; Shen, J. Semi-automatic extraction of ribbon roads form high resolution remotely sensed imagery by cooperation between angular texture signature and template matching. In Proceedings of the ISPRS Congress, Beijing, China, 3–11 July 2008; pp. 539–544.

24. Das, S.; Mirnalinee, T.T.; Varghese, K. Use of salient features for the design of a multistage framework to extract roads from high-resolution multispectral satellite images. *IEEE Trans. Geosci. Remote Sens.* **2011**, *49*, 3906–3931. [CrossRef]

25. Cheng, G.; Zhu, F.; Xiang, S.; Wang, Y.; Pan, C. Accurate urban road centerline extraction from VHR imagery via multiscale segmentation and tensor voting. *Neurocomputing* **2016**, *205*, 407–420. [CrossRef]

26. Cheng, G.; Wang, Y.; Gong, Y.; Zhu, F.; Pan, C.B.I. Urban road extraction via graph cuts based probability propagation. In Proceedings of the IEEE International Conference on Image Processing (ICIP), Paris, France, 27–30 October 2014; IEEE: New York, NY, USA, 2014; pp. 5072–5076.

27. Hu, X.; Zhang, Z.; Tao, C.V. A robust method for semi-automatic extraction of road centerlines using a piecewise parabolic model and least square template matching. *Photogramm. Eng. Remote Sens.* **2004**, *70*, 1393–1398. [CrossRef]

28. Miao, Z.; Wang, B.; Shi, W.; Zhang, H. A semi-automatic method for road centerline extraction from VHR images. *IEEE Geosci. Remote Sens. Lett.* **2014**, *11*, 1856–1860. [CrossRef]

29. Zhou, J.; Bischof, W.F.; Caelli, T. Robust and efficient road tracking in aerial images. In Proceedings of the Joint Workshop of ISPRS and DAGM (CMRT'05) on Object Extraction for 3D City Models, Road Databases and Traffic Monitoring—Concepts, Algorithms and Evaluation, Vienna, Austria, 29–30 August 2005; pp. 35–40.

30. Movaghati, S.; Moghaddamjoo, A.; Tavakoli, A. Road extraction from satellite images using particle filtering and extended Kalman filtering. *IEEE Trans. Geosci. Remote Sens.* **2010**, *48*, 2807–2817. [CrossRef]

31. Lv, Z.; Jia, Y.; Zhang, Q.; Chen, Y. An adaptive multifeature sparsity-based model for semiautomatic road extraction from high-resolution satellite images in urban areas. *IEEE Geosci. Remote Sens. Lett.* **2017**, *14*, 1238–1242. [CrossRef]

32. Dal Poz, A.P.; Gallis, R.A.B.; Silva, J.F.C.D.; Martins, E.F.O. Object-space road extraction in rural areas using stereoscopic aerial images. *IEEE Geosci. Remote Sens. Lett.* **2012**, *9*, 654–658. [CrossRef]

33. Yang, K.; Li, M.; Liu, Y.; Jiang, C. Multi-points fast marching: A novel method for road extraction. In Proceedings of the International Conference on Geoinformatics, Beijing, China, 18–20 June 2010; pp. 1–5.

34. Osher, S.; Sethian, J.A. Fronts propagating with curvature-dependent speed: Algorithms based on Hamilton-Jacobi formulations. *J. Comput. Phys.* **1988**, *79*, 12–49. [CrossRef]

35. Jbabdi, S.; Bellec, P.; Toro, R.; Daunizeau, J.; Pelegrini-Issac, M.; Benali, H. Accurate anisotropic fast marching for diffusion-based geodesic tractography. *Int. J. Biomed. Imaging* **2008**, *2008*, 2. [CrossRef] [PubMed]

36. Li, H.; Xue, Z.; Cui, K.; Wong, S. Diffusion tensor-based fast marching for modeling human brain connectivity network. *Comput. Med. Imaging Graph.* **2011**, *35*, 167–178. [CrossRef] [PubMed]

37. Barzohar, M.; Cooper, D.B. Automatic finding of main roads in aerial images by using geometric-stochastic models and estimation. *IEEE Trans. Pattern Anal. Mach. Intell.* **1996**, *18*, 707–721. [CrossRef]

38. Hinz, S.; Baumgartner, A. Automatic extraction of urban road networks from multi-view aerial imagery. *ISPRS-J. Photogramm. Remote Sens.* **2003**, *58*, 83–98. [CrossRef]

39. He, K.; Sun, J.; Tang, X. Guided Image Filtering. *IEEE Trans. Pattern Anal. Mach. Intell.* **2013**, *35*, 1397–1409. [CrossRef] [PubMed]

40. Xiang, S.; Nie, F.; Zhang, C. Learning a Mahalanobis distance metric for data clustering and classification. *Pattern Recognit.* **2008**, *41*, 3600–3612. [CrossRef]

41. Rosenfeld, A.; Pfaltz, J.L. Distance functions on digital pictures. *Pattern Recognit.* **1968**, *1*, 33–61. [CrossRef]

42. Bakker, W.H.; Schmidt, K.S. Hyperspectral edge filtering for measuring homogeneity of surface cover types. *ISPRS-J. Photogramm. Remote Sens.* **2002**, *56*, 246–256. [CrossRef]

43. Ahamada, I.; Flachaire, E. *Non-Parametric Econometrics*; Oxford University Press: Oxford, UK, 2010; ISBN 9780199578009.

44. Williams, D.J.; Shah, M. A Fast algorithm for active contours and curvature estimation. *CVGIP Image Underst.* **1992**, *55*, 14–26. [CrossRef]

45. Türetken, E.; Benmansour, F.; Fua, P. Automated reconstruction of tree structures using path classifiers and Mixed Integer Programming. In Proceedings of the 2012 IEEE Conference on Computer Vision and Pattern Recognition (CVPR), Providence, RI, USA, 16–21 June 2012; pp. 566–573.

46. VPLab Data. Available online: http://www.cse.iitm.ac.in/~vplab/satellite.html (accessed on 16 April 2017).

47. Wiedemann, C.; Heipke, C.; Mayer, H.; Jamet, O. Empirical evaluation of automatically extracted road axes. In *Empirical Evaluation Techniques in Computer Vision*, 1st ed.; Bowyer, K., Phillips, P.J., Eds.; IEEE CSP: Los Alamitos, CA, USA, 1998; pp. 172–187. ISBN 0818684011.

MDPI

St. Alban-Anlage 66

4052 Basel

Switzerland

Tel. +41 61 683 77 34

Fax +41 61 302 89 18

www.mdpi.com

Remote Sensing Editorial Office

E-mail: remotesensing@mdpi.com

www.mdpi.com/journal/remotesensing

www.ingramcontent.com/pod-product-compliance
Lightning Source LLC
Chambersburg PA
CBHW051725210326
41597CB00032B/5611